Rheinwerk
Computing

Have you been to our website?

For code downloads, print and e-book bundles,
extensive samples from all books, special deals,
and our blog, please visit us at:

www.rheinwerk-computing.com

Rheinwerk Computing

The Rheinwerk Computing series offers new and established professionals comprehensive guidance to enrich their skillsets and enhance their career prospects. Our publications are written by the leading experts in their fields. Each book is detailed and hands-on to help readers develop essential, practical skills that they can apply to their daily work.

Explore more of the Rheinwerk Computing library!

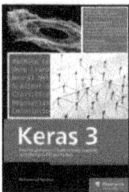

www.rheinwerk-computing.com

Bert Gollnick

PyTorch

The Practical Guide

Rheinwerk

Computing

Editors Meagan White
Acquisitions Editor Hareem Shafi
Translation Bert Gollnick
Copyeditor Doug McNair
Cover Design Abigail Damke
Photo Credits iStock: 129331108/© Niko_Cingaryuk; iStock 1426271120/© imaginima
Layout Design Vera Brauner
Production Abigail Damke, Hannah Lane
Typesetting III-satz, Germany
Printed and bound in Canada, on paper from sustainable sources

ISBN 978-1-4932-2786-0
1st edition 2026
1st German edition published 2026 by Rheinwerk Verlag, Bonn, Germany

© 2026 by:
Rheinwerk Publishing, Inc.
2 Heritage Drive, Suite 305
Quincy, MA 02171
USA
info@rheinwerk-publishing.com
+1.781.228.5070

Represented in the E.U. by:
Rheinwerk Verlag GmbH
Rheinwerkallee 4
53227 Bonn
Germany
service@rheinwerk-verlag.de
+49 (0) 228 42150-0

Library of Congress Cataloging-in-Publication Control Number: 2026001184

Contents at a Glance

Contents

4   Computer Vision   127

8 Time Series Forecasting

9 Language Models

10   Pretrained Networks and Fine-Tuning

Preface

Welcome to your journey into the world of deep learning with PyTorch—the most popular open-source deep learning framework! It provides a comprehensive library of tools for deep neural networks, which makes it a foundational platform for developing and training state-of-the-art deep learning models. In this preface, you'll find out what you can expect to learn from this comprehensive guide.

My goal goes beyond mere knowledge transfer. I want to empower you to use deep learning in a targeted way and fully prepare you to tackle and master the challenges and opportunities that this technology brings to the modern world. By the end of your journey through this book, you'll have achieved not only an understanding but also a mastery of the complex tools of artificial intelligence (AI).

Target Group

This book is aimed at a broad target group interested in deep learning with PyTorch, including students of computer science, mathematics, and related disciplines as well as professionals who want to expand their knowledge in this area. The book is written in such a way that a beginner in machine learning can work with it, but experienced professionals can also use it to improve their skills and knowledge.

The practical examples and detailed explanations will help you understand the concepts and techniques that are crucial for the development and application of deep learning. Whether you want to innovate in your field, start an academic project, or simply explore your fascination with AI, this book should be a valuable resource on your journey.

Requirements

Before we dive into the captivating world of deep learning, there are a few prerequisites you should keep in mind to ensure a smooth journey. First, you need to have a good understanding of Python programming. You should feel comfortable with the following:

- The creation of functions
- Working with different data structures like lists or dictionaries
- Manipulating these structures
- Writing loops (mainly for loops)
- Using packages like numpy and pandas

Ideally, but not necessarily, you should already have a basic understanding of machine learning concepts, such as training models with scikitlearn and working with datasets. Familiarity with basic statistics and linear algebra will also be an advantage as they underpin many AI algorithms. If these prerequisites sound like languages you speak, you're well equipped to master the challenges that lie ahead when training neural networks.

Structure of the Book

The book is intended as a practical guide for Python programmers who want to develop AI applications. The structure of the book follows a step-by-step approach, starting with an introduction to the basic concepts and progressing to more advanced topics. You're encouraged to go through the book in order, as some chapters build on knowledge from previous chapters—more on this in the next section.

Chapter 1 provides an introduction to deep learning, covering the basic building blocks such as neural networks, activation functions, and tensors to provide a solid theoretical foundation.

Chapter 2 leads you seamlessly to the creation and training of your first model, which will be a regression model.

Chapter 3 will expand your knowledge of training classification models with the aim of being able to predict two or more classes. You'll also learn where you need to make changes to adapt the model training to new tasks, such as other classifications.

Chapter 4 is the most extensive chapter, as there are an extremely large number of tasks and models in the field of computer vision. It begins with image classification using *convolutional neural networks (CNNs)*, which constitute a fundamental technique for many image processing tasks. It follows this up with more complex applications such as *style transfer*, in which the artistic and visual style of one image is transferred to another, as well as *object recognition* and *semantic segmentation*, which make it possible to localize and classify objects in images.

Chapter 5 begins with a basic introduction to the topic of recommendation systems and presents the different types of recommendation systems and their areas of application. It then covers the theoretical foundations of recommendation systems, including the various algorithms and techniques used.

Chapter 6 introduces the world of autoencoders and variational autoencoders (VAEs), two powerful architectures. After a basic introduction and an explanation of the theoretical principles, the focus is on the practical implementation using the Python-based framework PyTorch. Through detailed example code, you'll learn how to implement both simple autoencoders and more complex VAEs in PyTorch. We'll highlight the differences between and application areas of the two architectures.

Chapter 7 discusses graph neural networks (GNN), which you can use to model complex relationships in networked data. It covers both the theoretical foundations of GNNs and practical implementations of them with PyTorch, to help you use GNNs for a variety of applications.

Chapter 8 covers the prediction of time series, which is a central application area of recurrent neural networks (RNNs). RNNs, and in particular long-short term memory (LSTM), were developed to process sequential data. They have a "memory" that enables them to store information from past timesteps.

Chapter 9 introduces the world of natural language processing (NLP) with PyTorch. It covers the basic concepts and architectures required for understanding and developing large language models.

Chapter 10 deals with the concept of *fine-tuning*, in which pretrained models are adapted to specific tasks to improve performance and reduce training effort. It also discusses *parameter-efficient fine-tuning* (PEFT), a method that aims to minimize the number of trainable parameters and thus reduce memory requirements and computational costs. It also discusses other related techniques and considerations.

Chapter 11 introduces a new framework—PyTorch Lightning—since by then, you'll have acquired a basic understanding of working with PyTorch, which forms the basis for understanding neural networks. PyTorch Lightning builds on this foundation and simplifies and abstracts the work, especially the training process, so that you can focus more on the model and the data. Through example code, we'll demonstrate the practical application of PyTorch Lightning, focusing on the clear organization of the code and the automatic handling of training sequences. Finally, we'll discuss the implementation of *early stopping*, which is an important technique for avoiding overfitting and optimizing the training process.

Chapter 12 covers *model evaluation*, which is an essential task that's closely linked to model development. It begins with a basic introduction to the various metrics and methods used to evaluate the performance of models. The focus here is also on practical applications, with detailed coding examples illustrating the use of MLflow, Tensor-Board, and Weights & Biases (WandB). These tools make it possible to monitor the training process, visualize metrics, and evaluate models more efficiently.

Chapter 13 rounds out the book by dealing in detail with deploying a trained model, which is a key step in making trained deep learning models usable in real applications. It starts with a general introduction to the different aspects of model deployment, including challenges and best practices.

Also note that practical Python code examples are available for download to illustrate and consolidate the application of the concepts we discuss.

How to Use This Book

This book is intended both as a practical tool for enthusiasts, developers, and professionals and as a comprehensive guide for anyone who wants to learn about and control the power of AI.

Most chapters are stand-alone, and you can simply jump into them after you've gained an understanding of some of the basics. See Figure 1 for an overview of the chapters and their dependencies.

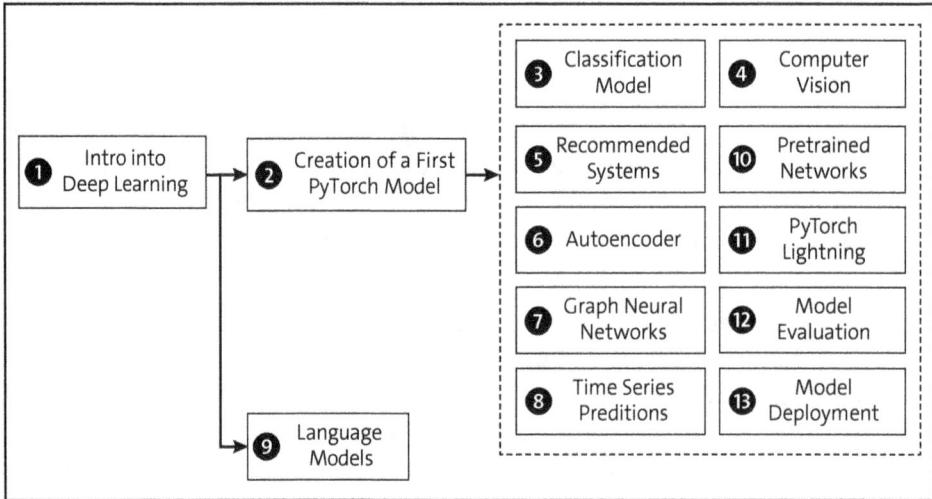

Figure 1 Chapter Dependencies

The information Chapter 1 is considered fundamental for understanding all subsequent chapters, and Chapter 2 lays the foundations for all the models you will train. So, you should read them first, and then, you'll be free to decide which topics interest you the most and read only the chapters on them if you wish. However, Chapter 9 is a special case. The models in it are handled somewhat differently because they place immense demands on the complexity of the network as well as the amount of data for training. So, you probably won't want to train a language model from scratch when working on Chapter 9—instead, it's best to use an existing language model to eliminate some of the requirements. That means you can jump into Chapter 9 as soon as you're done reading Chapter 1, which will familiarize you with the basics of deep learning.

Downloading Code and Additional Materials

You can download the code you'll use in this book so that you can reproduce the scripts yourself. You can obtain the material via GitHub at *https://github.com/DataScience-Hamburg/PyTorch_Book_Material*.

The following section describes how to download the material so that you can follow the Python scripts on your own system.

Preparing the System

Figure 2 shows the steps you should perform to set up the system so that you can work with the code material and run it on your local system.

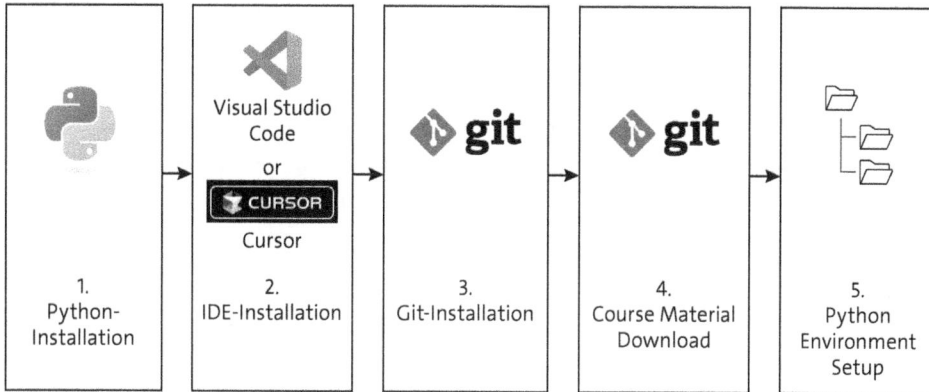

Figure 2 System Setup Steps

These steps may seem like a lot, but remember that this is only a one-off and should not take longer than 30 minutes.

Installing Python

Installing Python is very simple, and you can do it in different ways. I recommend downloading Python from the official website: *https://www.python.org/downloads/*. If you scroll down a bit on the download page, you'll find different versions.

Figure 3 shows a list of different versions of Python. The code in this book was developed based on Python version 3.12.7, so to avoid problems, you should use the same version. Please click on the **Download** button and follow the installation instructions. There is nothing special to note, and you can use the default settings.

On some systems, such as MacOS, Python is already preinstalled. You can type "python -version" in the command line to find out which Python version is installed.

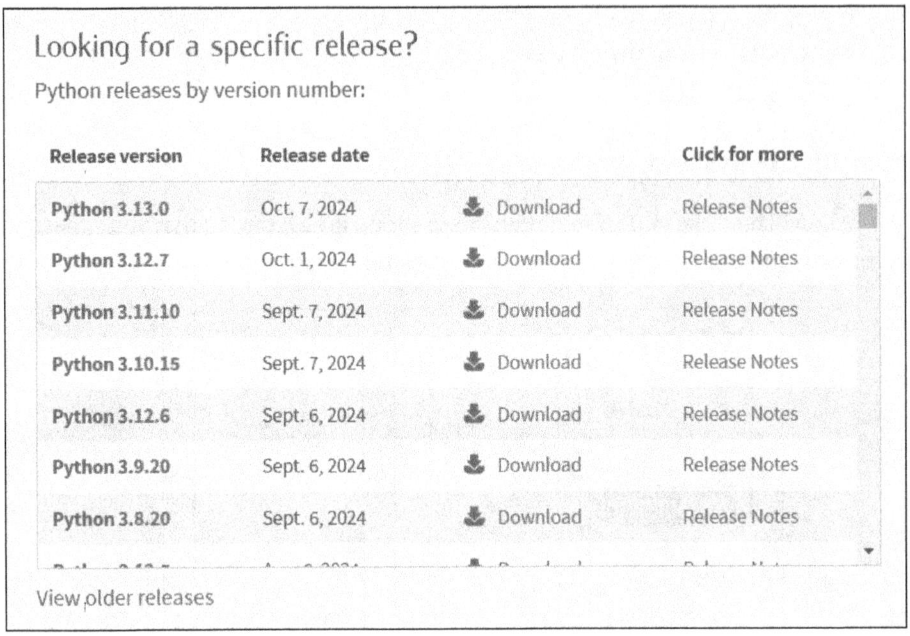

Figure 3 Python Versions

Installing an Integrated Development Environment

You need to have an *integrated development environment* (IDE) to efficiently write and manage code in Python projects. Popular options are as follows:

- **Visual Studio Code (VS Code)**
  This is a lightweight and highly customizable IDE that supports many different programming languages. That's one reason why it's one of my favorites. I can use VS Code for my Python, R, and Flutter projects, and I don't have to get used to a different IDE when I switch projects.

  VS Code is open source, and you can download it for free from *https://code.visualstudio.com/*. It also has a large community that provides thousands of extensions. I used it for many years and could hardly imagine switching—but recently, I did and switched to Cursor.

- **Cursor**
  This is a *fork* of VS Code, which means it's a very close relative. It has an AI-assisted code editor built in, and with it, you'll become much faster and more efficient at programming. It leverages the capabilities of generative AI, and you can get coding assistance via OpenAI's Codex, GPT, or Claude.

  Cursor offers features such as code completion, code explanation, debugging, refactoring, and interactive chat. There is a free version and a paid version, and you can find out more at *https://www.cursor.com/*. You'll also need to create an account to use the program.

- **PyCharm**

  This IDE focuses on Python, and it offers advanced features such as code completion and debugging. The Community Edition is free and has basic features, while the Professional version offers deeper integrations. You can find both versions on the developer website: *https://www.jetbrains.com/pycharm/download/*.

- **Windsurf Editor**

  This is a very new editor that claims to be the first agent-based IDE, and it's also a VSC fork. You can obtain it from *https://codeium.com/*, where you must register before you can use the program.

If you have a preference for and experience with one of them, I won't stop you. But if you're a beginner and looking for good advice, I recommend you start with Cursor.

Installing Git

Git is an indispensable version control tool that helps developers track changes in their code, collaborate with others, and keep a historical record of project development. It allows you to manage your codebase efficiently by enabling branching, merging, and rolling back changes. Git integrates seamlessly with platforms like GitHub, GitLab and Bitbucket, making remote collaboration easy. You'll need it to perform the next step, in which you'll source the course material, and you can download it at *https://git-scm.com/downloads*. Figure 4 shows the various download options.

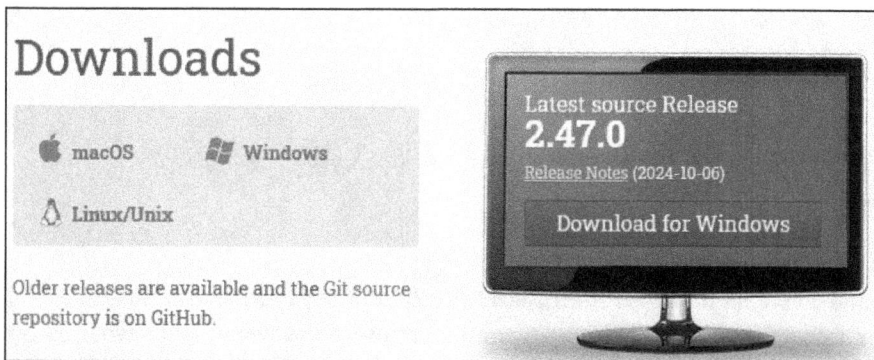

Figure 4 Downloading Git

Select the download that suits your operating system. There are many options you can choose from, but in all cases, you can use the default values to perform the installation.

Downloading Course Material

The sample code is hosted on GitHub, and you can find the materials at this link: *https://github.com/DataScienceHamburg/PyTorch_Book_Material*. When you open the page, you'll see all the materials as shown in Figure 5.

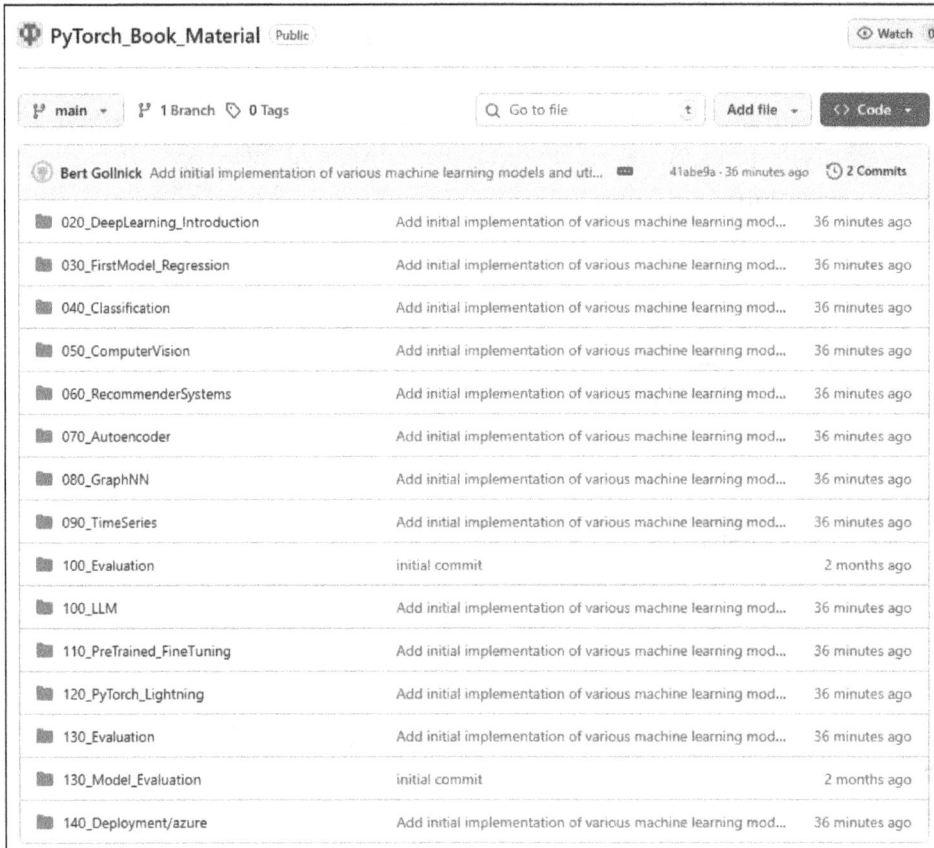

Figure 5 GitHub Repository for This Book

You can obtain the code in two ways by clicking on the green **Code** button:

- Download it as a ZIP-file.
- Clone the repository via Web-URL.

If you have no experience with Git, select the ZIP option, download the ZIP file, and unzip it into a folder of your choice on your computer. However, if you have used Git before, you can clone the repository by executing the following in the command line:

```
git clone
https://github.com/DataScienceHamburg/PyTorch_Buch_Material
```

This command creates a copy of the directory, which is saved on your computer.

Setting Up the Local Python Environment

A *Python environment* is an isolated workspace where you can install and manage dependencies for a specific project. It works like a *sandbox*—a protected and separate area—which ensures that the libraries and tools you use in a project won't conflict with others on your system. Figure 6 illustrates the concept of environments and why they are necessary.

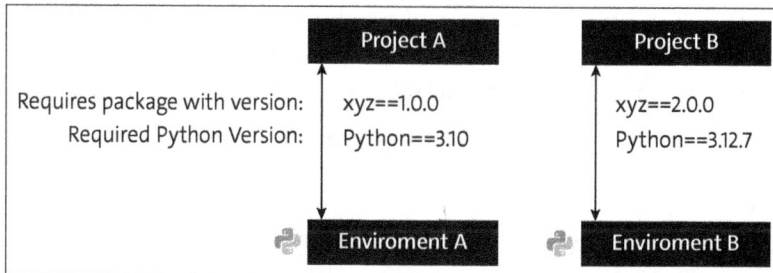

	Project A	Project B
Requires package with version:	xyz==1.0.0	xyz==2.0.0
Required Python Version:	Python==3.10	Python==3.12.7
	Enviroment A	Enviroment B

Figure 6 Python Environments

While Project A requires a specific package (xyz) in version 1.0.0 and a Python version of 3.10, you also need to work on another project that has completely different requirements. Project B needs the same package (xyz) in a newer version (2.0.0) and a different Python version of 3.12.7. If you worked with the same environment for all your projects, you would run into problems because of incompatibilities between the two projects.

The best practice is to set up a separate environment for each project you're working on. This will take up more space on your hard disk, but apart from that, you'll be on the safe side and avoid negative side effects between your projects.

Using a project environment has clear advantages, but now, we need to clarify which environments you can choose from. There are multiple tools for the management of Python environments.

- **virtualenv**
  This is a lightweight tool for managing Python environments. It allows you to install project-specific packages without affecting the global Python installation.

- **Poetry**
  This is a more modern dependency and environment manager. It simplifies package management by combining dependency specification, installation, and environment creation into a single tool. It's perfect for projects that require precise version tracking and easy deployment.

- **uv**
  This is a less commonly used but up-and-coming tool. It focuses on creating ultra-lightweight virtual environments, and thanks to its implementation in Rust, it's lightning fast compared to other environment tools.

My personal favorite is uv, and you can find many different options for installing uv on the developer page at *https://docs.astral.sh/uv/getting-started/installation/#stand-alone-installer*. I suggest that you install it with pip via the terminal (or the command line), as follows:

```
pip install uv
```

In the code folder, you'll find a file called *pyproject.toml*, which contains basic metadata information about the package as well as the dependencies (i.e., the packages and their versions as I installed them on my system while creating the materials). Next to the *pyproject.toml* file, you'll find the *uv.lock* file, which keeps track of the exact versions of all dependencies and their *transitive dependencies* (the dependencies of dependencies).

You can create the same environment by executing the following command in the terminal:

```
uv sync
```

This will download and install all dependencies at once in a *.venv* subfolder. If, for some reason, this doesn't work on your computer, don't worry. You can open *pyproject.toml*, look for the dependencies section, and install the packages listed there manually by simply executing the following:

```
pip install packagename
```

You can also do this by means of uv, as follows:

```
uv add packagename
```

As soon as you open a Python script, VSC (or Cursor) will recognize that an environment is available. You can determine which environment is being used via the horizontal bar at the bottom. Figure 7 depicts this bar, showing that the script is connected to the **'.venv': venv** Python environment and the corresponding version **3.12.7**.

```
Ln 8, Col 1    Spaces: 4    UTF-8    CRLF    {} Python    3.12.7 ('.venv': venv)    Cursor Tab    ⊘ Prettier    ○
```

Figure 7 Environment and Python Have Been Detected in VSC

This is the last step in our system preparation, and we're now ready to start.

Acknowledgements

I hope that the pages you have read so far have piqued your interest in the world of deep learning with PyTorch. It has been a fascinating journey writing this book—a journey that has taken me into the complex yet elegant depths of neural networks. However, such a journey is rarely a solitary affair. I would therefore like to take this moment to thank all the people who have accompanied and supported me along the way.

First and foremost, my deepest thanks go to my family. Their love and understanding have been the foundation that has carried me through the challenges of writing.

My beloved wife Lea's support and unwavering belief in me have been the beacon that has guided me when I've lost myself in the endless lines of PyTorch code. They have taken the long hours I've spent studying and trying to explain complex concepts with admirable equanimity (at least most of the time). Their strength and humor kept me motivated.

I would also like to thank my wonderful daughter, Elisa. Her lighthearted joy has often broken through the seriousness of my work and reminded me that life is more than just zeros and ones. Her understanding has been a precious gift.

To these two extraordinary people, I dedicate not only this book but the fruits of all the hours I put into creating this work on deep learning with PyTorch.

A big thank-you also goes to our extended family and my friends, who have always been there with words of encouragement. And of course, thanks to the entire open-source community, whose work and passion for PyTorch made this book possible.

This book isn't just the result of my work; it's a testament to the collective spirit of all those who have contributed in large or small ways to its completion.

Thank you.

Conventions Used in This Book

In this book, we've used certain typographical and formatting habits to enhance your learning and make the introduction it gives you to generative AI as clear and effective as possible. We've presented code blocks and programming examples in the following nonproportional fonts that make them stand out from the surrounding text so that you can easily recognize and follow the Python examples given:

```python
# sample code looks like this
my_welcome = "Welcome to the book"
```

```
I am a comment. Welcome to the book
```

You can also find the entire code and all files in the repository belonging to the book, and you can easily navigate to the source code files, which are displayed as *source_folder/source_file_name.py*.

To summarize important findings or provide additional context, we've strategically placed information boxes like this one in the chapters:

> **An Information Box**
> Here, we discuss some key concepts.

Links are underlined and italicized to allow you to quickly navigate to external resources that can give you deeper insight into the topics or access to datasets that are relevant to the exercises. They might look like this: *www.rheinwerk-computing.com.*

I hope these conventions make the book both user-friendly and beneficial to your learning journey into the world of generative AI.

Chapter 1
Introduction to Deep Learning

"We shape our tools, and thereafter, our tools shape us."
—*John Culkin*

These words are more relevant today than ever. Over the past decades, we've created computers and software to help us with complex tasks—but since the emergence of deep learning, we've been experiencing a new phase of this relationship. Today's tools can do much more than help us analyze large amounts of data—they can recognize patterns, evolve, and make decisions, often in a way that we ourselves can't fully comprehend.

This chapter introduces you to the basics of this fascinating technology. We'll find out how artificial neural networks work, how they are trained, and what potential they have. In Section 1.1, we'll start with the question of what deep learning is and how it can be differentiated. In Section 1.2, we'll look at the areas in which deep learning is used. In Section 1.3, you'll learn exactly how deep learning works, how model training works, and how you can use a model later in model inference. In Section 1.4, we'll talk about the history of AI, in Section 1.5, we'll discuss concepts such as the perceptron, and in Section 1.6 through Section 1.9, we'll discuss and the core components of AI. In Section 1.10, after all the theory, we'll conclude the chapter by working with tensors, which are the technical foundation of networks. We'll explain what tensors are, how to work with them, what a computational graph is, and how to optimize parameters in a graph.

1.1 What is Deep Learning?

Deep learning is closely related to other terms, and that often leads to confusion. Therefore, we've included Figure 1.1 to help you differentiate the terms.

The overarching term is *artificial intelligence* (AI), the best-known representative of which is machine learning. You may have already developed regression or classification models using statistical methods, which can also be classified as machine learning. You can use machine learning to develop algorithms that can learn from data and solve certain tasks without having been explicitly programmed to do so.

Figure 1.2 illustrates the difference between classic programming and machine learning. Whereas in traditional programming, a technical expert had to manually define rules that describe how the data is to be processed, in *machine learning*, the data (including the results) are passed to the algorithm and the algorithm can extract the rules independently, based on this information.

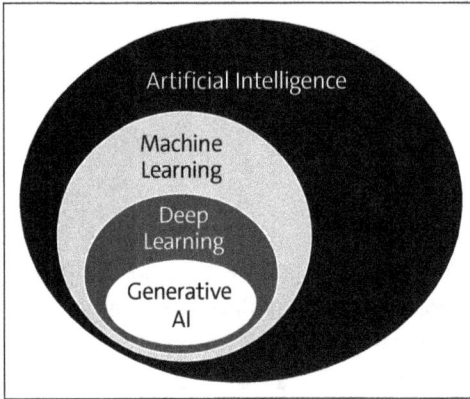

Figure 1.1 Relationships Among AI-Related Terms

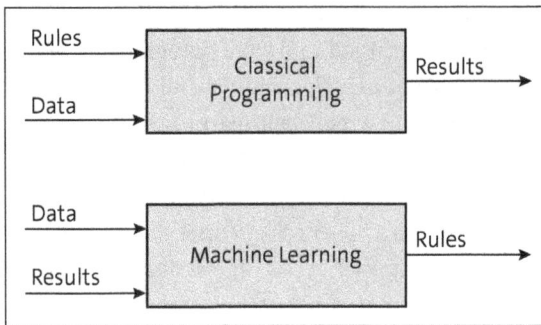

Figure 1.2 Difference Between Classic Programming and Machine Learning

Since the 2010s, however, deep learning has taken over the field. Deep learning belongs to the field of machine learning but works quite differently from traditional statistical methods in terms of technology. Deep learning uses artificial neural networks that attempt to replicate the human brain as computer-supported functions.

Then, in the 2020s, due to the success of large language models, other generative AI systems were added that can do things like taking text input and generating images or videos from them. This subarea is called *generative AI* because it deals with models that can generate (i.e., create) new things independently.

1.2   What Can You Use Deep Learning For?

Training data can be available in different forms, and that leads one to use different methods to train models. These methods and how they work are as follows:

■ **Supervised learning**
You train the model with labeled data. This means that each dataset contains a correct answer, which is called a *label* or *target variable*. The model learns by recognizing the relationship between the input data and the corresponding labels, and the goal is to find a function that correctly maps the input data to the output data. You can think of a neural network as a complex function.

■ **Unsupervised learning**
You train the model with data that isn't labeled. The model doesn't receive predefined answers. Instead, the goal is to find hidden patterns and structures in the data. Such models are often used to group similar data points (in a process known as *clustering*) or reduce dimensionality, as in dimension reduction or autoencoders.

■ **Reinforcement learning**
This is a type of learning in which an agent learns through interaction with its environment. The agent makes autonomous decisions or performs actions for which it receives rewards or punishments, and its goal is to develop a strategy that maximizes its cumulative reward. You'll use reinforcement learning for tasks in which the model must make a series of decisions, such as in robotics, autonomous driving, and playing video games.

■ **Semi-supervised learning**
This is an approach that combines the advantages of supervised and unsupervised learning. You'll use it when only a small amount of labeled data is available but a large amount of unlabeled data is available. You'll use special algorithms to create pseudo-labels and train a model with a combination of the originally labeled and pseudo-labeled data. This approach can be helpful when labeling data is very time-consuming and expensive, such as in image recognition or speech recognition.

Figure 1.3 shows the tasks for which you can use deep learning. The learning method is shown horizontally, and the two types of target variable are plotted vertically.

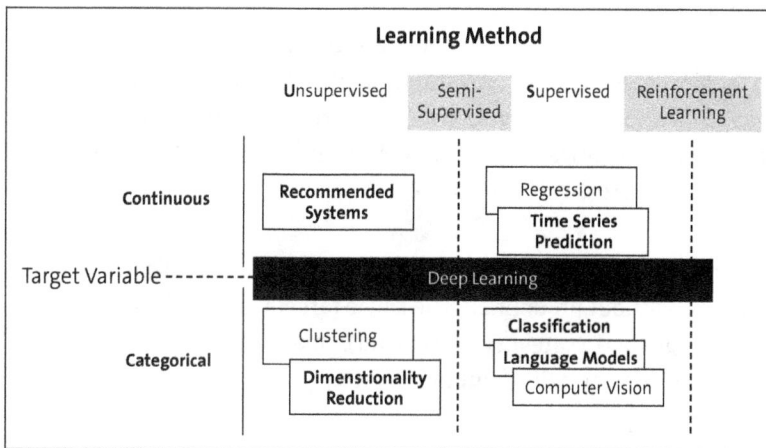

Figure 1.3 Areas of Application for Deep Learning

The target variable can be continuous or categorical, and the various task fields of deep learning result from this combination. The book follows this structure, so that many chapters correspond exactly to these task areas.

We explain the individual areas of responsibility in more detail in the following list:

- **Regression**
  This is a supervised learning task in which the model predicts a continuous numerical value and learns the relationship between the input data and the continuous target value. Examples of regression include the prediction of house prices, share prices, and physical parameters such as temperatures.

- **Time series prediction**
  This is a special type of regression that is also part of supervised learning. The model predicts future values based on historical, time-ordered data, and the target variable is continuous. Areas of application include the prediction of electricity consumption, sales figures, and weather trends.

- **Classification**
  This is a supervised learning task in which the model categorizes input data into one of several predefined classes. Using labeled data, the model learns to identify the characteristics that distinguish one class from another. Classic examples are the recognition of e-mails as "spam" or "not spam" and the classification of patients into "healthy" and "sick" categories.

- **Language models**
  Natural language processing models are part of supervised classification, specifically self-supervised learning. In this approach, the models generate their own "labels" from the unlabeled input data. The models learn the probability that a certain word sequence will occur and uses that to generate or understand text. They can categorize sentences or text excerpts, as is the case with sentiment analysis, for example. They can therefore assign text such as "I like the product" to the "positive" class.

- **Models for computer vision**
  Computer vision is a broad field that mostly belongs to supervised learning and deals with the automatic processing and understanding of images and videos. The tasks include *image classification* (classification of the entire image into a specific class), *object recognition* (e.g., categorization of objects in an image), and image segmentation. The target variable here is categorical.

- **Clustering**
  This is a task of unsupervised learning, and its aim is to classify similar data points into groups (clusters) without predefined labels. The target variable is categorical, as each data point is assigned to a category (the cluster). One application example is the segmentation of customers into different target groups.

- **Dimensionality reduction**
  This is a task of unsupervised learning in which the number of features (dimensions)

in a dataset is reduced to facilitate data visualization and improve the computational efficiency of models without losing important information. The target variable is categorical here, as it represents the affiliation of the data points to a reduced space.

- **Recommender systems**
  This can be considered a nonsupervised, semisupervised, or even supervised learning field. These systems analyze user behavior to find patterns and recommend products, content, or services to the user. The target variable can be continuous if, for example, the rating (e.g., 1 to 5 stars) is predicted, but often, there is no target variable and the model analyzes which products are similar or have been purchased by similar users.

This list provides a good overview of the types of problems you can tackle with deep learning. Now, let's turn to the question of how deep learning works.

1.3 How Does Deep Learning Work?

Deep learning is based on training a model in a first step, and the trained model is then deployed and used. This step is called *model inference*. The most complicated part is model training, which we'll look at now.

1.3.1 Model Training

The diagram in Figure 1.4 illustrates the training process for a neural network. It starts with the training data, which contains descriptive features (often called *independent features*) that are abbreviated here as a capital X (as is usual in statistics), as well as the actual target values (often called *dependent features*) that are usually abbreviated as Y.

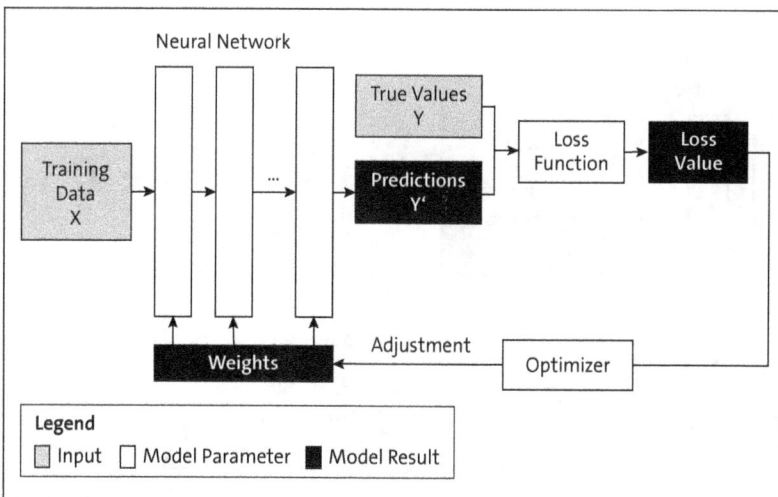

Figure 1.4 Model Training

An example of such training data would be a dataset created from past transactions on the real estate market. The target value in this case would be the sales price of the house, and the descriptive features would be things like the age of the house, the living space, the condition, etc.

The independent features are fed into the neural network, which consists of several layers. In each layer, there is a clearly defined number of neurons with specific weights that represent the strength of the connections between the neurons. The independent features are processed in the network, and the predictions (Y') are delivered.

These predictions are compared with the actual values (Y), and this comparison is made using a loss function. The loss function measures the deviation between the predictions and the actual values, resulting in a loss value. An optimizer uses this loss value to determine how the weights must be adjusted to minimize the error.

This process is repeated in loops, in which the training data is "shown" again to the now "smarter" model, in which the weights are continuously updated. The aim is for the model to make better and more accurate predictions, and as soon as it does, the training can be stopped. The trained model is checked using *test data* (i.e., data that the model hasn't seen before) to ensure that it delivers good results with more than just the training data. It can then prove itself in practice by making predictions for "new" data.

1.3.2 Model Inference

Figure 1.5 shows how the previously trained model is applied to make predictions for unknown data. This model inference process uses the model architecture with *frozen* weights, meaning that the model weights are no longer adjusted in this step—nor should they be. The test data, which has never been seen by the model, is used to make predictions. These predictions are used with the actual, real values in the model evaluation to check the quality of the model.

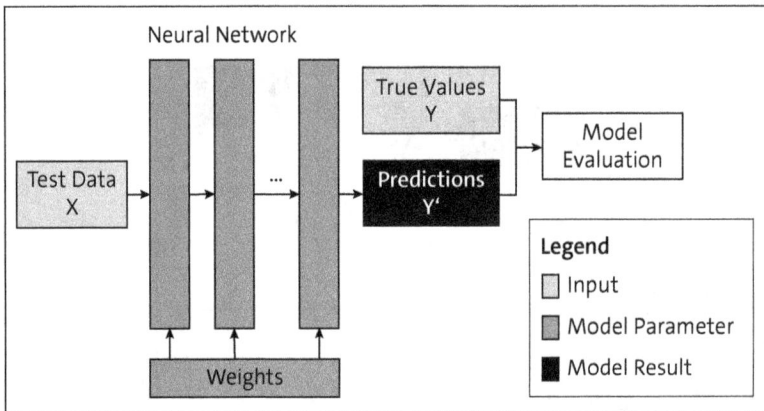

Figure 1.5 Model Inference

1.4   Historical Development

There have been several highs and lows during the development of AI, and the most important milestones during the evolution of AI are shown in Figure 1.6.

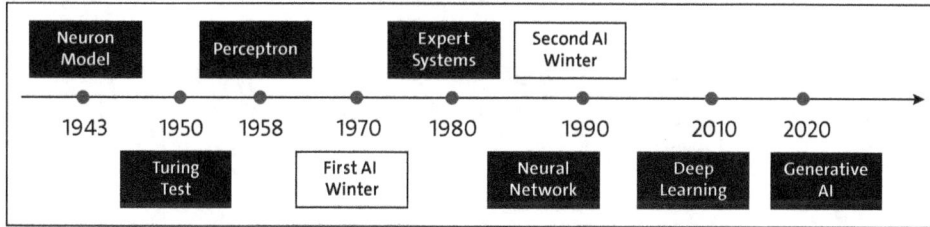

Figure 1.6 AI Evolution

The evolution of AI began with the McCulloch and Pitts developing the first neuron models in the 1940s. Alan Turing is regarded as one of the founding fathers of AI, and it was he who developed the *Turing test* (named after him) to assess a machine's ability to imitate human behavior.

The first *perceptron* (which we present in the next section) was developed in 1958, and researchers focused on developing expert systems in the 1970s. Such systems constitute an early form of AI that encodes human expert knowledge in a specific field to solve complex problems or make decisions. However, these systems ultimately failed to gain acceptance, and that led to the *first AI winter*, which was a period of little progress, low funding, and increasing skepticism. The reasons for this included inflated expectations for AI that could not be met. Then, in the 1980s, interest in neural networks began to increase again—but the expert systems in use at the time were unable to establish themselves, and that led to the *second AI winter* at the end of the 1980s.

It was the 2010s before we saw the beginning of the age of *deep learning*—an AI summer that continues to this day. It was initially driven by better computing power, the availability of large amounts of data (thanks to the internet), and very good and publicly accessible (open-source) frameworks.

The two dominant frameworks in the field of deep learning are *TensorFlow* and *PyTorchError! Bookmark not defined.*. TensorFlow was published by Google in 2015, and PyTorch was first made available to the world by Facebook (now Meta) in 2017.

Generative AI was then added in the 2020s. Since ChatGPT was published in November 2022, new developments in the field of AI have been coming thick and fast. Hardly a day goes by without the appearance of new and better models, more tools, and technological breakthroughs in the field.

1.5   Perceptrons

A *perceptron*, as shown in Figure 1.7, is the simplest form of an artificial neuron and is therefore the basic building block of neural networks. It was developed by Frank Rosenblatt in 1958 and was one of the first algorithms to put the idea of neural learning into practice.

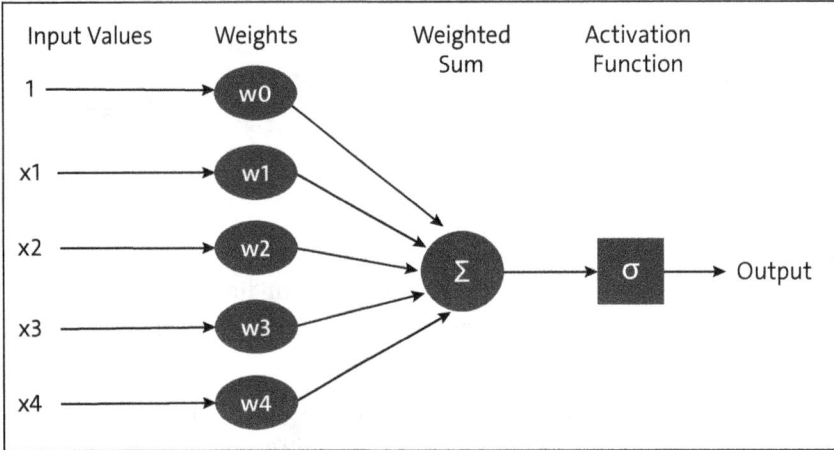

Figure 1.7 Structure of Perceptron

A perceptron processes several inputs, which are multiplied by specific weights that represent the importance or relevance of each input. All weighted inputs are summed up and passed through an activation function that decides whether the neuron "fires" or not. In the simplest case, the activation function outputs a 1 or 0 based on whether the sum exceeds a certain value. The perceptron laid the foundation for the development of deeper and more complex neural networks, which consist of many layers of perceptron-like neurons and form the basis of deep learning today.

In the following sections, let's look at the core components of neural networks, starting with their general structure and layers.

1.6   Network Structure and Layers

Figure 1.8 shows the basic structure of a neural network, which consists of several interconnected layers. The data is passed from the input layer through one or more hidden layers before arriving at the output layer.

The usual layer types are as follows:

- **Input layer**
  This type of layer receives the raw data, with each circle representing a neuron that corresponds to a feature in the dataset.

- **Neurons**
  These are all connected to the neurons in the next layer, which is referred to as a hidden layer.

- **Hidden layer**
  This is where the actual processing takes place. The number of hidden layers and the number of neurons in the hidden layers can vary. There are also the following different types of hidden layers that are closely related to the task the model is supposed to solve:

  - **Convolutional layer**
    This is used in computer vision.

  - **Recurrent layer**
    This is for processing sequential data in time series.

  - **Dense layer**
    This is the traditional hidden layer and is also referred to as a *fully connected layer*. Each neuron in a dense layer is connected to every neuron in the previous layer.

- **Output layer**
  This is where the information is finally forwarded to, and it provides the result, such as the probability of belonging to a class or a numerical value.

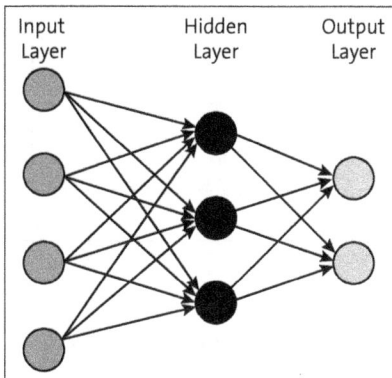

Figure 1.8 Basic Network Structure

The connections among the neurons are represented by arrows, and each connection has a weight that determines the strength of the connection and is adjusted during the training process.

Now, we come to the next important components—the activation functions.

1.7 Activation Functions

The output of a neuron can be controlled with the help of *activation functions*, which give the network the ability to recognize nonlinear patterns. Such nonlinearities are extremely important because without them, the entire neural network would function like a single linear model, which would severely limit its ability to solve complex problems.

An activation function decides, based on the weighted sum of the inputs, whether and to what extent a neuron is activated. Put simply, it introduces a nonlinearity into the calculation. In the following sections, we'll take a closer look at a few of the most important activation functions, which are shown in Figure 1.9. We'll also cover the softmax activation function, which isn't shown in the figure.

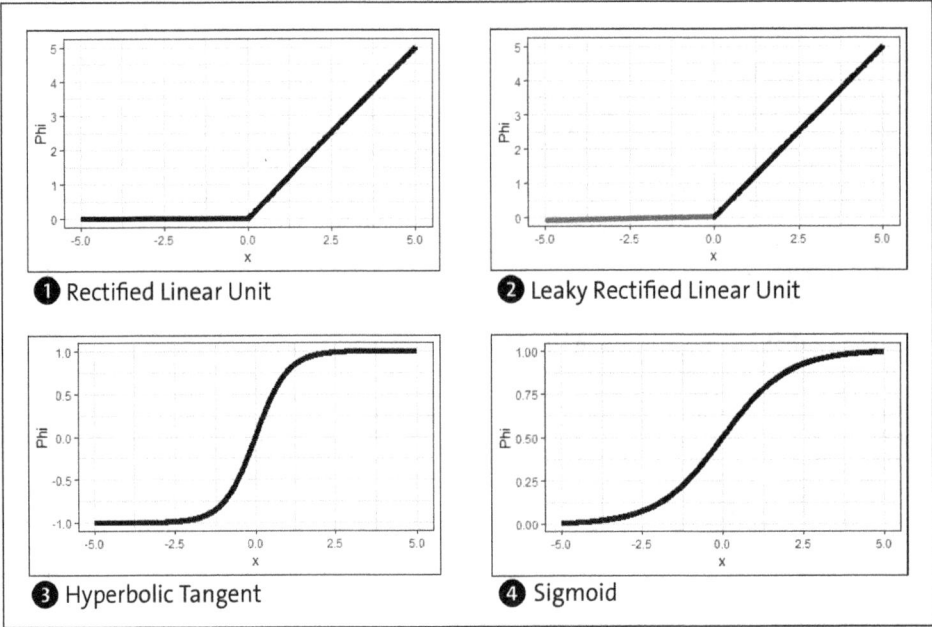

Figure 1.9 Activation Functions

Basically, the function processes the weighted sum and returns a corresponding output.

1.7.1 Rectified Linear Units

The *rectified linear unit* (ReLU) function shown in Figure 1.9 ❶ returns a value of zero for weighted sums below zero and returns the actual value if the value is positive. The mathematical instruction is simple:

$$f(x) = max(0, x)$$

A slight adaptation is the leaky rectified linear unit (Leaky ReLU), which is shown in Figure 1.9 ❷. For the negative range, a value close to zero is returned instead of a constant zero, as follows:

$$f(x) = \begin{cases} x, & if\ x > 0 \\ ax, & if\ x \leq 0 \end{cases}$$

The value of α is therefore very small, for example, 0.01. The biggest advantage is that the problem of "dying neurons" is prevented. This means that the output of the neurons is permanently zero, which means that the gradient also becomes zero and they no longer receive weight updates during training (i.e., they effectively become inactive).

In the normal ReLU function, neurons with negative input values no longer change their weight. This means that they contribute nothing to the learning process, so in special cases, *Leaky ReLU* can make training more robust and accelerate convergence.

1.7.2 Hyperbolic Tangents

The function shown in Figure 1.9 ❸ is the hyperbolic tangent (tanh). It scales the output of a neuron to the value range from –1 to +1. It's an extended version of the sigmoid function, which we'll get to know in a moment.

It has an S-shaped curve, and its outputs are symmetrically centered on the zero point. The main disadvantage of tanh is the *vanishing gradient problem*. This means that for very large or very small input values, the gradient of the function is almost zero, which means that the weights of the neurons in the earlier layers are updated very slowly or not at all during the backpropagation process.

Nowadays, tanh is rarely used; the very similar-looking sigmoid function is much more common.

1.7.3 Sigmoid

Figure 1.9 ❹ shows the sigmoid function, which also has an S-shaped curve. It scales the output values to the range from 0 to 1. Mathematically, it can be described by the following function:

$$f(x) = \frac{1}{1 + e^{-x}}$$

We'll encounter the sigmoid function again in the output layer of binary classification because it scales the output to the range from 0 to 1, and therefore, the result can be interpreted as the probability of belonging to one of two classes.

1.7.4 Softmax

The softmax activation function is mainly used in the output layer of neural networks. Its main purpose is to convert a series of arbitrary real numbers into a probability distribution in which the sum of all output values is 1 or 100%. This property is very valuable for multiclass classification (i.e., the division of the prediction into several classes, in which each class is assigned a certain percentage value). Figure 1.10 shows how the softmax activation function works.

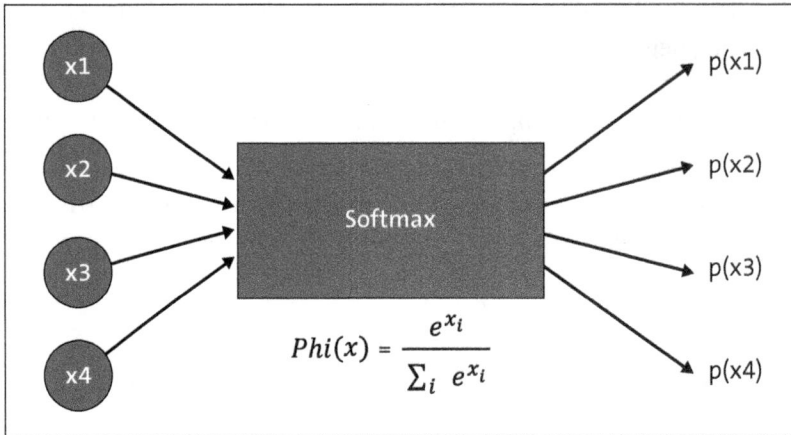

Figure 1.10 Softmax Activation Function

The neurons of the output layer are shown on the left. Their respective weight values of x1 to x4 are activated using softmax, resulting in a probability value for each neuron. The class with the highest probability is then usually selected as the class prediction.

1.8 Loss Functions

The *loss function* is another fundamental component in deep learning. It measures the error between the model prediction and the actual values. It takes the output of the neural network and the actual (correct) value as inputs and returns a single, nonnegative numerical value that reflects the quality of the prediction.

Figure 1.11 uses an example from the field of regression to illustrate what the loss function does. Observed values (depicted here as red dots) are shown in a two-dimensional coordinate system. The aim of the model is to determine a function that best describes the points, which is shown here as the black line.

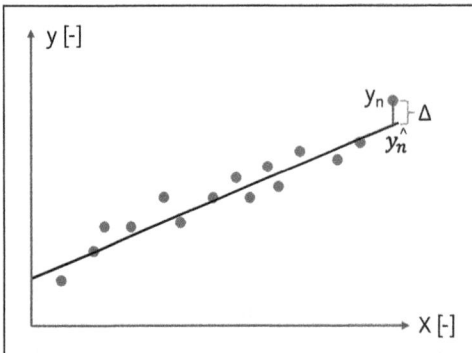

Figure 1.11 Loss Function

1

There are an infinite number of ways to draw straight lines "through the points," but only one minimizes a loss. For the model shown, the result of the prediction is the horizontal point y_n. There is an error Δ between these two points, such an error occurs for each individual point, and the errors are sometimes larger and sometimes smaller. The task of the loss function is to accumulate all errors and condense them into a single value. A low loss value means that the prediction of the model is very accurate, while a high value indicates a large error.

The main goal during training is to adjust the weights so that the loss value is minimized. It's crucial for you to choose the correct loss function, and which one you should choose depends on the specific task the model will solve. The following sections present some loss functions for regression problems, binary classification, and multiclass classification.

1.8.1 Regression Problems

The *mean squared error* (MSE) is often used for regression problems (i.e., when numerical values are to be predicted), as follows:

$$MSE = \sum_{i=1}^{N}(y_i - \hat{y_i})^2$$

The errors between the prediction and the actual value are determined, squared, and added up across all values. With this loss function, individual large errors have a very large influence on the total error, as the error is squared in each case.

This behavior isn't always desirable, and for this reason, you may want to use other loss functions such as the *mean absolute error* (MAE) and the *mean bias error* (MBE). The MAE measures the average size of the prediction error without taking the sign into account, while the MBE shows the systematic bias of the model by averaging the errors while retaining the sign.

1.8.2 Binary Classification

Binary classification normally uses the *binary cross-entropy loss* (BCE loss) function, which measures the performance of a classification model whose output is a probability between 0 and 1. The BCE "penalizes" a model more if the predicted probability is far from the actual value. Figure 1.12 shows how high the loss is for different predictions when the true value is 1.

If the prediction in this case is also 1, then the error is 0. The further away the prediction is from this, the exponentially greater the error. The situation becomes extreme with predictions that are very close to 0, where the error increases extremely sharply.

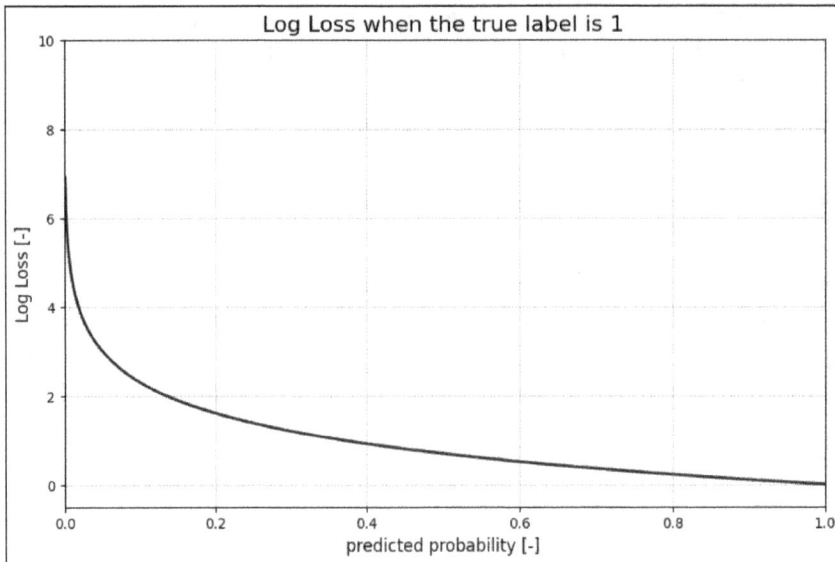

Figure 1.12 Cross-Entropy Loss

1.8.3 Multiclass Classification

Categorical cross-entropy loss (CCE loss) is usually used for multiclass classification. It's a generalization of BCE for more than two classes, and it's usually used in conjunction with the softmax activation function in the output layer. The softmax function transforms the logits (raw outputs) into a probability distribution, which ensures that all predicted class probabilities sum to one. CCE loss then quantifies the dissimilarity between the predicted probability distribution and the true one-hot encoded label. In this way, it penalizes predictions that are confident but wrong.

1.9 Optimizers and Updating Parameters

Now that we've defined the loss function, the next critical step is to minimize it, which requires us to update the model parameters. This process is governed by the optimizer, which is the engine that drives the learning. In this section, you'll learn about what optimizers are and how they update parameters.

1.9.1 Optimizers

An *optimizer* is a special algorithm that adjusts the weights of a neural network during training to minimize the error (loss). It's the real brain behind the learning process because it finds a quick way to adjust the parameters so that error minimization is exactly what happens. This is where the delta rule comes into play. The *delta rule* is a fundamental algorithm that describes how the weights in a network are gradually adjusted to minimize the difference (the so-called delta) between the desired output *y* and the actual output. To do this, it calculates the gradients in the backpropagation process.

After the training data has been sent forward through the entire network (through a forward pass and forward propagation) and passed through the layers and activations, backpropagation starts at the end and goes backward through the network from the output layer. It uses *gradients* (also known as *slopes*) for this, and you can imagine a gradient as a directional arrow in an error mountain. It indicates in which direction and how steeply the weights must be adjusted to effectively minimize the overall error and thus improve the network.

First, the gradients are calculated for the parameters of the last layer. They show how the total error changes if a certain weight in this layer changes minimally. Then, the gradients are propagated backward, step by step. Each layer uses the gradients of the following layer to calculate its own gradients.

1.9.2   Updating Parameters

Once all gradients have been determined, the optimizer uses that information to adjust the weights of the network and therefore adjust the network so that it makes smaller errors in the next run. Figure 1.13 shows an example of a gradient descent method.

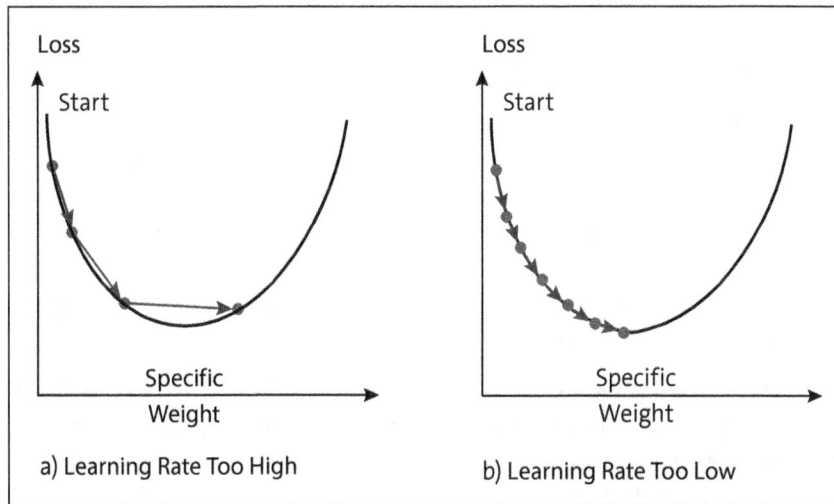

Figure 1.13 Gradient Descent

The weights are randomly initialized at the beginning of the training process, and this usually results in relatively high losses. Then, the optimizer determines how the weights need to be adjusted to achieve a slightly lower loss. The new weight ensures a small loss in the next training loop, but that's still not optimal. The optimal solution is only reached after a certain number of training runs.

How long does the training take to reach the optimal solution? It depends on many parameters, which are called *hyperparameters*. One decisive parameter here is the *learning rate*, which influences how much the weights of the network are adjusted in each training step. Figure 1.14 shows two extreme examples—for a learning rate that is too high and one that is too low.

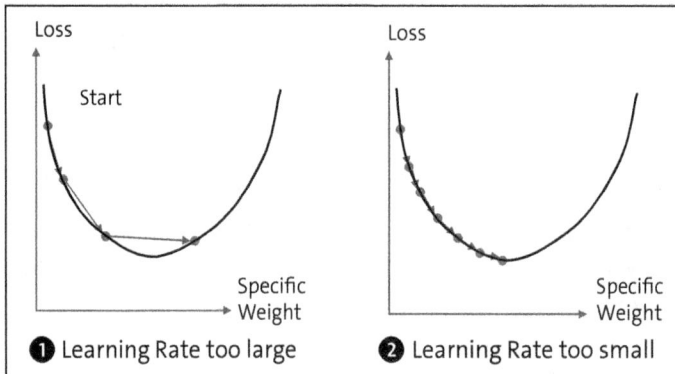

Figure 1.14 Influence of Learning Rate on Loss Development

If the learning rate is too high, as shown in ❶, large adjustments will be made to the weights in each training step. However, there is a risk that the optimizer will "skip" the global minimum of the loss function and the model will not converge well.

Conversely, if the learning rate is too low, as in ❷, the training will creep slowly toward the goal because the weights are only adjusted in homeopathic doses. The training can also get stuck at a local minimum.

You need to choose the right learning rate to implement good model training. You must set it carefully to find a balance between fast convergence and avoid unstable training. Nowadays, making the choice is easier because many optimizers use adaptive learning rates that dynamically adjust during training—for example, by making larger weight changes at the beginning of training when you are far from optimal and reducing the adjustments later when you are closer to optimal.

In terms of theory, we now know all the fundamental elements, and what's missing is the practical implementation. To do this, we need to look at the mapping of the data and the model, so the next section describes the data and the model parameters with the mathematical construct of the tensor.

1.10   Tensor Handling

Deep learning is closely related to the mathematical concept of tensors. In this section, you'll learn what tensors are and how to create them, and then, you'll implement a simple neural network and train it.

1.10.1   What are Tensors?

A *tensor* is essentially a generalization of scalars, vectors, and matrices. Figure 1.15 illustrates the concept of tensors with a few examples.

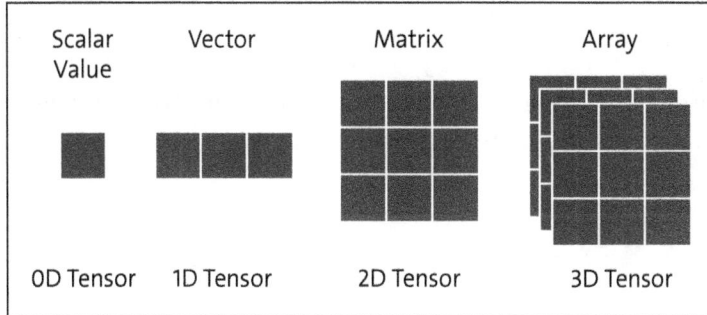

| Scalar Value | Vector | Matrix | Array |

0D Tensor   1D Tensor   2D Tensor   3D Tensor

Figure 1.15 Tensors of Various Orders

A *scalar* is a single numerical value, and it's also referred to as a *zero-order tensor* (0D tensor). A *vector* is a list of numerical values, and it's also referred to as a *first-order tensor* (1D tensor). The concept can be extended to *matrices* (2D tensors), *arrays* (3D tensors), and N-dimensional tensors with a total of N dimensions.

The size of the tensors reflects the number of elements per dimension. A tensor with the form (3, 4) has three rows and four columns. A tensor consists of many elements that all have the same data type, such as float32 or int64. In the following section, you'll learn how to create tensors.

1.10.2   Coding: Tensor Creation and Attributes

You can find the complete script for this coding exercise in *020_DeepLearning_Introduction\working_w_tensors.py*.

To start, you import the required torch and numpy packages, as follows:

```
import torch
import numpy as np
```

You can easily create tensors, for example, from a list:

```
my_list = [1, 2, 3, 4, 5]
my_tensor1 = torch.tensor(my_list)
my_tensor1
```

```
tensor([1, 2, 3, 4, 5])
```

When preparing the data, you may need to convert it from numpy arrays. You can perform this conversion of arrays into tensors with torch.from_numpy(), as follows:

```
my_data = np.array([[1, 2, 3], [4, 5, 6]])
my_tensor2 = torch.from_numpy(my_data)
my_tensor2
```

```
tensor([[1, 2, 3],
        [4, 5, 6]], dtype=torch.int32)
```

In many cases, only the dimension of the result tensor is specified and the entire tensor is filled with zeros via torch.zeros(), for example, like this:

```
shape = (3, 2)
my_tensor3 = torch.zeros(shape)
my_tensor3
```

```
tensor([[0., 0.],
        [0., 0.],
        [0., 0.]])
```

Alternatively, the tensor may be randomly initialized with random numbers via torch.rand(). This happens, for example, when the weights of a network are initialized, as follows:

```
#%% tensor from random values
my_tensor4 = torch.rand(shape)
my_tensor4
```

```
tensor([[0.2589, 0.5477],
        [0.6280, 0.0358],
        [0.9364, 0.5995]])
```

The similarities to NumPy arrays are striking. You can use the shape attribute to determine the dimension of the tensor, and each tensor consists of elements of the same data type, which you can display by using dtype.

However, a key difference is that PyTorch tensors can track gradients (using the autograd feature), a capability that NumPy does not have. With requires_grad, you can find out whether the tensor has a gradient or a gradient calculation.

The tensor is created on the CPU (central processing unit) by default. However, if it's to be part of a model training, you're better off performing the calculation on the graphics processing unit (GPU). You can determine the device where the tensor is located by using the device attribute.

It's also a good idea to determine the shape and type of the tensor, as well as its requires_grad property and the device where the tensor is located, as follows:

```
print(f"Shape of my_tensor4: {my_tensor4.shape}")
print(f"Type of my_tensor4: {my_tensor4.dtype}")
print(f"Device of my_tensor4: {my_tensor4.device}")
print(f"Gradient of my_tensor4: {my_tensor4.requires_grad}")
```

Shape of my_tensor4: torch.Size([3, 2])
Type of my_tensor4: torch.float32
Device of my_tensor4: cpu
Gradient of my_tensor4: False

If you create a tensor manually, its `requires_grad` attribute will be deactivated, as you can see in the following. You can also change this.

```
#%% set gradient to true
my_tensor4.requires_grad = True
print(f"Gradient of my_tensor4: {my_tensor4.requires_grad}")
```

Gradient of my_tensor4: True

You can access individual elements of the tensor (in a process called *indexing*) or sections of it (in a process called *slicing*) in exactly the same way you would with NumPy:

```
print("first row: ", my_tensor4[0])
print("first column: ", my_tensor4[:, 0])
print("last column: ", my_tensor4[:, -1])
```

first row: tensor([0.2589, 0.5477], grad_fn=<SelectBackward0>)
first column: tensor([0.2589, 0.6280, 0.9364], grad_fn=<SelectBackward0>)
last column: tensor([0.5477, 0.0358, 0.5995], grad_fn=<SelectBackward0>)

1.10.3   Coding: The Calculation Graph and Training

In this example, we'll determine model parameters iteratively to help you complete your understanding of the basics of model training. You'll work with very simple data that you'll create yourself. As Figure 1.16 shows, these are measuring points that represent a linear function. The model will approximate a straight line that best describes the points. The data has only one independent variable (their X values) and the resulting dependent y variable.

The connection is simple, as shown here:

$$y_{true} = w * X_{true} + b$$

The model parameters are the slope value w and the offset value b. The "real" values are $w = 3$ and $b = 2$, and the model should determine these two values.

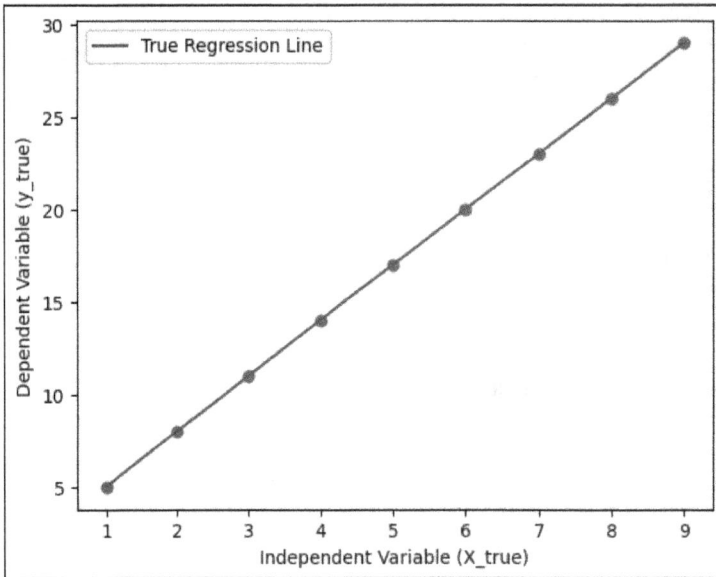

Figure 1.16 Training Data

You can find the complete script at *O20_DeepLearning_Introduction\automatic_diff.py*.

First, you load the required packages, as follows:

```
import torch
import matplotlib.pyplot as plt
```

You use torch for the modeling and matplotlib for the visualization of results.

Then, you manually create the X_true and y_true data, with X_true as your input tensor and y_true as your target variable, as follows:

```
X_true = torch.arange(1, 10, dtype=torch.float32).unsqueeze(1)
y_true = 3 * X_true + 2
```

The connection shown in Figure 1.16 has been created with the code shown in Listing 1.1.

```
plt.scatter(X_true, y_true)
plt.plot(X_true, 3 * X_true + 2, 'r-', label='True Regression Line')
plt.xlabel('Independent Variable (X_true)')
plt.ylabel('Dependent Variable (y_true)')
plt.legend()
plt.show()
```

Listing 1.1 Calculation Data

The data is processed in a calculation graph, which is shown in Figure 1.17. This graph illustrates how the data interacts with functions and parameters.

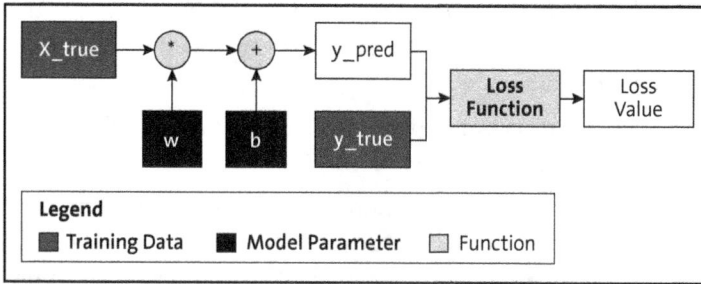

Figure 1.17 Calculation Graph

The independent characteristics X_true are multiplied by the slope value w, and then, the bias term b is added. The data is shifted from left to right by these operations, in a process known as a *forward pass*. The model prediction y_pred follows from this calculation.

The prediction is then compared with the actual value y_true in the loss function, which determines a loss value from the difference between the values.

Now, you need the model parameters w and b, both of which are initialized randomly. You should set the requires_grad parameter to True because that's the only way you'll be able to calculate the gradients later.

```
w = torch.randn(1, requires_grad=True)
b = torch.randn(1, requires_grad=True)
print(f "Weights 'w' (randomly initialized): {w.item():.4f}")
print(f "Bias 'b' (randomly initialized): {b.item():.4f}")
```

Weights 'w' (randomly initialized): -0.9095
Bias 'b' (randomly initialized): -0.0187

Later, you'll iteratively improve the model parameters, and during that process, you'll need to know the LEARNING_RATE and the number of EPOCHS. The EPOCHS parameter determines how many iterations are carried out, and the LEARNING_RATE describes how quickly parameter changes are made, as follows:

```
LEARNING_RATE = 0.01
EPOCHS = 1000
```

1. In Listing 1.2, you now come to the adjustment of the parameters in an iterative process. This process comes very close to the later training of neural networks, as you can see in the steps involved:

2. You create two empty lists for the model parameters (w_list and b_list) and another one for the loss's loss_list.

3. The actual training takes place in a loop that runs through a specified number of EPOCHS, And within the loop, you generate the y_pred predictions in a forward pass.

4. You determine the losses using mean squared error. To do this, you calculate the square distance between the prediction and the actual target value for all training data and determine the mean value.

5. You calculate the gradients in the backward pass. This step is crucial for the subsequent determination of the w and b parameters.

6. You update the parameters in the opposite direction of the gradient to minimize the loss. You perform these calculations within the torch.no_grad() scope to ensure that the updates don't become part of the computation graph. You then add the parameters of the current iteration to the corresponding list so that you can visualize progress afterward.

7. After updating the parameters, you must reset the gradients to zero or they will accumulate in the next iteration. This is the last required step in the training loop.

8. Optionally, you can output progress to the console.

```python
# 1. Initialization of parameter lists
w_list = []
b_list = []
loss_list = []
# 2. Training loop
for epoch in range(EPOCHS):

    # 3. Forward Pass
    y_pred = w * X_true + b

    # 4. Loss Calculation
    loss = torch.mean((y_pred - y_true)**2)

    # 5. Backward Pass: calculate gradients
    loss.backward()

    # 6. Parameter-Update
    with torch.no_grad():
        w -= LEARNING_RATE * w.grad
        b -= LEARNING_RATE * b.grad
        w_list.append(w.item())
        b_list.append(b.item())
        loss_list.append(loss.item())
    # Reset gradients
    w.grad.zero_()
    b.grad.zero_()

    # output progress (optional)
    if (epoch + 1) % 10 == 0:
```

```
print(f"Epoch [{epoch+1}/{EPOCHS}],
        Loss: {loss.item():.4f},
        w: {w.item():.4f}, b: {b.item():.4f}")
```

```
Epoch [10/1000], Loss: 0.2662, w: 3.1779, b: 0.8803
Epoch [20/1000], Loss: 0.2451, w: 3.1708, b: 0.925
...
Epoch [990/1000], Loss: 0.0001, w: 3.0032, b: 1.9802
Epoch [1000/1000], Loss: 0.0001, w: 3.0030, b: 1.9810
```

Listing 1.2 Iterative Adjustment of Model Parameters

When you analyze the losses and model parameters, you'll see that the parameters have come closer to the actual parameters and that the loss has decreased over time. Figure 1.18 shows the weights, bias values, and the losses over the epochs.

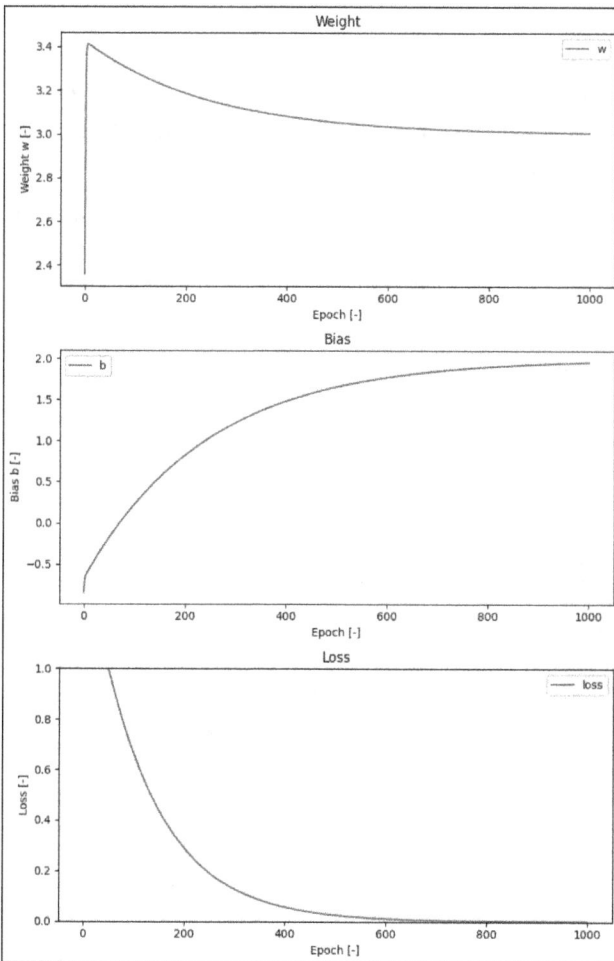

Figure 1.18 Weights and Model Parameters During Training

At this point, the model parameters have come very close to the actual values and the loss has decreased over time. This is the desired behavior and represents good training.

1.11   Summary

In this chapter, you've learned the basics of neural networks. We began by looking at the definition of deep learning and how it differs from other areas of AI. Then, we looked at the diverse areas of application of deep learning, and you learned how deep learning works by looking at the process from model creation to training and model inference.

A historical review covered the development of AI from the perceptron to today's complex networks, and we examined the perceptron in more detail.

You've dived deeply into the core components of a neural network, and you now know how a network is structured, how the neurons in the different layers are connected, and how the data flows through them.

You've also learned that activation functions play an important role in changing the data, and we also looked at loss functions that measure the quality of the network during training. Optimizers are essential for successful training as they specify how the parameters of the model need to be adjusted to improve the results.

At the end of the chapter, we looked at tensors, which are used for data representation and calculation in neural networks. We discussed what tensors are and how to work with them, and we taught you about calculation graphs, which you can use to efficiently optimize parameters.

In the next chapter, you'll train your first neural network with PyTorch.

Chapter 2
Creating Your First PyTorch Model

"Any sufficiently advanced technology is indistinguishable from magic."
—Arthur C. Clarke

For me, when I trained my first deep learning model, I felt exactly the same—like magic. You might feel the same way.

In this chapter, we lay the foundation for all subsequent chapters by teaching you how to train PyTorch models. First, we'll train a model from scratch, implementing almost every step by hand. This will help you gain a better understanding of how PyTorch works.

After that, we'll successively use more features of PyTorch to define our scripts as modularly as possible. Ideally, our script should ultimately be so modular that we can use a different dataset and train the model without having to make adjustments at various points in the code.

In Section 2.1, we'll familiarize ourselves with the data set and prepare it. In Section 2.2, we'll train our model, which we'll then improve. In Section 2.3, we'll learn how to define a model class, and in Section 2.4, we'll introduce the concept of batches. In Section 2.5, we'll abstract the dataset using the Dataset and DataLoader classes. In Section 2.6, we'll address the question of how to save and load models and their model weights so that we can avoid having to retrain the model again. In Section 2.7, we'll conclude the chapter by turning to data sampling, which is ultimately about ensuring that the model learns to generalize in order to predict not only the training data well but (in the best case) to predict any data that it has never seen before.

In general, this chapter is less about regression and more about all the trappings of model training. But before we get into our first model training, we need to familiarize you with the data set for which you'll be training the model.

2.1 Data Preparation

We'll work with a data set from Kaggle (*www.kaggle.com*), which is an online community with a special focus on data analysts and data scientists. With this platform, you can explore datasets, perform analyses, and learn from others who have already worked with the data. It's a very valuable source of knowledge.

Specifically, we'll work with the "Social Anxiety Dataset" (*https://www.kaggle.com/datasets/natezhang123/social-anxiety-dataset*), which contains more than 10,000 samples of people who have different levels of social anxiety. Each person's anxiety level is rated on a point scale from 1 to 10, and this is the target (or dependent) variable that the model will ultimately predict. Figure 2.1 shows a section of the dataset.

	Age	Gender	Occupation	Sleep Hours	Physical Activity (hrs/week)	Caffeine Intake (mg/day)	Alcohol Consumption (drinks/week)	Smoking	Family History of Anxiety	Stress Level (1-10)	Heart Rate (bpm)	Breathing Rate (breaths/min)	Sweating Level (1-5)	Dizziness	Medication	Therapy Sessions (per month)	Recent Major Life Event	Diet Quality (1-10)	Anxiety Level (1-10)
0	29	Female	Artist	6.0	2.7	181	10	Yes	No	10	114	14	4	No	Yes	1	Yes	7	5.0
1	46	Other	Nurse	6.2	5.7	200	8	Yes	Yes	1	67	23	2	Yes	No	2	No	8	3.0
2	64	Male	Other	5.0	3.7	117	4	No	Yes	1	91	28	3	No	No	1	Yes	1	1.0
3	20	Female	Scientist	5.8	2.8	360	6	Yes	No	4	86	17	3	No	No	0	No	1	2.0

Figure 2.1 Section of Dataset

You need to prepare the dataset, and before you can do that, you need to learn about different feature types (which we'll cover in Section 2.1.1) and different data types (which we'll cover in Section 2.1.2). Often, you'll need to reshape the data, especially to transform numerical data into categorical data (which we'll cover in Section 2.1.3). An important aspect of data preparation is that you need to make yourself familiar with the data in a process called exploratory data analysis (which we'll cover in Section 2.1.4). Finally, we'll cover data scaling (in Section 2.1.5), which is an important step for ensuring a stable model training.

2.1.1 Feature Types

Our dataset contains various *independent variables*, including demographic characteristics such as age, gender, and occupation. Other characteristics can fall into the areas of general health, mental indicators, and mental health.

Independent and Dependent Features

The terms *independent features* and *dependent features* refer to the roles that variables (columns) play in a dataset. This concept is primarily used in supervised learning.

Independent features are also called *input variables*, predictors, or *characteristics*, and they are the inputs for ML models. It's assumed that independent features are the causes of or influencing factors for the dependent variable. In statistics, *dependent features* are also called *target variables*, *output variables*, or *labels*, and they are the output values that are ultimately predicted by the model.

The ML model thus learns the relationships or patterns that exist between the independent and dependent variables, and after it has been trained, it can use this knowledge to predict future values of the dependent variable based on new values of the independent features.

At the beginning of each script, we load all the required packages and classes as shown in Listing 2.1. As previously mentioned, the dataset comes from Kaggle, and we can import it directly using the kagglehub in-house package. We'll import the data as a pandas dataframe, and we'll need the numpy package to later convert the data from a dataframe into a numpy array.

We'll get into the topic of scaling the data later. At this point, we load the standard scaler from the sklearn package. We always use the os package when we want to interact with operating system functions, and we use seaborn and matplotlib to visualize the data and results.

```
#%% packages
import numpy as np
import pandas as pd
import kagglehub
import os
from sklearn.preprocessing import StandardScaler
import seaborn as sns
import matplotlib.pyplot as plt
```

Listing 2.1 Data Preparation: Package Import

Kaggle provides us with an easy way to import data into Python by using the kagglehub package. In Listing 2.2, we only have to load the dataset via its id. During loading, the dataset is copied to the hard disk and the folder is returned. The file is saved in the folder, and we can then load it directly via pd.read_csv(). At that point, we will have successfully loaded the data and created the anxiety dataframe.

```
#%% Download latest version
path = kagglehub.dataset_download("natezhang123/social-anxiety-dataset")
print("Path to dataset files:", path)
#%% Data import
anxiety_file = os.path.join(path, 'enhanced_anxiety_dataset.csv')
anxiety = pd.read_csv(anxiety_file)
```

Path to dataset files: C:\Users\BertGollnick\.cache\kagglehub\datasets\nate-zhang123\social-anxiety-dataset\versions\2

Listing 2.2 Data Preparation: Data Import (Source: 030_FirstModel_Regression\DataPrep.py)

Now, let's take a look at which columns and how many rows and columns the dataset has, as shown in Listing 2.3.

```
print(f"anxiety.columns: {anxiety.columns}")
print(f"anxiety.shape: {anxiety.shape}")
```

```
anxiety.columns: Index(['Age', 'Gender', 'Occupation', 'Sleep Hours',
      'Physical Activity (hrs/week)', 'Caffeine Intake (mg/day)',
      'Alcohol Consumption (drinks/week)', 'Smoking',
      'Family History of Anxiety', 'Stress Level (1-10)',
      'Heart Rate (bpm)', 'Breathing Rate (breaths/min)',
      'Sweating Level (1-5)', 'Dizziness', 'Medication',
      'Therapy Sessions (per month)', 'Recent Major Life Event',
      'Diet Quality (1-10)', 'Anxiety Level (1-10)'],
    dtype='object')
anxiety.shape: (11000, 19)
```

Listing 2.3 Data Preparation: Checking Dataset Columns and Dimensions

The dataset comprises a total of 11,000 samples and 19 features, some of which contain text rather than numerical information. This is the case with the "Smoking" feature, for example, which has two states: "Yes" and "No."

2.1.2   Data Types

At this point, let's consider what types of data there are. There are generally two main data types: numerical data and categorical data, as follows:

- **Numerical data**
  This is also known as *quantitative data* or *metric data*, and it consists of numbers that can be measured.

- **Categorical data**
  This is also called *qualitative data* or *nominal data*, and it describes qualities or categories (like gender or occupation) that can't be measured or counted in the conventional sense. Categorical data can be further subdivided into *nominal data*, which are unordered (e.g., favorite colors) and *ordinal data*, which are categories with a natural order (e.g., academic degrees, educational qualifications).

Since PyTorch can only process numerical data, you must convert all features that contain categorical information into numerical information. You can do this with one-hot encoding.

2.1.3   One-Hot Encoding

One-hot encoding is a special technique used in machine learning to convert categorical data into a numerical format, which is the only way the data can be processed by algorithms. How does one-hot encoding work? We can illustrate the underlying concept with an example. Imagine that the favorite-color column has been entered into a data set about people (see Table 2.1).

Person	favorite_color
Bob	Yellow
Stuart	Green
Kevin	Red
Gru	Green

Table 2.1 Sample Table in Long Format

With one-hot encoding, all unique values recorded are displayed as a single column. After applying one-hot encoding, the favorite_color column is converted into as many columns as there are unique values. In our example, there are three unique values: [yellow, green, red]. These are used to create the favorite_color_yellow, favorite_color_green, and favorite_color_red columns.

These columns only contain binary information, so the numerical value is 1 if it corresponds to the favorite color and 0 if it doesn't. For each person, the 1 is then entered into the column corresponding to the favorite color. Therefore, Table 2.1 looks like Table 2.2 after one-hot encoding.

Person	favorite_color_yellow	favorite_color_green	favorite_color_red
Bob	1	0	0
Stuart	0	1	0
Kevin	0	0	1
Gru	0	1	0

Table 2.2 Sample Table in One-Hot-Encoded Format

You can even omit a column without losing information because the column is then implicitly derived from the other columns—specifically, if there are only the colors yellow, green and red and each person has exactly one favorite color.

In this form, the information is now represented numerically and is therefore suitable for use in most ML algorithms. Another advantage is that there's no implicit order. Imagine the original colors had been encoded numerically (e.g., yellow = 1, green = 2, red = 3). Then, the original form would have formally met the requirements of ML algorithms since the information would have been encoded numerically. But the algorithm would have also implicitly "assumed" an order of the colors in which green would have counted twice as much as yellow and red would have counted three times as much as yellow, which makes no sense. You can avoid such problems with one-hot encoding.

One clear disadvantage of one-hot encoding is that the number of dimensions increases. Especially when there are many different characteristics, this is reflected in many new features, which is associated with increased training time for the model and the *curse of dimensionality*—the fact that problems occur when the number of features is large compared to the number of data points.

Now, let's apply this newly learned technique to our data. Thankfully, the developers of the pandas package have made our work here very easy so that we can create one-hot coding with the pd.get_dummies method.

Listing 2.4 illustrates how one-hot encoding is implemented. In addition to the anxiety data set, several other parameters are passed. The drop_first parameter ensures that the first encoded column is omitted and dummy variables are obtained.

```
anxiety_dummies = pd.get_dummies(anxiety, drop_first=True, dtype=int)
anxiety_dummies.head()
#%% df shape
anxiety_dummies.shape
```

(11000, 31)

Listing 2.4 Data Preparation: One-Hot Encoding (Source: 030_FirstModel_Regression\Data-Prep.py)

By using this technique, we've increased the number of columns from 19 to 31. Now, we can look at the context of the data to improve our understanding.

2.1.4 Exploratory Data Analysis

In our next example, we'll look at how sleep behavior affects anxiety disorders. The corresponding code is shown in Listing 2.5.

```
sns.regplot(x='Sleep Hours', y='Anxiety Level (1-10)', data=anxiety_dummies,
color='blue', line_kws={'color': 'red'})
# add a title
plt.title('Sleep Hours vs Anxiety Level')
# add x title
plt.xlabel('Sleep Hours')
# add y title
plt.ylabel('Anxiety Level')
```

Listing 2.5 Data Preparation: Data Visualization (Source: 030_FirstModel_Regression\Data-Prep.py)

This results in the correlation shown in Figure 2.2. The data points are shown as a dot plot (with blue dots). In addition, the linear correlation between the two variables is shown as a red line.

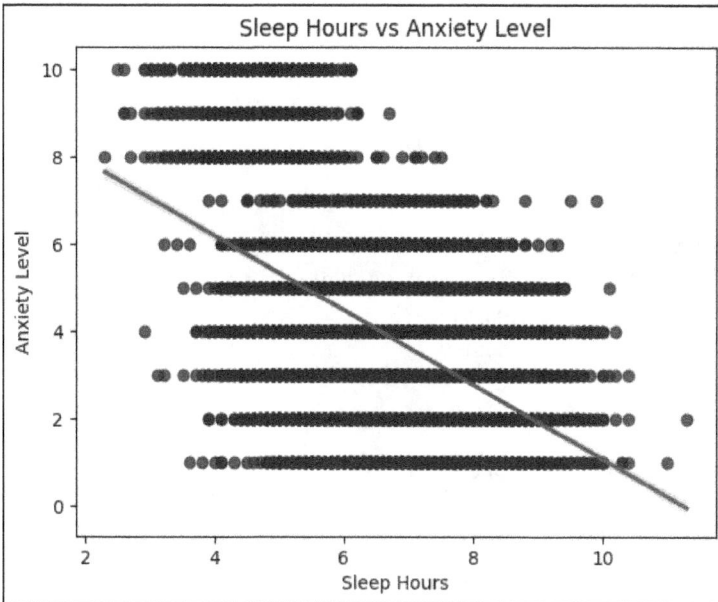

Figure 2.2 Connection Between Sleep and Anxiety Disorder

The correlation is quite clear: anxiety levels increase as hours of sleep decrease. This is just one possible connection—we have a total of 30 independent features that we could look at.

To get a quick overview, we can determine and display the correlation between the independent features and the target variable in a correlation matrix as a heat map. A *heat map* is a diagram form in which the categorical information is coded as color values. The linear correlations among all variables are determined and can then be visualized as color values.

Listing 2.6 illustrates how the correlations are determined. For the sake of clarity, only numerical features are analyzed. The filtered pandas numerical_features data frame has the corr() method, which we can used to determine the linear correlation among all features. With N columns, this results in a corr correlation matrix with the dimensions N x N.

```
#%% check correlation
# Select only numerical features for correlation analysis
numerical_features = anxiety.select_dtypes(include=['int64', 'float64'])
corr = numerical_features.corr()
```

Listing 2.6 Data Preparation: Calculation of Correlation Coefficients

Listing 2.7 shows how we can now visualize these correlations with sns.heatmap. As the matrix is symmetrical, we just need to look at the upper or lower triangle, and we can implement this by using a mask, which is then passed as a parameter to the heatmap.

This mask consists of N x N Boolean values, and it specifies which values are to be displayed.

```
# Create mask for upper triangle
mask = np.triu(np.ones_like(corr, dtype=bool))

# Plot correlation heatmap
sns.heatmap(corr, annot=False, cmap='coolwarm', vmin=-1, vmax=1, mask=mask)
plt.title('Correlation Heatmap (Numerical Features Only)', fontsize=10)
plt.xticks(rotation=45, ha='right', fontsize=8)
plt.yticks(rotation=0, ha='right', fontsize=8)
plt.tight_layout()
plt.show()
```

Listing 2.7 Data Preparation: Visualization of Correlations

Our visualization of the numerical features is shown in Figure 2.3, where the color coding ranges from –1 (blue), to 0 (gray), to +1 (red).

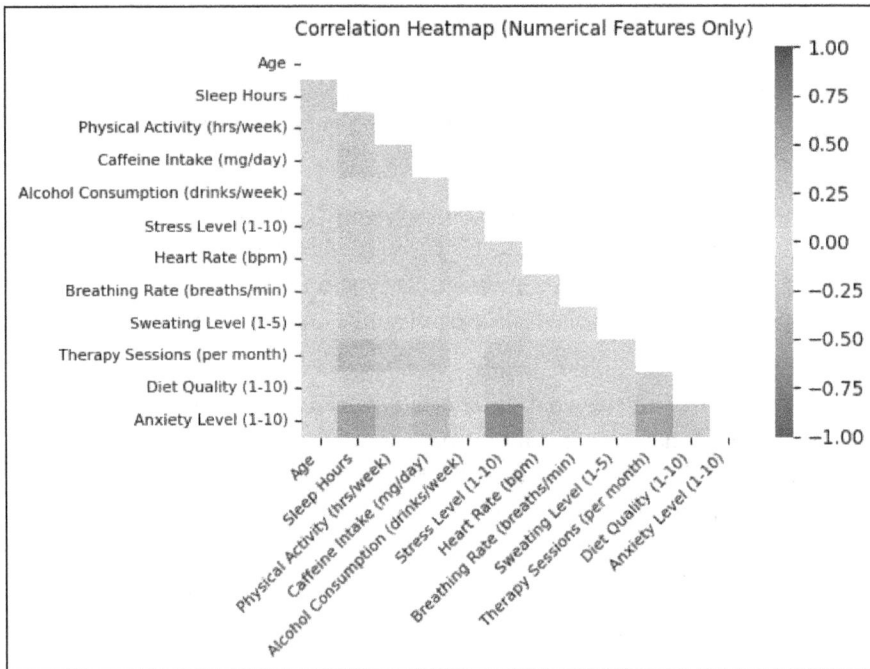

Figure 2.3 Correlation of Numerical Features

In the figure, a correlation coefficient of +1 represents the *maximum positive correlation*, which means that an increasing value of one feature is accompanied by an increasing value of the other feature. On the other hand, we can't say here that the rising value of one feature causes or results in the rising value of the other feature—that would mean that there is *causality* between the two variables. For now, it only means that there is a correlation, and we can't determine whether this correlation is causal on this basis.

Conversely, a correlation coefficient of −1 represents a *perfect-negative correlation*, which means that an increasing value of one feature is accompanied by a decreasing value of the other feature.

We're particularly interested in the correlations between our Anxiety Level target (1–10) and the descriptive features. These are shown in the last line in the figure, where it becomes clear that the Anxiety Level is strongly correlated with Sleep Hours and Stress Level.

Up to this point, we've stored the data in a pandas data frame. We now need to do two things: separate the data into independent and dependent features and then to convert it into NumPy arrays (i.e., pure number matrices). Both steps are combined in Listing 2.8. The independent features are stored in object X, and the dependent features are stored in object y. This terminology comes from mathematics and contradicts naming conventions in Python, especially in the case of the capital X, but since the terms are so common, I will follow the statistical convention at this point.

The independent features correspond to all features of the anxiety_dummies dataset, except for the column with the target variable. In contrast to this is the independent feature y, in which only the target variable is stored.

Finally, we check the output by visualizing the sizes of the objects.

```
#%% convert data to numpy array
X = np.array(anxiety_dummies.drop(
    columns=['Anxiety Level (1-10)']),
    dtype=np.float32)
y = np.array(anxiety_dummies[['Anxiety Level (1-10)']],
    dtype=np.float32)
print(f"X shape: {X.shape}, y shape: {y.shape}")
```

X shape: (11000, 30), y shape: (11000, 1)

Listing 2.8 Data Preparation: Data Conversion into NumPy Arrays (Source: 030_FirstModel_Regression\DataPrep.py)

Of the 31 original columns, we've now transferred 30 to object X and 1 to object y.

2.1.5 Data Scaling

The next step involves scaling the data. Here, we should first look at why this step is necessary at all. Data scaling plays a decisive role in the training of many models. Why? It's because raw data that varies greatly in its values can lead to problems during training.

Also, large values can cause gradients to "explode" during the backpropagation process, and that would cause the training to become unstable and even fail completely. Conversely, very small values could lead to disappearing gradients (see Chapter 1), which could also make learning unstable. So, we scale the data with the aim of transforming the values of the input features into a similar value range.

There are various ways to do this. One common method is *min-max scaling*, in which the data is usually scaled in the value range from 0 to 1. Another approach is *standardization*, which involves transforming the data so that it fluctuates around a mean value of 0 and has a standard deviation of 1. It's also important to make the scaling consistent in order to achieve comparable results.

You should only calculate the scaling parameters (the mean values and standard deviations) on the training data and only then apply them to the validation and test dataset. In this way, you can avoid data leakage. We'll come back to these aspects in Section 2.7, which is on the topic of data splitting.

We can carry out the scaling (in our case, the standardization) by using the Standard-Scaler class. First, we create an instance of the class, and then, we transfer the data to the fit_transform method so that the parameters are determined and the standardization is carried out. The final object X will contain the standardized data, as follows:

```
#%% normalize data
scaler = StandardScaler()
X = scaler.fit_transform(X)
```

At this point, we've prepared our data sufficiently and are ready to train our first model.

2.2   Model Creation

In our first model, we'll implement many details ourselves as this will help us understand the model better. For example, we'll determine the predictions of the model by using matrix multiplication, implement the model parameters ourselves, and adjust the model parameters independently. The trained model parameters, slopes, and offsets denote the two most important learnable parameters within a neuron or a linear transformation.

Later, we'll hand over these tasks more and more to the PyTorch framework. If we were to do this from the outset, many aspects of model training would remain black boxes that we wouldn't fully understand.

Finally, we'll train a model to predict the anxiety level (y) based on a variety of independent features, using the following formula:

$$y = w1 * X1 + w2 * X2 + ... + w30 * X30 + b$$

We'll start by importing and preparing the data in Section 2.2.1. Then, we'll train the model in Section 2.2.2 and evaluate the training progress in Section 2.2.3. Finally, we'll check the model predictions in Section 2.2.4.

2.2.1 Data Import

We start as usual by importing the packages in Listing 2.9. Since we're building directly on the data preparation from the previous section, we import the independent features X and the dependent feature y directly from the Dataprep script. For the creation of tensors, we also load NumPy and torch, and for visualization, we load seaborn and matplotlib. Finally, we use the value to evaluate the model, and we therefore load the r2_score function from sklearn.

```
#%% packages
from DataPrep import X, y
import torch
import numpy as np
import seaborn as sns
import matplotlib.pyplot as plt
from sklearn.metrics import r2_score
```

Listing 2.9 Model Creation: Package Import (Source: 030_FirstModel_Regression\00_LinReg-FromScratch.py)

PyTorch only works with tensors, so we first convert the NumPy array into tensors with torch.from_numpy, as follows:

```
X_tensor = torch.from_numpy(X.astype(np.float32))
y_tensor = torch.from_numpy(y.astype(np.float32))  # Ensure y is float32
```

Now that we have the data in shape, we can get started. Our regression model is ultimately described by a bias parameter and a slope parameter (slope or weight). For each feature, there is a slope parameter and a total of one bias parameter.

We need to initialize the w (weight) and b (bias) terms first, and we can implement this with torch.zeros. We should also set the requires_grad parameter to True because that's the only way to enable automatic backpropagation and training of the model.

```
# Initialize weights with smaller values to prevent exploding gradients
w = torch.zeros(X.shape[1], 1, requires_grad=True, dtype=torch.float32)
b = torch.zeros(1, requires_grad=True, dtype=torch.float32)
print(f"w shape: {w.shape}, b shape: {b.shape}")
```

w shape: torch.Size([30, 1]), b shape: torch.Size([1])

2.2.2   Model Training

The training process is influenced by a number of parameters, and the most important ones are the number of epochs and the learning rate. Before we start the training, let's take a closer look at these two important parameters.

Let's start with the epochs. Our training dataset has 11,000 samples, and these are usually transferred to the model in smaller chunks called *batches*. We'll come back to this concept later. Once all the samples have been used once to adjust the weights of the model, the epoch is complete, and the process is then repeated so that the model can "see" the same data many times to learn from it. Typically, a model is trained for several epochs and the patterns in the data are captured better with each successive epoch.

Now that we have an understanding of the concept of epochs, let's clarify what the learning rate is all about. We can imagine model training as the search for the deepest point in an unknown valley. Say that a hiker descending from a mountaintop is blindfolded and must slowly feel his way down. They can decide whether their steps should be long or short, and the length of each step corresponds to the learning rate.

With long stride lengths (high learning rates), the hiker quickly covers a large area, but they could also quickly pass the lowest point and start climbing up the neighboring mountain. Conversely, they could choose to take very small steps (which correspond to low learning rates) in order to move very carefully. In that case, the hiker would have a high probability of finding the lowest point exactly and not going beyond it, but they could take a relatively long time to get there. However, they could also get stuck in a small, shallow puddle on the valley floor. The hiker could conclude that he has reached the deepest point, stop moving, and get stuck, even though the deepest point of the valley is still further away. The technical term for such an issue is a *local minimum*, which is in strong contrast to the deepest point (*global minimum*).

The learning rate therefore defines how quickly or carefully the blindfolded hiker explores the valley of the error to find the optimum point.

Now, we can really get started and define our two parameters:

```
EPOCHS = 100
LEARNING_RATE = 0.01
```

This brings us to the actual core of the model training—the *training loop*, which is implemented in Listing 2.10. The data is shown to the model 100 times, and this is implemented with a for loop via the EPOCHS. We can then assess how well the model is learning by studying the losses, which are extracted in each epoch and added to the loss_list.

The loop always runs through the same steps, as follows:

1. The predictions are generated in the *forward pass*. Here, the independent features are multiplied by the model weights and the activation functions are applied.

2. These predictions are compared with the correct results, and the loss is calculated. There are various *loss functions* for this (as we learned in the previous chapter), and for regression models, the *mean squared error loss* (MSE loss) is a good choice.

3. Now, we can calculate all the gradients by executing `loss.backward()`.

4. These gradients are now used to update the model weights. The learning rate is multiplied by the gradients, and this correction is subtracted from the previous model weight.

5. Before the next epoch starts, we must reset the gradients to zero to prevent them from adding up and distorting the result.

6. The loss value of the current epoch is added to the total list of all losses.

7. To check the model training, we output the current epoch and the current loss.

```python
loss_list = []
for epoch in range(EPOCHS):
    # 1. Forward pass
    y_predict = torch.matmul(X_tensor, w) + b

    # 2. Calculate loss (MSE)
    loss = torch.nn.functional.mse_loss(y_predict, y_tensor)

    # 3. Backward pass
    loss.backward()

    # 4. Update weights and biases
    with torch.no_grad():
        w -= LEARNING_RATE * w.grad
        b -= LEARNING_RATE * b.grad
        # 5. Zero gradients after using them
        w.grad.zero_()
        b.grad.zero_()

    # 6. Store loss for plotting
    loss_list.append(loss.item())

    # 7. Print loss for this epoch
    print(f"Epoch {epoch}, Loss: {loss.item():.4f}")
```

```
Epoch 0, Loss: 19.9446
Epoch 1, Loss: 19.0938
...
Epoch 98, Loss: 1.5858
Epoch 99, Loss: 1.5732
```

Listing 2.10 Model Creation: Training Loop (Source: 030_FirstModel_Regression\00_LinReg-FromScratch.py)

This shows us how the training is progressing, and we can see that the losses are getting smaller.

2.2.3 Model Evaluation

Now, we can visualize the training losses again in a graphic. The corresponding code is shown in Listing 2.11, and the losses stored in the loss_list are shown as a line diagram above the number of EPOCHS. To do this, we use seaborn with the sns.lineplot function.

```
#%% plot loss
sns.lineplot(x=range(EPOCHS), y=loss_list)
plt.title('Loss over Epochs')
plt.xlabel('Epoch [-]')
plt.ylabel('Loss [-]')
```

Listing 2.11 Model Creation: Loss Visualization (Source: 030_FirstModel_Regression\00_Lin-RegFromScratch.py)

Figure 2.4 shows the result of the model training. The loss decreases continuously with each subsequent epoch, but you can also see here that although the model has fewer losses, the losses are asymptotically approaching a limit. We'll return to the question of the optimum training duration at a later point.

The picture rarely looks as "clean" as it does here. There's usually more fluctuation (i.e., periods when losses rise again slightly for a short time before returning to the longer-term falling trend).

Next, let's take a closer look at the model weights (the slope values [w] and offset value [b]), as follows:

```
#%% check results
print(f"Weights: {w.detach().numpy().flatten()}, Bias: {b.item()}")
```

```
Weights: [-0.09916524 -0.5292779  -0.16298231  0.34272027  0.06862636
...
Bias: 3.408252716064453
```

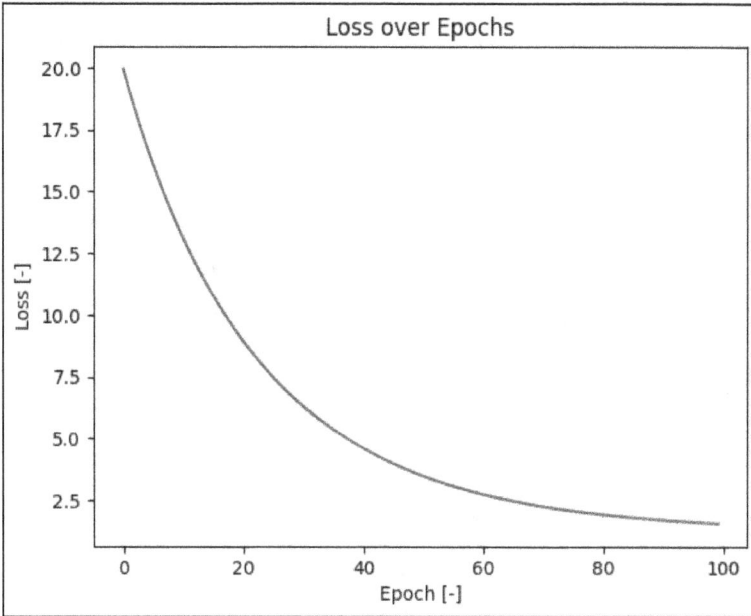

Figure 2.4 Our First Model: Losses and Epochs

2.2.4   Model Inference

Finally, in this simple case, we can use these values and perform the calculation based on the formula of the regression we provided at the start of Section 2.2. We achieve this by multiplying the independent features by the model weights. We need to perform the calculation within the scope of torch.no_grad() to prevent gradients from being calculated. It's also important to note that we're performing *model inference* (i.e., testing the model) here, not model training. In model inference, we don't want to perform any operations that could influence the network but should not be part of the training process. The positive side effects are that this saves resources (such as memory and computing time) and ensures that certain operations are not incorrectly included in the gradient calculation.

```
# %%
with torch.no_grad():
    y_pred = (torch.matmul(X_tensor, w) + b).detach().numpy().flatten()
```

We've calculated the predictions y_pred and displayed them in connection with the real values y, and Listing 2.12 shows the corresponding code. We use the sns.regplot() function to create a scatterplot with a superimposed regression line. The data points are displayed in blue with a transparency value of 0.1, and that value ensures that the points are displayed in a bluer color in areas where many values lie on top of each other and the points in areas with very few points are displayed in a faint blue color. In addition, the regression line is displayed as a red line.

```
# %% visualize correlation
sns.regplot(x=y_pred, y=y, color='red',
            scatter_kws={'s': 10,
                         'color': 'blue',
                         'alpha': 0.1})
plt.title('Predicted Anxiety Level vs Actual Anxiety Level')
plt.xlabel('Predicted Anxiety Level [-]')
plt.ylabel('Actual Anxiety Level [-]')
```

Listing 2.12 Model Creation: Correlation Visualization (Source: 030_FirstModel_Regression\ 00_LinRegFromScratch.py)

The result is a correlation diagram and can be seen in Figure 2.5. The actual anxiety level is plotted above the predicted anxiety level.

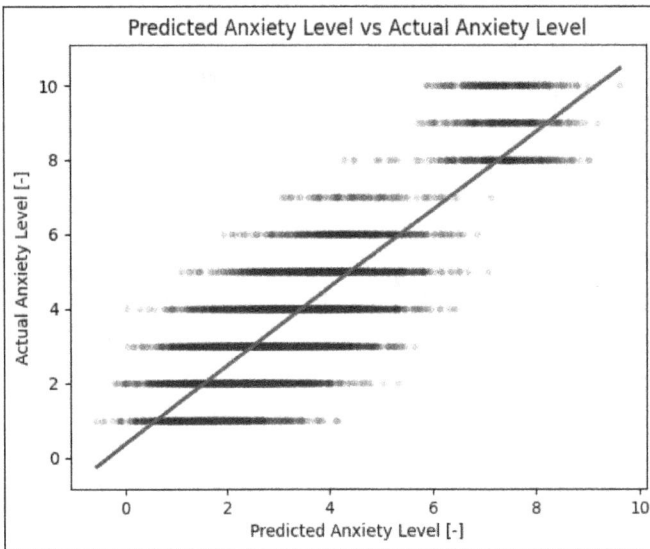

Figure 2.5 Our First Model: Predicted and Actual Anxiety Level

The correlation is positive, and on average, an anxiety level of 5 is also predicted as 5. But of course, there is scatter in the data, so in some cases, values between 1 and 7 are predicted. This illustration gives you a good overview of the areas in which the model works well and where improvements may still be necessary.

However, you'll often want to compare different models with each other, and this is easier if you summarize the model quality as a single numerical value. In the field of regression models, the R^2 value is a frequently used measure. The R^2 value (also known as the *coefficient of determination*) is a statistical indicator that shows how well the independent features in a regression model match the variance (or dispersion) of the dependent variable. The value range is generally between 0 and 1 or 0% and 100%. We can understand the extreme values as follows:

- $R^2 = 0$

  The trained model can't explain the dependent variable at all, and there's no linear relationship between the independent variables and the dependent variable. This means that while there may well be a relationship between the variables, it's simply nonlinear.

- $R^2 = 1$

  The model is perfectly able to explain the entire variability of the dependent variable. Note that this is almost never the case in practice, as there are always measurement inaccuracies, random errors, or other independent variables that weren't considered in the model. As a result, you should always treat a very high value as a red flag that may indicate overfitting of the model.

If the R^2 value is 0.75, for example, you can interpret it as meaning that 75% of the variance of the dependent variable can be explained by the independent variables contained in the model. The remaining 25% of the variance is due to other factors not included in the model or random errors.

In general, a higher R^2 value reflects a better fit of the model to the data.

It's also extremely important to note at this point that a high R^2 value doesn't necessarily mean that there is a causal relationship, meaning the independent variables do not automatically have a causal influence on the dependent variable. The high value merely shows a strong statistical correlation between the variables.

The calculation is performed using the r2_score function from sklearn, the real and predicted values are passed to the function, and a single numerical value is obtained, as follows:

```
r2 = r2_score(y_true=y,
              y_pred=y_pred)
print(f"R-squared: {r2:.2f}")
```

R-squared: 0.65

Our first model achieves an R^2 value of 0.65, and now, we can use this value as a benchmark to compare our model with other models.

Whether an R^2 value is considered good or bad depends heavily on the context. In certain cases, an R^2 of 0.98 is considered poor, and in other cases, an R^2 of 0.4 is considered very good. But for now, we're satisfied with the result, and we want to further improve our model training by using PyTorch's capabilities to make our code more modular and therefore more versatile.

In the next section, we'll learn how to define the model in a separate class and how to embed the optimizer in the training.

2.3 The Model Class and the Optimizer

Defining the model in a separate class makes the code much easier to adapt later, and if we want to use the model later, we also need to be able to save and load it very easily at that time. For these reasons, it helps to save the model in a separate class.

To do this, we start by importing the required packages again as shown in Listing 2.13.

```
#%% packages
from DataPrep import X, y
import torch
import numpy as np
import seaborn as sns
import matplotlib.pyplot as plt
from sklearn.metrics import r2_score
```

Listing 2.13 Model Class: Package Import (Source: 030_FirstModel_Regression\10_ModelClass.py)

It's also a best practice to define the constant training parameters, which are referred to as the hyperparameters and are bundled at the beginning of the script. After we've loaded the packages, we need to define the hyperparameters as follows:

```
EPOCHS = 100
LEARNING_RATE = 0.1
```

Hyperparameters

Hyperparameters are constants (i.e., variables that are not changed during the program run). They are defined before the training process and used to influence it, they have a major influence on model training, and they are defined manually by the model developers (us) on the basis of certain optimization procedures or best practices (empirical values).

Hyperparameters control the behavior of the learning process and have an enormous influence on the subsequent efficiency and performance of the model. The most frequently used hyperparameters are the learning rate, batch size, and number of epochs. The parameters depend heavily on the data used and the model selected, so hyperparameter optimization is an important step during model training.

We need to convert the data into tensors because all objects must be available as tensors during training, and it's also important to adjust the data type if necessary. We'll usually use Float32 here, as follows:

```
X_tensor = torch.from_numpy(X)
y_tensor = torch.from_numpy(y.astype(np.float32))
```

At this point, we come to the *model class*, which is a class defined by us that is derived from the torch.nn.Module class. Together with the optimizer, the model class represents the blueprint for a model. On this basis, we'll later create a model instance that we'll use during training.

The model class must have two methods: __init__() and forward():

- **__init__**
  This is called once during the creation of a model instance. Ideally, we want to create a model class that we can use flexibly and therefore adapt to the dataset or other parameters. For this reason, we pass the input_size and output_size parameters to the __init__() method, which allows us to adapt the model to the number of independent features in the dataset and the number of features to be predicted. Usually, we'll create the network layers we want to use at this point as well. In our example, we're using a linear layer that directly connects the input features to the next layer—which in this case is the output layer.

- **forward()**
  This method describes how the network is structured and how data flows through the network. In addition to the class-specific self parameter, this function receives the x parameter, which is sometimes referred to as input or data. Here, x refers to what is entered into the model, which is the data that the model will process. The function then describes how this data is further processed (i.e., how it's passed from layer to layer). In this simple example, we only use a linear layer with self.linear(x) and the result then overwrites the value of x and is returned as the result of the function.

Listing 2.14 shows the definition of the model class.

```
#%% Model class
class LinearRegression(torch.nn.Module):
    def __init__(self, input_size, output_size):
        super(LinearRegression, self).__init__()
        self.linear = torch.nn.Linear(input_size, output_size)

    def forward(self, x):
        x = self.linear(x)
        return x
```

Listing 2.14 Model Creation: Model Definition (Source: 030_FirstModel_Regression\10_ModelClass.py)

Once we've created the model class, we can create an instance of the class. To do this, we pass the dimensions that result from our dataset. We can determine the number of independent features by using the number of columns in the dataset with X.shape[1]. As only one target value is to be predicted, the output_size is 1, as follows:

```
#%% Model instance
model = LinearRegression(input_size=X.shape[1],
                         output_size=1)
```

In addition to the model instance, we need the optimizer and the loss function as important elements of the network training. We'll use Adam as the optimizer at this point, and the most common loss function for regression tasks is the MSE loss and is called via torch.nn.MSELoss():

```
#%% Optimizer and loss function
optimizer = torch.optim.Adam(model.parameters(), lr=LEARNING_RATE)
loss_fn = torch.nn.MSELoss()
```

At this point, we have prepared everything and can start the training. Listing 2.15 shows how we first initialize the loss_list as an empty list. During the iteration over the epoch, we go through the same steps as before:

1. In forward pass predictions, we create y_predict.
2. We calculate loss list loss based on y_predict and actual y_tensor values. It's important to make sure that the dimensions of the two objects are always identical.
3. In the backward pass, we determine all gradients with loss.backward().
4. Finally, we update model weights with optimizer.step().
5. To avoid a distortion of the result, we need to use optimizer.zero_grad() to reset gradients to zero.
6. (Optionally) For later evaluation of how the model training continuously improves, we can add the losses of the current loss.item() epoch to the total list of losses. To temporarily store the actual calculated value, we can call the item() method, which will return a numerical value rather than a tensor.

```
loss_list = []
for epoch in range(EPOCHS):
    # 1. Forward pass
    y_predict = model(X_tensor)

    # 2. Calculate loss (MSE)
    loss = loss_fn(y_predict.squeeze(), y_tensor)

    # 3. Backward pass
    loss.backward()

    # 4. Update weights and biases
    optimizer.step()

    # 5. zero gradients
```

```
optimizer.zero_grad()

# 6. Store loss for plotting
loss_list.append(loss.item())

# 7. Print loss for this epoch
print(f"Epoch {epoch}, Loss: {loss.item():.4f}")
```

Listing 2.15 Model Creation: Training Loop (Source: 030_FirstModel_Regression\10_ModelClass.py)

Once we've trained the model, we can use the losses to get an idea of how well the model training went. For this purpose, we can plot losses over the epochs as shown in Listing 2.16.

```
sns.lineplot(x=range(EPOCHS), y=loss_list)
plt.title('Loss over Epochs')
plt.xlabel('Epoch [-]')
plt.ylabel('Loss [-]')
```

Listing 2.16 Model Creation: Loss Visualization (Source: 030_FirstModel_Regression\10_ModelClass.py)

The result of the training can be seen in Figure 2.6. The losses decrease sharply in the initial epochs and then level off around epoch 40. The model can still learn, but it benefits less from further runs, as can be seen from the flattening of the losses.

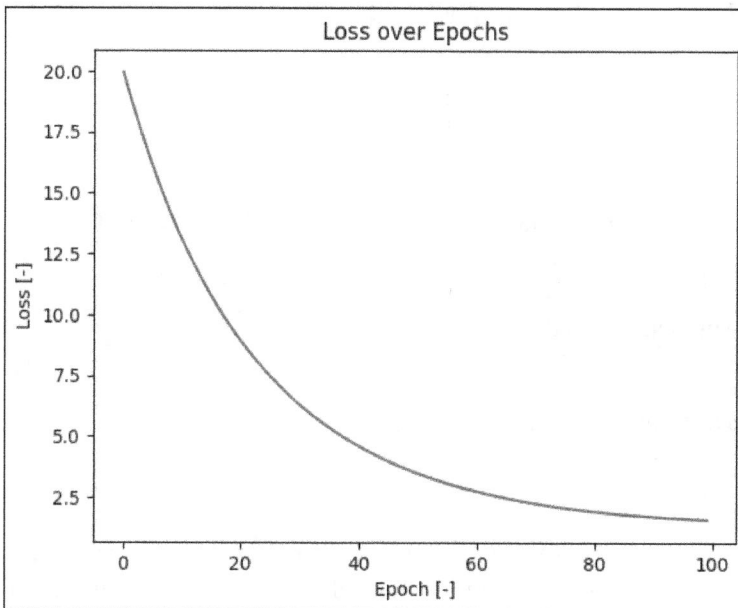

Figure 2.6 Model Class and Optimizer: Development of Losses

Now that we've introduced the model class and the optimizer, we can improve our system even further. One improvement concerns the transfer of training data to the model during training. If the data is not passed in its entirety but in smaller chunks, the chunks are called batches, and we'll deal with them in detail in the following section.

2.4   Batches

Batches are integral parts of any training and can also influence the performance of the training, and they are also hyperparameters that can be used for training. As batches can also influence the results, you'll often investigate different batch sizes and their influence.

In this section, we'll clarify what batches are, what advantages they offer, which batch sizes you should use, and how to implement them (using a practical example).

2.4.1   What Are Batches?

First, we should clarify what batches are exactly. The basic concept is illustrated in Figure 2.7. Up to now, we've transferred the entire dataset to the model in each epoch, but when training with batches, we divide the entire dataset into smaller chunks that we then transfer to the model.

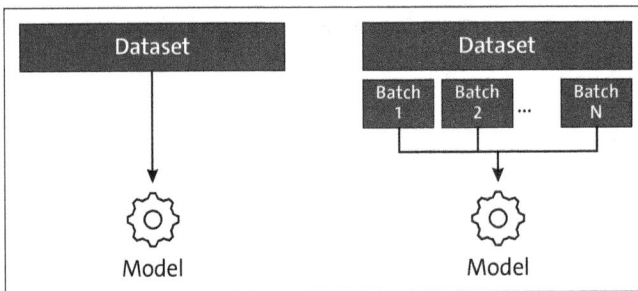

Figure 2.7 Batches: Left = Situation Without Batches; Right = Dataset Separation into Different Batches

Let's look at why we use batches in the next section.

2.4.2   Advantages of Using Batches

We use batches for the following reasons:

- **Memory**
  There are often technical factors that require the use of batches. Due to the size of the dataset, it may not be possible to process the whole thing in one go because it doesn't

fit completely into the RAM or graphics card memory. This is often the case with large image or video datasets, for example.

- **Parallelization**
  We can accelerate training by using batches and then running them in parallel on modern hardware such as GPUs.

- **Improved learning**
  During training, it's important for the model to learn *generalization*, meaning it should learn to not only predict the "known" training data well but also make good predictions for completely unknown data. Using smaller and medium batch sizes can help slow down model overfitting. The stronger fluctuations in the gradients can bring the model out of local minima and thus lead to a better generalization capability on unknown data.

Let's assume that we want to use batches. The next question is: which batch size should we use?

2.4.3 Optimal Batch Size

There's no universal best batch size—it strongly depends on parameters such as the dataset used, the model, and the available hardware resources. Nevertheless, there are some best practice recommendations, as follows:

- The batch sizes that are typically used are powers of 2 (e.g., 16, 32, 64).
- As a rule of thumb, start with moderate batch sizes such as 32 or 64 and increase them gradually if computing resources allow and it improves model performance.
- If the model is more complex, you may need to use smaller batch sizes.
- There are side effects related to batch size and learning rate. Larger batch sizes can benefit from higher learning rates, so it may be important to optimize both parameters.
- In the field of computer vision, such as for classifying images, batch sizes between 32 and 512 are often used.
- In speech processing, batch sizes from 8 to 256 are used. Smaller batches are preferred, especially for frequently used transformer models, as the models themselves have high memory requirements.
- In the end, you usually must test several batch sizes to determine the optimal batch size for your use case.

In the following section, we look at how to practically embed batches in training.

2.4.4 Coding: Implementation of Batches

In Listing 2.17, the packages and the prepared data *X*, *y* from our previous script data preparation are loaded and then converted into tensors X_tensor and y_tensor.

```
import numpy as np
from sklearn.preprocessing import StandardScaler
import seaborn as sns
import matplotlib.pyplot as plt
import kagglehub
import torch
from sklearn.metrics import r2_score
from DataPrep import X, y


#%% convert to tensor
X_tensor = torch.from_numpy(X.astype(np.float32))
y_tensor = torch.from_numpy(y.astype(np.float32))  # Ensure y is float32
```

Listing 2.17 Using Batches: Data Preparation (Source: 030_FirstModel_Regression\ 20_Batches.py)

Training is influenced by several hyperparameters, including EPOCHS, LEARNING_RATE, and the newly added BATCH_SIZE, as follows:

```
#%% Hyperparameters
EPOCHS = 100
LEARNING_RATE = 0.01
BATCH_SIZE = 512
```

In Listing 2.18, the model class, optimizer, and loss function are defined. There are no adjustments to previous scripts here.

```
#%% Model class
class LinearRegression(torch.nn.Module):
    def __init__(self, input_size, output_size):
        super(LinearRegression, self).__init__()
        self.linear = torch.nn.Linear(input_size, output_size)

    def forward(self, x):
        x = self.linear(x)
        return x

#%% Model instance
model = LinearRegression(X.shape[1], 1)
```

```
#%% Loss function
criterion = torch.nn.MSELoss()

#%% Optimizer
optimizer = torch.optim.Adam(model.parameters(), lr=LEARNING_RATE)
```

Listing 2.18 Using Batches: Model Class, Loss Function, and Optimizer
(Source: 030_FirstModel_Regression\20_Batches.py)

Listing 2.19 shows the training loop with the necessary adjustments for processing batches. The decisive factor is that there is now an additional loop in the epoch loop for processing the individual batches (comment 1). The dataset is iterated through in constant steps of the size of the BATCH_SIZE.

The next adjustment (comment 2) concerns the data. We must extract the corresponding independent X_batch features and dependent y_batch features for the respective batch items. From then on, the training follows the usual pattern.

```
loss_list = []
for epoch in range(EPOCHS):
    epoch_loss = 0
    for i in range(0, len(X_tensor), BATCH_SIZE):    # (1)
        # (2) get batch
        X_batch = X_tensor[i:i+BATCH_SIZE]
        y_batch = y_tensor[i:i+BATCH_SIZE].unsqueeze(1)

        # forward pass
        y_predict = model(X_batch)

        # calculate loss
        loss = criterion(y_predict, y_batch)

        # backward pass
        loss.backward()

        # update weights and biases
        optimizer.step()

        # zero gradients
        optimizer.zero_grad()

        # Store loss for plotting
        epoch_loss += loss.item()
```

```
    # Print loss for this epoch
    print(f"Epoch {epoch}, Loss: {epoch_loss/len(X_tensor):.4f}")
    loss_list.append(epoch_loss)
```

Listing 2.19 Using Batches: Training Loop (Source: 030_FirstModel_Regression\
20_Batches.py)

We've successfully integrated the concept of batches into the training, but there's still room for improvement. For example, we currently access the data directly within the training loop and must "manually" take care of iterating in batches. Other auxiliary functions that can help us here are Dataset and DataLoader. We'll look at these concepts in the next section.

2.5   Coding: Implementation of Dataset and DataLoader

Dataset and DataLoader are PyTorch classes that are used as abstractions of the data. Dataset provides you with an interface for accessing individual data in the dataset. This ensures that the logic for loading, preprocessing, and retrieving the data is encapsulated, which makes it possible to separate the data access logic from the model training. Also, this separation of the logic makes our scripts much more flexible and easier to adapt to new data. The DataLoader then wraps the Dataset and provides an iterable over the dataset, among other features like batch handling, shuffling, and multiprocess data loading.

As we've already mentioned, consistent access to the data and the separation of the data from the model logic are the main advantages of this concept. It also makes it possible to easily customize specific data structures. Adaptation can also include data transformations such as *data augmentation*, which involves artificially increasing the number and variety of training data. We'll come back to this topic in Chapter 4. For now, let's look at how to implement these classes.

We'll extend the script we've created so far. The Dataset and DataLoader classes are provided via torch, and the other packages have already been covered and are listed in Listing 2.20.

```
#%% packages
import numpy as np
import pandas as pd
import os
import torch
from torch.utils.data import Dataset, DataLoader
from sklearn.metrics import r2_score
from DataPrep import X, y
import seaborn as sns
```

```
#%% Hyperparameters
EPOCHS = 50
LEARNING_RATE = 0.1
BATCH_SIZE = 512
```

Listing 2.20 Dataset and DataLoader: Data Preparation (Source: 030_FirstModel_Regression\
30_DatasetDataLoaders.py)

First, we implement our own `AnxietyDataset` dataset class. Our custom class inherits from the `Dataset` class, and we must define three methods in this class:

- **__init__()**
  This constructor of the class is called when a new instance of the class is created. It's used to load file paths and metadata that are needed for later access. Transformations are also applied to the data here.

- **__len__()**
  This method determines the total number of data items, and the `DataLoader` uses this method to determine the number of data points in the dataset.

- **__getitem__()**
  This is the most important of the three methods. It's called to retrieve individual data points using an index.

Look at the practical example in Listing 2.21. In the __init__ method, the independent features X and the dependent feature y are passed, and the corresponding properties of the class are created. We can then access these via `self.X` and `self.y`. At this point, the data type of y is converted to float using `astype()` and the __len__ method is used to determine the number of data records via the `len` function.

In addition to the `self` class object, the __getitem__ function requires an `idx` index, which is used to access individual data points. This method returns the independent and dependent features at the position of the index.

```
#%% Dataset class
class AnxietyDataset(Dataset):
    def __init__(self, X, y):
        self.X = X
        self.y = y.astype(np.float32)

    def __len__(self):
        return len(self.X)

    def __getitem__(self, idx):
        return self.X[idx], self.y[idx]
```

Listing 2.21 Dataset and DataLoader: Dataset Class (Source: 030_FirstModel_Regression\
30_DatasetDataLoaders.py)

Dataset and DataLoader always occur in pairs. DataLoader is the partner of the dataset, which ensures that the model is always optimally supplied with finished data.

The DataLoader takes care of the batching. In addition, you can easily shuffle the data before each training epoch to prevent the model from always seeing the data in the same order and thus adapting to certain patterns that may only arise due to the order of the data. This can make the model more robust and achieve better generalizability.

The DataLoader can also be configured so that the data is loaded in the background. This ensures that the fast GPU is always fully utilized. The loading and preprocessing of the data (e.g., *augmentation, normalization, batch compilation*) usually takes place on the slower CPU and is therefore often slower than the actual model calculation. This can turn the CPU into a bottleneck.

We previously had to integrate many of these steps manually into the training code, but we can now simply delegate them to the DataLoader. We can perform the actual creation of the model instances of our dataset with little effort, and we can create the dataset instance with our previously defined AnxietyDataset class.

We can obtain the dataloader instance using the DataLoader class, which receives the dataset as the most important parameter. Other important parameters are the BATCH_ SIZE and the shuffle parameter, which allows the data to be shuffled.

```
#%% DataLoader
dataset = AnxietyDataset(X, y)
dataloader = DataLoader(dataset, batch_size=BATCH_SIZE, shuffle=True)
```

Listing 2.22 shows the model class, the loss function, and the optimizer. There are no surprises or adjustments here, compared to previous implementations.

```
#%% Model class
class LinearRegression(torch.nn.Module):
    def __init__(self, input_size, output_size):
        super(LinearRegression, self).__init__()
        self.linear = torch.nn.Linear(input_size, output_size)

    def forward(self, x):
        x = self.linear(x)
        return x

#%% Model instance
model = LinearRegression(X.shape[1], 1)

#%% Loss function
loss_fun = torch.nn.MSELoss()
```

```
#%% Optimizer
optimizer = torch.optim.Adam(model.parameters(), lr=LEARNING_RATE)
```

Listing 2.22 Dataset and DataLoader: Model Class, Loss Function, and Optimizer (Source: 030_FirstModel_Regression\30_DatasetDataLoaders.py)

In Listing 2.23, you can see the training loop and one of the major advantages of using the approach with Dataset and DataLoader. There is no longer any direct access to the original dataset. Instead, X_batch and y_batch are extracted directly from the dataloader.

```
#%%
loss_list = []
for epoch in range(EPOCHS):
    epoch_loss = 0
    for i, (X_batch, y_batch) in enumerate(dataloader):
        # forward pass
        y_predict = model(X_batch)

        # calculate loss
        loss = loss_fun(y_predict, y_batch.reshape(-1, 1))

        # backward pass
        loss.backward()

        # update weights and biases
        optimizer.step()

        # zero gradients
        optimizer.zero_grad()

        # Store loss for plotting
        epoch_loss += loss.item()

    # Print loss for this epoch
    print(f"Epoch {epoch}, Loss: {epoch_loss}")
    loss_list.append(epoch_loss)
```

Listing 2.23 Dataset and DataLoader: Model Training (Source: 030_FirstModel_Regression\30_DatasetDataLoaders.py)

This approach is advantageous in that we don't need to make any changes to the training loop when adjusting the dataset. We'll also encounter the code presented here with the outer loop for the epochs, the inner loop for the batches, and the individual elements for model training very often in this form.

At this point, we've familiarized ourselves with the various elements of model training, and what we're still missing is the ability to save and load models. We don't want to have to retrain a model every time we use it, so instead, we want to load and deploy trained networks with little effort.

2.6 Loading and Saving a Model

To save the trained model, we must first familiarize ourselves with the process for doing so. Only the model weights are saved, and Figure 2.8 illustrates this process using an analogy. Imagine that you've built a house and now want to rebuild it at a different location. You can't move the whole house to the new location, so instead, you must take the house apart brick by brick and then formulate a detailed construction plan that describes how to rebuild it.

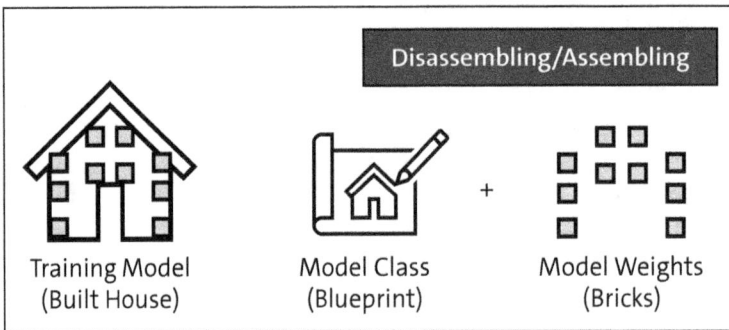

Figure 2.8 Procedure for Saving and Loading Model

The procedure for saving and loading a model is very similar. The construction plan is already known—it's the previously created model class—and the bricks in our analogy correspond to the model parameters. The two are saved separately and linked together again when loading.

2.6.1 Saving Model Parameters

We have access to the model parameters via the state_dict() method of the model. This is an ordered dictionary in which tuples reflect the respective layer name and the associated model weights, as follows:

```
model.state_dict()
```

```
OrderedDict([('linear.weight',
            tensor([[-1.3479e-01, -5.1887e-01, -1.8175e-01, ... ]])),
            ('linear.bias', tensor([3.9327]))])
```

We can then save this state dictionary in a file using `torch.save()`. We usually use the *.pt* or *.pth* file extension for such a weight file, and the code shows how the model weights are saved in the *Model1.pth* file:

```
torch.save(model.state_dict(), 'models/Model1.pth')
```

Now that we know how to save models, let's look at how we can restore them.

2.6.2   Loading a Model

This step has two stages. In the first step, we create an instance of the model class then load the model weights into the model. We can save the model class in a separate file. Listing 2.24 shows how to define your own model class.

```
import torch


#%% Model class
class LinearRegression(torch.nn.Module):
    def __init__(self, input_size, output_size):
        super(LinearRegression, self).__init__()
        self.linear = torch.nn.Linear(input_size, output_size)


    def forward(self, x):
        x = self.linear(x)
        return x
```

Listing 2.24 Model Saving and Loading: Model Class (Source: 030_FirstModel_Regression\models\Model1.py)

Now, let's move on to the actual script, which is based on this model class. In Listing 2.25, we start by loading all packages. Our model class is outsourced to a `models/Model1.py` script so that we can load the `LinearRegression` model class from there. We then create an instance of the model.

```
import torch
from models.Model1 import LinearRegression
import seaborn as sns
import matplotlib.pyplot as plt
# %% create model instance
model = LinearRegression(37, 1)
```

Listing 2.25 Model Saving and Loading: Packages and Model Instance (Source: 030_FirstModel_Regression\40_ModelLoading.py)

We want to use the code in Listing 2.26 to determine whether the model weights have been loaded successfully. To do this, we create a show_model_parameter function that displays the model weights as a histogram.

```
def show_model_parameters(model):
    params = []
    for param in model.parameters():
        params.append(param.detach().numpy().flatten())
    g = sns.histplot(params, kde=True)
    # add title
    g.set_title('Model Parameter Distribution')
    g.set_xlabel('Parameter Value')
    g.set_ylabel('Frequency')
    return g
```

Listing 2.26 Model Saving and Loading: Visualization Function (Source: 030_FirstModel_Regression\40_ModelLoading.py)

We then use the following function to show the model weights directly after instantiating the model and then after loading the model weights, as depicted in Figure 2.9:

```
show_model_parameters(model)
#%% load model weights
model.load_state_dict(torch.load('models/Model1.pth'))
```

<All keys matched successfully>
```
show_model_parameters(model)
```

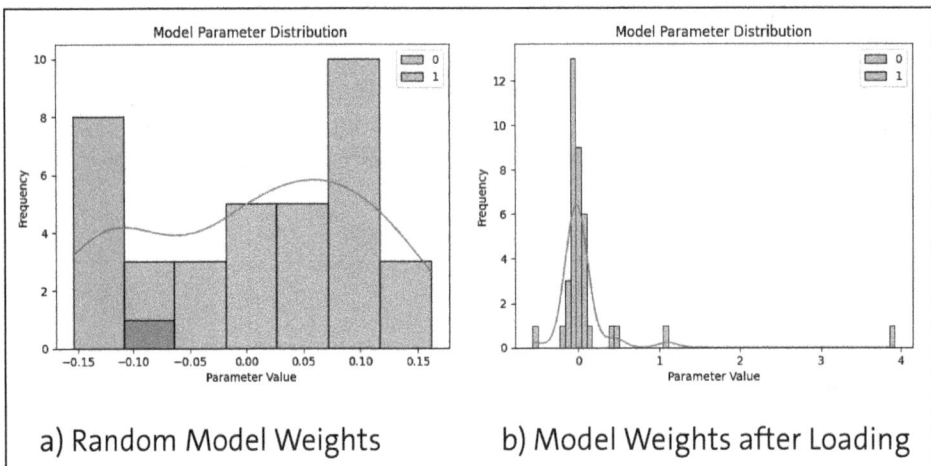

a) Random Model Weights   b) Model Weights after Loading

Figure 2.9 Model Weights: Left = Random Weights, Right = Weights after Loading

On the left is the state immediately after the model instance is created, when the weights are all around zero. On the right, you can see that the model weights cover a much larger range and the bias value (in gray) is almost four. The loading of the model parameters has therefore worked.

This brings us to the next extension of our training, where we'll learn how to split data and why you should do it.

2.7 Data Sampling

In this section, we'll look at the next concept that is important for model training: data sampling. First, you'll first learn exactly what data sampling means and why you need this approach. Then, you'll learn how to implement it.

2.7.1 What Is Data Sampling?

Data sampling is the process of selecting data points from a larger dataset. Instead of training the model based on the complete dataset, only a subset is used for training and another is used for validating the data. This is the concept of *train-test split*, as illustrated in Figure 2.10.

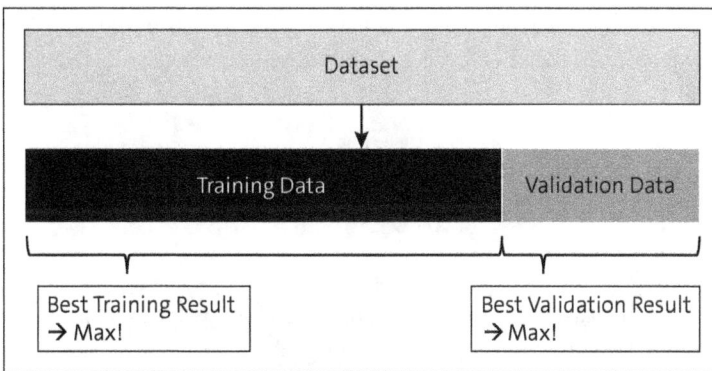

Figure 2.10 Illustration of Train-Test-Split Concept

The original data is divided into two (or sometimes three) separate parts, as follows:

- **Training dataset**
  We use this, the largest part of the dataset, to train the model.

- **Validation dataset**
  We use the validation dataset, which is usually the smaller part of the dataset, to evaluate the performance of the trained model with unseen data and to check how well it generalizes. The model should not only work well on the training data but above all on new, as yet unknown data.

■ **Testing dataset**
Sometimes, there's also a third dataset, called the test dataset. We use this dataset only at the very end to evaluate the final, unbiased performance of the trained and validated model. The model must never "see" this data during training and validation. Figure 2.10 also shows that it's advantageous to make the training dataset as large as possible, but this becomes problematic because the same applies to the test (or validation) dataset. How can we overcome this conflict of objectives and find the optimum ratio of training to validation data? That's the subject of the following info box.

Optimal Distribution Ratio of Training to Validation Data

There's no universally valid "perfect" ratio, as it depends on various factors. The exact structure depends on whether there's a two-way split (between training and validation data) or a three-way split (among training, validation, and test data). As a rough guide, you can assume a split of 80% training data and 20% validation data or 70% training data, 15% validation data, and 15% test data.

In the following, we assume a two-way split. The following factors influence distribution:

■ **Size of the dataset**
An important parameter is the size of the dataset. The larger the dataset, the larger the training dataset can be in percentage terms because the absolute number of data points in the validation dataset will still be large enough to derive statistically significant statements.

■ **Model complexity**
The more complex a model is, the more training data it tends to require in order to avoid overfitting. Validation data is very important here for monitoring the generalization capability.

■ **Time series data**
Time series data is a special case because it must not be randomly split to maintain the chronological order. We address this topic in Chapter 8.

■ **Imbalanced data**
Imbalanced data is (rather obviously) also referred to as *imbalanced datasets*, which are data for classification problems in which certain classes are strongly underrepresented. We must therefore take care to ensure that the class distributions in the various datasets are similar.

The cross-validation presented in the next section is one way of dispensing with data sampling.

2.7.2 Cross-Validation

Cross-validation is a highly recommended technique, especially for small datasets. You'll mainly use this method to determine stable model performance and extend the data sampling described earlier. You'll proceed as follows:

1. Divide the entire dataset into equal parts (folds) of size K.

2. Train the model K times, using a different fold as the validation dataset in each iteration and using the remaining $K–1$ folds as the training dataset.

3. Determine the metrics for evaluating model performance for each model.

4. Average these metrics to obtain a robust assessment of the overall model performance.

The process of splitting the data is illustrated in Figure 2.11.

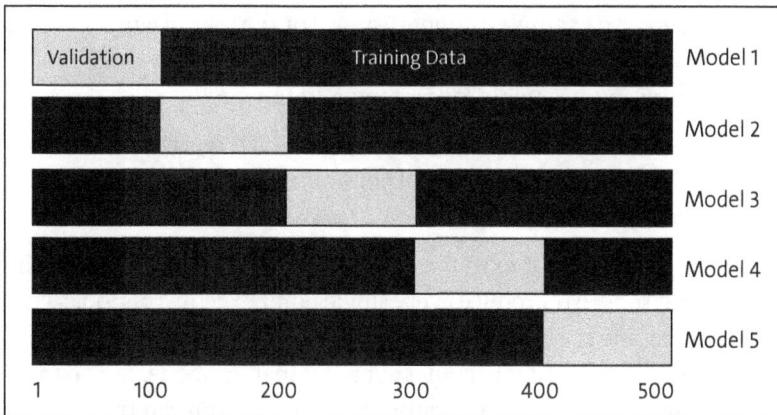

Figure 2.11 Cross-Validation

Normally, we don't define the validation data "all in one piece" but select it at random, but we show it all at once here to simplify matters. The usual values for the number of folds are 5 or 10. This method has the advantage that each data point is used for both training and validation.

One disadvantage, of course, is that the computational effort increases significantly. Whereas previously, we only had to train and validate one model, now, we must train and validate the model K times. For this latter reason, we don't use this method further in this book, but it's important for you to know how to use it so you can if you need to.

2.7.3 Why Is Data Sampling Needed?

The aim of data sampling is to ensure the model's ability to generalize. This helps it avoid overfitting, which is what happens when a model learns the training data too precisely. It should understand the basic correlations, but in addition, it will learn the noise and other specific characteristics by heart.

Figure 2.12 shows examples of an underadjusted and an overadjusted model.

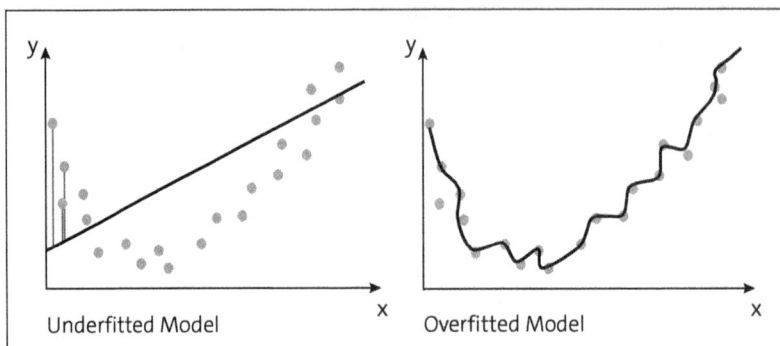

Figure 2.12 Underfitted and Overfitted Models

The left-hand panel presents an example of underfitting. For the sake of simplicity, it's a regression problem with an independent variable X and the dependent variable y. The point cloud is slightly noisy, but it roughly follows a quadratic trend.

A linear model was trained on the left-hand partial image, which reflects the relationship via a straight line. The errors of the individual points are shown as vertical lines (in red), and the errors are relatively high.

The right-hand panel shows the other extreme, in which the model is adapted to the training data in a way that tries to reproduce every single point exactly. This looks very good with the training data, and the error between the model (the blue line) and the data point is minimal. However, in practice, it becomes apparent that the model can't handle data it hasn't seen before and therefore generalizes very poorly. You can think of the model like a student who simply memorizes answers without understanding the underlying context. The student can answer all the old questions perfectly, but as soon as a question is slightly modified, they can't answer the question.

Therefore, a good model finds a happy medium between underadjustment and overadjustment.

2.7.4   Coding: Separation of Training and Validation Data

In this example, we again expand our existing code to include training and validation data. Listing 2.27 shows all the required packages, and the data is imported from the DataPrep script.

```
import numpy as np
import pandas as pd
import os
import matplotlib.pyplot as plt
```

```
import torch
from torch.utils.data import Dataset, DataLoader
from DataPrep import X, y
from sklearn.model_selection import train_test_split
from sklearn.preprocessing import StandardScaler
from sklearn.metrics import r2_score
import seaborn as sns
```

Listing 2.27 Train-Test-Split: Packages and Data Import (Source: 030_FirstModel_Regression\ 50_DataSplitting.py)

The hyperparameters found in Listing 2.28 include the maximum number of EPOCHS, the LEARNING_RATE, and the BATCH_SIZE.

```
EPOCHS = 20
LEARNING_RATE = 0.1
BATCH_SIZE = 512
```

Listing 2.28 Train-Test-Split: Hyperparameters (Source: 030_FirstModel_Regression\50_ DataSplitting.py)

Now, we come to the data split approach we have discussed. In Listing 2.29, we split the data into training data and validation data by using the train_test_split function. We use 80% for the training data and 20% for the test data.

```
#%% split data
X_train, X_val, y_train, y_val = train_test_split(X, y,
    test_size=0.2,
    random_state=42)
```

Listing 2.29 Train-Test-Split: Data Split (source: 030_FirstModel_Regression\ 50_DataSplitting.py)

It's common and almost always advisable to scale the data, and Listing 2.30 shows how we do it. We scale the independent training features X_train, and we also determine the scaling parameters in this step and save them in the scaler object.

An interesting detail here is that we adjust the validation data X_val based on the scaling parameters of the training data. The background to this is that we must assume we know only the training data and its parameters. If you were to scale the validation data to your own distribution values, it would imply that you have information about the validation data that you should not have in practice. The impact of this may usually be small, but we suggest following this best practice. Obtain the adjusted validation data by applying the scaler's transform method to the data.

```
scaler = StandardScaler()
X_train = scaler.fit_transform(X_train)
X_val = scaler.transform(X_val)
```

Listing 2.30 Train-Test-Split: Data Scaling (Source: 030_FirstModel_Regression\
50_DataSplitting.py)

There are no changes to the dataset class shown in Listing 2.31, compared with previous sections.

```
#%% Dataset class
class AnxietyDataset(Dataset):
    def __init__(self, X, y):
        self.X = torch.from_numpy(X.astype(np.float32))
        self.y = torch.from_numpy(y.astype(np.float32))

    def __len__(self):
        return len(self.X)

    def __getitem__(self, idx):
        return self.X[idx], self.y[idx]
```

Listing 2.31 Train-Test-Split: Dataset Class (Source: 030_FirstModel_Regression\
50_DataSplitting.py)

The DataLoader is where things get exciting again. As there are two datasets, we instantiate two corresponding DataLoaders (train_dataloader and val_dataloader) in Listing 2.32.

```
#%% DataLoader
train_dataset = AnxietyDataset(X_train, y_train)
train_dataloader = DataLoader(train_dataset,
                              batch_size=BATCH_SIZE,
                              shuffle=True)

val_dataset = AnxietyDataset(X_val, y_val)
val_dataloader = DataLoader(val_dataset,
                            batch_size=BATCH_SIZE,
                            shuffle=False)
```

Listing 2.32 Train-Test-Split: DataLoader (Source: 030_FirstModel_Regression\
50_DataSplitting.py)

In Listing 2.33, we create the model class and thus the model instance.

```
#%% Model class
class LinearRegression(torch.nn.Module):
    def __init__(self, input_size, output_size):
        super(LinearRegression, self).__init__()
        self.linear = torch.nn.Linear(input_size, output_size)

    def forward(self, x):
        x = self.linear(x)
        return x

#%% Model instance
model = LinearRegression(input_size=train_dataset.X.shape[1],
                         output_size=1)
```

Listing 2.33 Train-Test-Split: Model Class and Model Instance
(Source: 030_FirstModel_Regression\50_DataSplitting.py)

As can be seen in Listing 2.34, the loss function is MSELoss and the optimizer is based on Adam.

```
#%% Loss function
loss_fun = torch.nn.MSELoss()

#%% Optimizer
optimizer = torch.optim.Adam(model.parameters(), lr=LEARNING_RATE)
```

Listing 2.34 Train-Test-Split: Loss Function and Optimizer (Source: 030_FirstModel_Regression\50_DataSplitting.py)

Now, we're ready to train the model in Listing 2.35. The key differences from previous trainings are as follows:

1. We consider the two different datasets by tracking their losses separately. For this purpose, we create the epoch_loss_train and epoch_loss_val lists.

2. At the end of each epoch, we determine the losses of the validation data. We use the torch.no_grad() scope for this, and we calculate the prediction with a forward pass and then determine the loss value.

```
# (1) Empty lists initialized
loss_train_list, loss_val_list = [], []
for epoch in range(EPOCHS):
    epoch_loss_train = 0
    epoch_loss_val = 0
    for i, (X_train_batch, y_train_batch) in enumerate(train_dataloader):
        # get batch
```

```
        # forward pass
        y_pred_train = model(X_train_batch)

        # calculate loss
        loss_train = loss_fun(y_pred_train, y_train_batch.reshape(-1,
1)).mean()

        # backward pass
        loss_train.backward()

        # update weights and biases
        optimizer.step()

        # zero gradients
        optimizer.zero_grad()

        # Store loss for plotting
        epoch_loss_train += loss_train.item()

    # (2) evaluate on test set
    with torch.no_grad():
        for X_val_batch, y_val_batch in val_dataloader:
            y_pred_val = model(X_val_batch)
            loss_val = loss_fun(y_pred_val,
                            y_val_batch.reshape(-1, 1)).mean()
            epoch_loss_val += loss_val.item()
    # Store the losses for plotting
    loss_train_list.append(epoch_loss_train / len(train_dataloader))
    loss_val_list.append(epoch_loss_val / len(val_dataloader))

    # Print loss for this epoch
    print(f"Epoch {epoch}, Train Loss: {epoch_loss_train}, Test Loss: {loss_
val.item()}")
```

Listing 2.35 Train-Test-Split: Model Training (Source: 030_FirstModel_Regression\
50_DataSplitting.py)

In Listing 2.36, you can now see the visualization of the losses based on the training and validation losses. We use most of the code to scale the data so that the losses are limited to the range between 0 and 1. Otherwise, the losses could be so different that they would be difficult to recognize.

```
#%% plot loss
# Convert to numpy arrays
loss_train_arr = np.array(loss_train_list)
loss_val_arr = np.array(loss_val_list)

# Train loss: scale independently
train_min = loss_train_arr.min()
train_max = loss_train_arr.max()
train_range = train_max - train_min if train_max > train_min else 1
loss_train_scaled = (loss_train_arr - train_min) / train_range

# Val loss: scale independently
val_min = loss_val_arr.min()
val_max = loss_val_arr.max()
val_range = val_max - val_min if val_max > val_min else 1
loss_val_scaled = (loss_val_arr - val_min) / val_range

sns.lineplot(x=range(EPOCHS), y=loss_train_scaled, color='blue', label='Train')
sns.lineplot(x=range(EPOCHS), y=loss_val_scaled, color='red', label='Valida-
tion')
plt.title('Losses over Epochs: Train (blue) vs. Validation (red)')
plt.xlabel('Epoch [-]')
plt.ylabel('Loss [-]')
plt.legend()
plt.show()
```

Listing 2.36 Train-Test-Split: Visualization of Training and Validation Losses
(Source: 030_FirstModel_Regression\50_DataSplitting.py)

The results of this effort are shown in Figure 2.13. The training (blue) and validation (red) losses follow the same trend. Initially, they decrease very sharply, and then, they converge asymptotically towards zero.

The fact that both curves are almost congruent is rather unusual and is due to the specific dataset with which we trained the model.

At this point, we've covered everything related to what data sampling is and how to implement it.

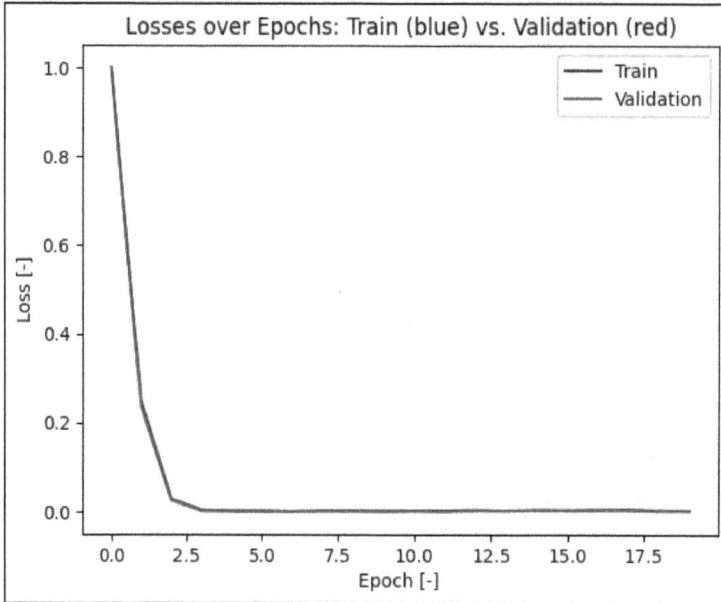

Figure 2.13 Training and Validation Losses

2.8 Summary

This chapter introduced the training of deep learning models using PyTorch. After you learned how to prepare the data in Section 2.1, we began the actual training in Section 2.2 with data import, model training, and model evaluation. This first model training delivered a functional model but still left plenty of room for improvement, and we devoted ourselves to that in the subsequent sections. We introduced the model class and the optimizer in Section 2.3, and we dealt with batches, what they are, and how and why to implement them in Section 2.4. We introduced another abstraction in Section 2.5 when we learned about Dataset and DataLoader. With these concepts, we can separate the data from the actual model training, which makes our code more modular and therefore easier to extend and maintain.

Since we usually train models to use them afterwards, we need to know how to save them and then load them again, and that was the focus of Section 2.6. Finally, we divided the data into training and validation data in Section 2.7. This step, called data sampling, ensures that the model learns to generalize to work well not only with the training data but also with unknown data.

With this knowledge in hand, we're now equipped to learn about other model architectures. The concepts we learned in this chapter will guide us through the rest of the book.

Chapter 3
Classification Models

"Science is the systematic classification of experience."
—George Henry Lewes

Science is about learning from experience, observations, and experiments. We record these systematically to help us learn from them. We can find, analyze, and classify patterns in the data, and in this respect, there is a great deal of similarity between science and machine learning.

Sorting or classifying things is one of the most fundamental and widely used tasks in machine learning. We encounter classification when assessing whether an email is spam or not, diagnosing diseases, and recognizing objects in images, to name just a few examples.

In this chapter, we'll first look at different classification types in Section 3.1. There are various methods for checking the results of a classification algorithm, and we can display the predictions with the actual class assignments by using a confusion matrix, which we'll explore in Section 3.2. A confusion matrix is useful for evaluating the results of a model, but the ROC curve is more suitable for comparing several models, and we'll discuss it in detail in Section 3.3.

We'll then begin the practical implementation of classification models. We'll get to know an implementation of binary classification that deals with the prediction of two states in Section 3.4. The implementation of multiclass classification, in which the model learns to distinguish among three or more different categories, is somewhat broader, and we'll explore it in Section 3.5.

We'll start with an introduction to the different classification types.

3.1 Classification Types

The simplest way to perform classification is to divide the data into two different classes. This is referred to as *binary classification*. Imagine that you're dealing with the classification of pictures. The possible classes into which you can divide the images "tree" and "house," so the model only recognizes these two categories. Examples of binary classification are shown in Figure 3.1.

Figure 3.1 Binary Classification

As models don't work with terms such as "tree" and "house" but do work with numerical target values, we must code these categories. In the example, "tree" is always coded with class 0 and "house" is always coded with class 1.

This works very well for images in which only one of the two categories is displayed, but it becomes problematic as soon as several categories are displayed in the image. This can be seen in Figure 3.1 in the image on the right, which shows a tree and a house. The creator of the dataset has the problem of having to choose one of the two categories and the model, which can only predict one of the two classes.

The solution here could be multilabel classification. In this case, we can assign each image to several classes. For example, we can code the image shown on the right as [0, 1], which means "both the tree class tree and the house class are in the image."

This concept is, of course, not limited to two different categories. We can define any number of categories. This can be seen clearly in Figure 3.2, in which another class, "Street," is coded. We're now dealing with three different classes: tree = 0, house = 1, and street = 2. Of course, the principle isn't limited to three classes—we can easily extend it to N classes.

Figure 3.2 Multiclass and Multilabel Classification

If we assign each image to exactly one class, we'll be dealing with multiclass classification again. If, on the other hand, we can assign each image to one or more classes, then that's called *multilabel classification*.

3.2   Confusion Matrix

Before we train an initial model, we need to know how to distinguish a good classifier from a bad one. This is where the *confusion matrix* comes into play. We use it to compare predictions with real values, and it basically has four different fields. As an example, let's take a binary classification model that has been trained to predict tsunamis. We can classify the predictions as follows:

- **True positive (TP)**
  The model didn't predict a tsunami and was right.

- **True negative (TN)**
  The model correctly predicted that a certain event was not a tsunami, and it actually represents a situation in which no tsunami occurred.

- **False positive (FP)**
  The model predicted that a tsunami would occur, and it didn't. This is also referred to as a *false alarm*.

- **False negative (FN)**
  The model predicted that there would be no tsunamis, overlooking a situation in which such an extreme event actually occurred.

We can assign all predictions to one of the four classes, and all values will be aggregated and displayed in a matrix. This matrix, shown in Figure 3.3, is the confusion matrix.

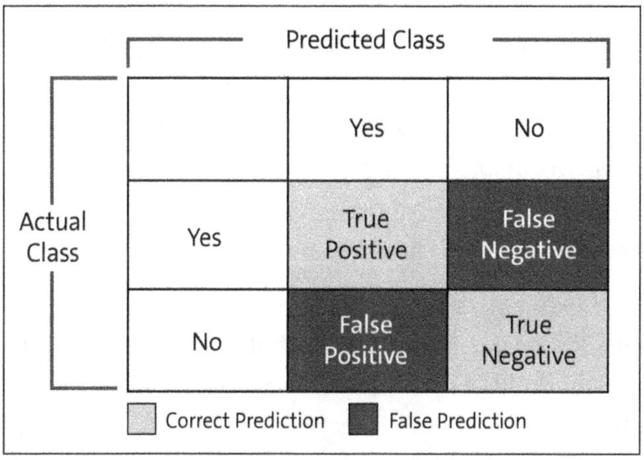

Figure 3.3 Confusion Matrix

In this matrix, the values of the actual and predicted classes are divided into the four fields, and we can then use these four values to derive further measures, such as accuracy, precision, and sensitivity.

In the case of *accuracy*, for example, all correct predictions (*TP* + *TN*) are set in relation to all predictions. The formula reads accordingly:

$$Accuracy = \frac{TP + TN}{TP + FP + FN + TN}$$

Precision is calculated as the ratio of true positive predictions to the total number of positive predictions. A high precision value (close to 1) means that the model produces only a few FPs when it makes a positive classification. The positive predictions of the model are therefore very reliable:

$$Precision = \frac{TP}{TP + FP}$$

Recall is calculated as the ratio of true positive predictions to the total number of actual positive cases. A high recall value (close to 1) means that the model makes few FN predictions, which means that it only overlooks a few of the cases that are actually positive:

$$Recall = \frac{TP}{TP + FN}$$

We've neglected one detail up to this point: the threshold value. For example, the model provides us with a probability that a tsunami will or won't occur, but only when we apply a threshold value can we derive a clear prediction class. If we code the classes as 0 (no tsunami will occur) and 1 (a tsunami will occur), we can derive classes from the prediction probabilities.

Let's assume that the prediction value is 0.65. The class assignment will depend on the threshold value, as follows:

- If the threshold value is 0.5, then every value above 0.5 is classified as class 1 and every value below 0.5 is classified as 0. In our example, class 1 would be predicted.

- This would change if the threshold value were 0.8. In that case, the predicted value of 0.65 would be below the threshold value and class 0 would be predicted.

Figure 3.4 shows vertically the predictions of 10 data points for the two classes (0 or 1). A threshold value is shown with a dashed line Above the threshold value, the points are assigned to class 1, and below the threshold value, the points are assigned to class 0. The actual classes (0 and 1) are plotted on the horizontal axis.

In the table at the bottom of Figure 3.4, the actual class, the class predicted by the model, and the assigned group resulting from the two values are entered for each data point.

Class 1 was predicted for data point 1, although the actual class is 0. There was therefore an incorrect assignment—specifically, an FP.

In the next step, the points of all groups are counted and entered into the confusion matrix. The result is shown in Figure 3.5.

Point Number	1	2	3	4	5	6	7	8	9	10
Actual Class	0	0	0	0	0	0	1	1	1	1
Predicted Class	1	1	1	0	0	0	1	1	0	0
Assigned Group	FP	FP	FP	TN	TN	TN	TP	TP	FN	FN

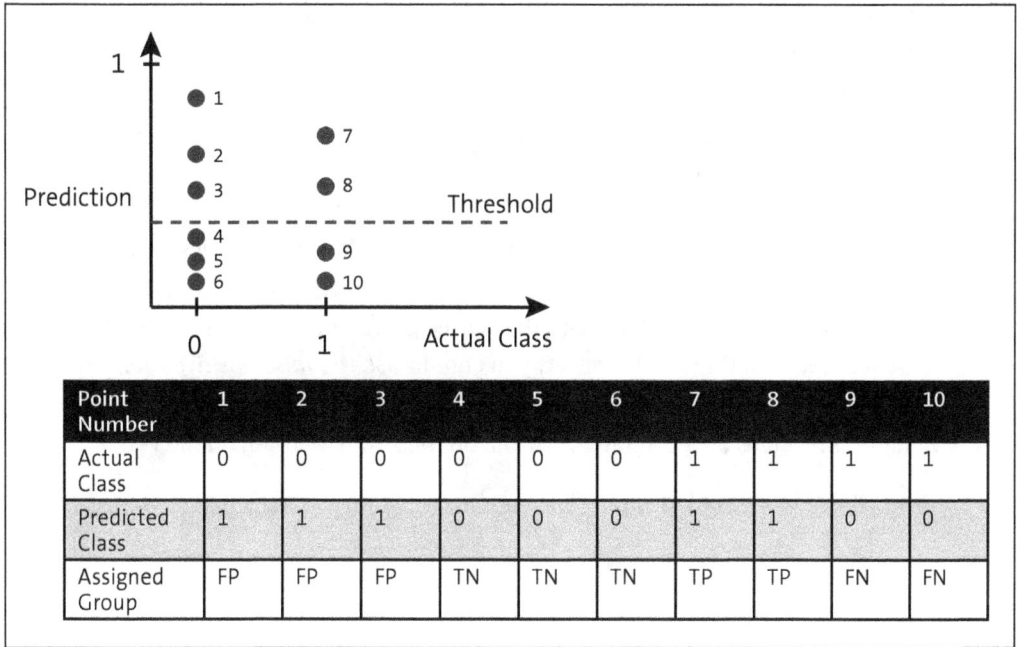

Figure 3.4 Model Predictions and Classes

Figure 3.5 Confusion Matrix for Our Example

The confusion matrix gives us a snapshot of the model, but with the ROC curve, we can generalize the concept.

3.3   Receiver Operator Characteristic Curve

The receiver operating characteristic curve (ROC curve) is based on a concept that was developed during the Second World War in an effort to detect enemy vessels such as submarines. The concept was later adopted in many other scientific disciplines, such as psychology, medicine, and the prediction of extreme weather events. Ultimately, the concept was also adopted in machine learning to assess the quality of models.

In the previous section, we learned that the threshold value has a decisive influence on the confusion matrix. Each change in the threshold value leads to a different confusion matrix. In this section, we'll see that the ROC curve shows us how well our model works across all possible threshold values. You only need to perform the following steps:

1. Determine the sensitivity, which is also known as the *true positive rate* (TPR), and the specificity, which is also known as the *true negative rate* (TNR), for various threshold values between 0 and 1. However, you should use the *false positive rate*, which we abbreviate as 1 – specificity (we'll use this abbreviation again later).

2. Plot both values on a diagram, with FPR on the x-axis and TPR on the y-axis.

One result is illustrated in Figure 3.6.

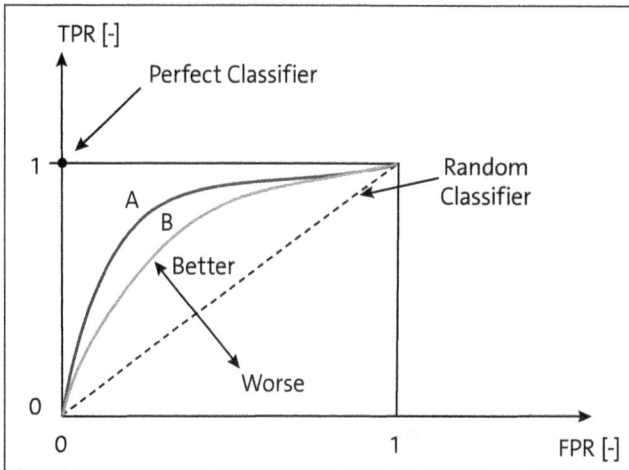

Figure 3.6 ROC Curve

The perfect classification model has an FPR of 0 and a TPR of 1, so it occupies the top left corner. The opposite is a model that has learned nothing and makes predictions completely at random (i.e., it's no better than guessing). Such a model would be exactly on the diagonal. Practical models are located somewhere in this area of tension, with better models being shifted to the top left. In our example, model A shows a consistently better performance than model B.

The concept of the ROC curve can be condensed into a single numerical value: the area under curve (AUC), which is the area under the ROC curve. It summarizes the performance of the model in a single scalar value between 0 and 1. The AUC is the probability that the model will rate a randomly selected positive case higher than a randomly selected negative case. At a value of 1, the model always makes the right decision, whereas at a value of < 0.5, the model performs worse than a model based purely on chance.

3.4 Coding: Binary Classification

In our first example, we'll train a model that predicts whether network traffic is normal behavior or a hacker attack (meaning it's a model that performs intrusion detection). We'll use the "Intrusion Detection Logs" dataset from Kaggle for this purpose. We'll have to prepare the data first, and then, we'll train the model and evaluate the model results.

3.4.1 Data Preparation

The entire script for this example can be found in *040_Classification\data_prep_binary.py*. Listing 3.1 shows us the required packages, and we'll load the data directly from Kaggle using `kagglehub`. The data is available as a CSV file, so we'll use `pandas` to load and modify the data. We'll also separate the data into training and test data by using functions from the `sklearn` package. Finally, we'll use the `matplotlib` and `seaborn` packages for visualization.

```
#%% packages
import kagglehub
import os
import pandas as pd
import numpy as np
from sklearn.model_selection import train_test_split
import matplotlib.pyplot as plt
import seaborn as sns
```

Listing 3.1 Binary Classification: Import Packages

Let's start by importing the data as shown in Listing 3.2. To do this, we can use the kagglehub package directly by passing the ID of the dataset to the `dataset_download` function. This ensures that the data is downloaded to the local hard disk—namely, to a folder whose path is stored in the path variable. This path contains the *Network_logs.csv* file, which we load with `pd.read_csv`. The file has 8,846 data points and 10 features.

```
#%% data import
path = kagglehub.dataset_download("developerghost/intrusion-detection-logs-nor-
mal-bot-scan")

print("Path to dataset files:", path)
```

```
file_path = os.path.join(path, "Network_logs.csv")
df = pd.read_csv(file_path)
```

(8846, 10)

Listing 3.2 Binary Classification: Import Data

First, let's look at the dataset. Figure 3.7 shows the first four lines of it.

	Source_IP	Destination_IP	Port	Request_Type	Protocol	Payload_Size	User_Agent	Status	Intrusion	Scan_Type
0	192.168.142.55	42.156.67.167	80	FTP	UDP	2369	curl/7.68.0	Success	0	Normal
1	53.39.165.18	94.60.242.119	135	SMTP	UDP	1536	Wget/1.20.3	Failure	1	BotAttack
2	192.168.127.91	7.10.192.3	21	SMTP	TCP	1183	Wget/1.20.3	Success	0	Normal
3	192.168.30.40	130.169.82.211	25	HTTPS	TCP	666	Mozilla/5.0	Success	0	Normal

Figure 3.7 Binary Classification: Dataset

Now, let's take a closer look at the features we're dealing with:

```
df.columns
```

```
Index(['Source_IP', 'Destination_IP', 'Port', 'Request_Type', 'Protocol',
    'Payload_Size', 'User_Agent', 'Status', 'Intrusion', 'Scan_Type'],
    dtype='object')
```

The IP address from which a packet is sent (Source_IP) and the IP address to which the packet is sent (Destination_IP) are recorded, and the data record also contains information on which port and which format (Request_Type) was used. This refers to the network protocols (e.g. FTP, SMTP, HTTP), which define the rules according to which computers exchange data via the internet.

The port is a virtual address within a computer that enables the operating system to assign the incoming request to a specific application or service.

The size of the parcel is recorded via the Payload_Size, and the Status feature tells us whether the packet was sent successfully. Finally, we have the Intrusion target value, which is binary coded. A value of 0 means that no attack has taken place, whereas a value of 1 represents an attack.

We can safely disregard some of the features because they are not suitable for drawing conclusions about an attack. These include Source_IP and Destination_IP. An attacker would certainly not launch a reproducible attack from the same source IP and would very probably not always attack only certain destination IP addresses. We can therefore delete these two features with df.drop. The Scan_Type feature breaks down our target variable more precisely, and we'll come back to that in Section 3.5, when we will no longer only differentiate between two categories (normal network traffic and attacks) but will also want to predict further classes.

In our current script, we can delete the feature, as follows:

```
#%% drop features that are not useful for the analysis
df = df.drop(columns=["Source_IP", "Destination_IP", "Scan_Type"])
```

All data must be available as numerical information if we want to train a model, so let's look at the data types with which the features are coded, as follows:

```
df.dtypes
```

```
Port            int64
Request_Type    object
Protocol        object
Payload_Size    int64
User_Agent      object
Status          object
Intrusion       int64
dtype: object
```

Only some of the features are available as integers; many of them have the Object data type. Therefore, we must convert these features into numerical data by using one-hot encoding. We can use the pd.get_dummies method for this, and it processes the dataset and converts the features as follows:

```
#%% treat categorical variables
df_cat = pd.get_dummies(df, drop_first=True, dtype=int)
```

The df_cat.shape property shows that 7 more columns have been created, so we now have 17 features. The first return value corresponds to the number of rows, and the second value corresponds to the number of columns:

```
df_cat.shape
```

```
(8846, 17)
```

Now, you should familiarize yourself with the target variable because you should generally not assume that the different categories are evenly distributed. For the visualization in Figure 3.8, we can use seaborn with the countplot function, as follows:

```
#%% visualize the distribution of the target variable
sns.countplot(x="Intrusion", data=df_cat)
plt.title("Target variable imbalance")
```

Figure 3.8 shows that there is significantly more "normal" behavior (indicated by class 0) in the network traffic than attacks (indicated by class 1). This is reassuring, but we'll have to take this into account later if we want to check the quality of the model.

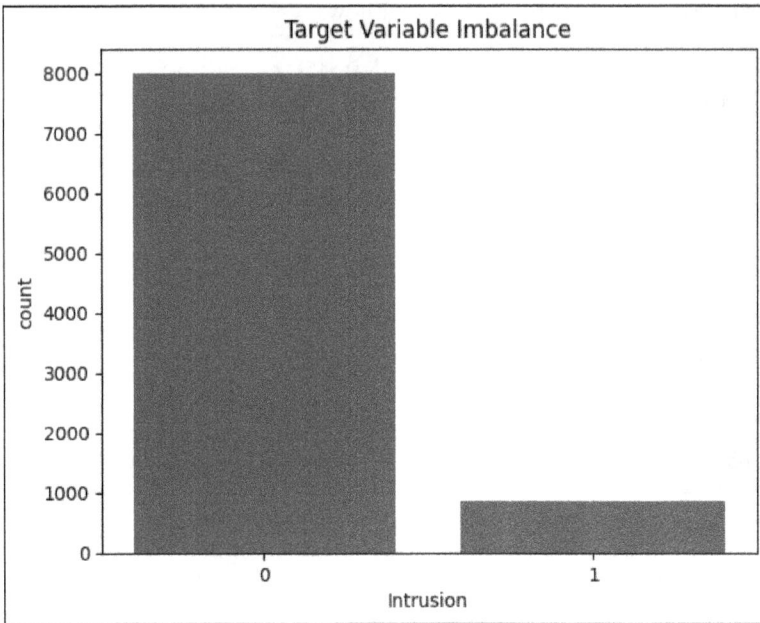

Figure 3.8 Binary Classification: Imbalance of Target Variables

Now, we can look at the correlations (i.e., the relationships) within the features to recognize which features are related to each other. To do this, we first determine the correlation matrix:

```
# Create correlation matrix
corr_matrix = df_cat.corr()
```

The code in Listing 3.3 generates a correlation matrix that is visualized as a heat map.

```
# Create heat map
plt.figure(figsize=(12, 8))
mask = np.triu(np.ones_like(corr_matrix), k=1).T
sns.heatmap(corr_matrix, annot=True, cmap='coolwarm', center=0, fmt='.2f',
mask=mask)
plt.title('Correlation matrix of variables')
plt.tight_layout()
```

Listing 3.3 Binary Classification: Correlation Matrix Visualization

Figure 3.9 shows the correlation matrix for all variables. All features are shown as rows and columns, and for each combination of features, the correlation value shows how strong the linear relationship between the two variables is. The larger the value, the stronger the correlation.

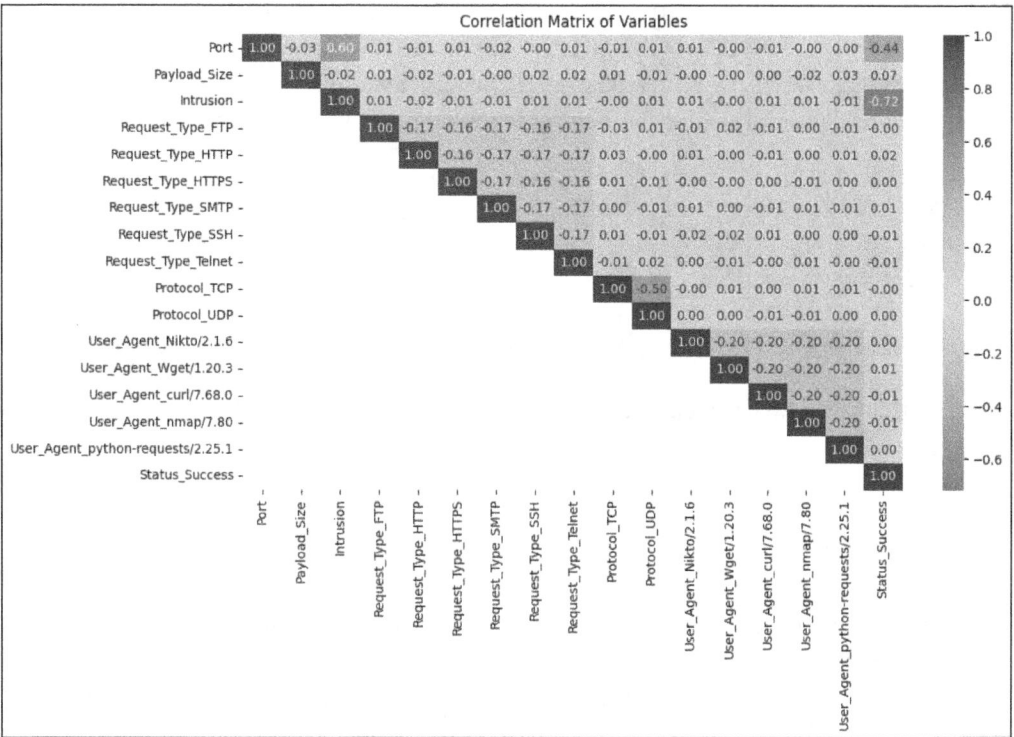

Figure 3.9 Binary Classification: Correlation Matrix

Above all, we're interested in the connection with the Intrusion target variable, so we can concentrate on line 3. All correlations between the independent variables and the target variable can be seen there.

The strongest correlation is with the Status_Success variable, which has a strong negative value that we can interpret to mean that unsuccessful parcel deliveries are more likely to indicate an attack.

As usual, we'll separate the independent (X) and dependent (y) features from each other, and we'll also use the drop method again. We'll remove the Intrusion column from the independent features and store it in y as a dependent feature to act as a target variable, as follows:

```
#%% separate independent and dependent variables
X = df_cat.drop(columns=["Intrusion"])
y = df_cat["Intrusion"]
print(f"X shape: {X.shape}, y shape: {y.shape}")
```

X shape: (8846, 16), y shape: (8846,)

At this point, we can split the data into training and test data by using train_test_split. In this case, we use 90% of the data for training and 10% for testing. The stratify parameter ensures that the ratio of classes in the target variable y in the generated training and test datasets remains the same as in the original overall dataset:

```
#%% split data into training and testing sets
X_train, X_test, y_train, y_test = train_test_split(X, y, test_size=0.1, ran-
dom_state=42, stratify=y)
print(f"X_train shape: {X_train.shape}, X_test shape: {X_test.shape}")
```

X_train shape: (7961, 16), X_test shape: (885, 16)

At this point, we have reached the end of the data preparation and can start creating the model.

3.4.2 Modeling

Now, we come to the core of our classification: creating and training the model. The final script can be found in *040_Classification\model_binary.py*. We start by loading the required packages using the code in Listing 3.4, and we load the DataLoader and a class called TensorDataset (a simplified form of the previously used Dataset class) from the torch package.

In addition, we import packages for evaluating the model result (sklearn) and for visualization (seaborn and matplotlib). Finally, we need the split training and test data, which is provided via the previously created data_prep_binary script.

```
from torch.utils.data import DataLoader, TensorDataset
import torch
import torch.nn as nn
import torch.optim as optim
import seaborn as sns
import numpy as np
from sklearn.metrics import confusion_matrix, accuracy_score, roc_curve, auc
from sklearn.dummy import DummyClassifier
import matplotlib.pyplot as plt
import torch
from data_prep_binary import X_train, X_test, y_train, y_test
```

Listing 3.4 Binary Classification: Import Packages for Modeling

The model and the training are controlled via the hyperparameters listed in Listing 3.5, in which the model is parameterized via the number of nodes in the HIDDEN_SIZE hidden layer and the number of OUTPUT_SIZE output nodes. The actual model training is parameterized via the BATCH_SIZE, the LEARNING_RATE, and the number of EPOCHS. The DEVICE variable is created to be able to execute the code flexibly on the GPU (if available) or (if not) on the CPU.

```
#%% Hyperparameters
BATCH_SIZE = 32
LEARNING_RATE = 0.0005
EPOCHS = 40
HIDDEN_SIZE = 4
OUTPUT_SIZE = 1
DEVICE = torch.device("cuda" if torch.cuda.is_available() else "cpu")
```

Listing 3.5 Binary Classification: Hyperparameter Definition

Now, let's move on to creating the train_dataset and test_dataset datasets. For this, we use an instance of the TensorDataset class, which is very easy to use. The independent and dependent variables are passed to it, and if no further preprocessing steps are required, we can use this simple class instead of the dataset class:

```
# %% create a custom dataset class
train_dataset = TensorDataset(torch.FloatTensor(X_train.values), torch.Float-
Tensor(y_train.values))
test_dataset = TensorDataset(torch.FloatTensor(X_test.values), torch.FloatTen-
sor(y_test.values))
```

To create the DataLoader, we call the class of the same name and pass the BATCH_SIZE and the shuffle parameter for random sampling, as follows:

```
# %% create a data loader
train_loader = DataLoader(train_dataset, batch_size=BATCH_SIZE, shuffle=True)
test_loader = DataLoader(test_dataset, batch_size=BATCH_SIZE, shuffle=False)
```

Now, let's move on to creating the model. We need to start by gaining an understanding of the activation of the output layer and the loss model used. Figure 3.10 shows what options we have here in the context of binary classification. We already know that the output layer will have a node that will give us the probability of class membership, but how should we activate this output layer? We have the following two options, and both are directly linked to the loss model used:

- We can activate the output layer with *sigmoid* and use *BCELoss* as the loss function.
- We can dispense with sigmoid activation if we use the *BCEWithLogitsLoss* loss function, which already has sigmoid activation integrated. If we were to use sigmoid in addition to this loss function in the network, we would be activating twice with sigmoid, which could lead to instability and poor learning progress.

In general, we recommend that you use the latter approach (without explicit sigmoid activation) as your standard approach.

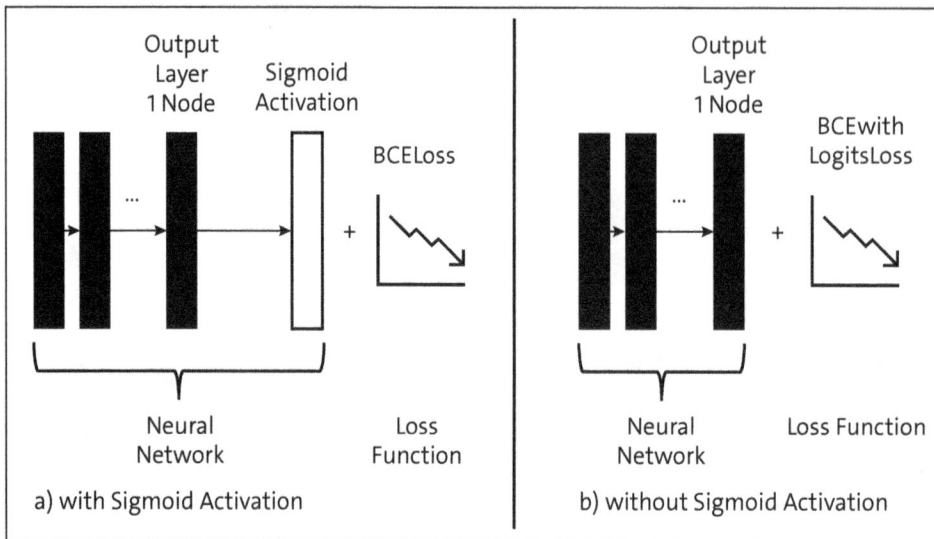

Figure 3.10 Binary Classification: Output Layer and Loss Function

In Listing 3.6, we create the model class. The model consists of its hidden layer with hidden_size nodes. The model parameters of the hidden layer are activated with ReLU, and Dropout is used for regularization. The output layer isn't explicitly activated, as the optimizer takes care of this and activates the data internally with Sigmoid.

```
class BinaryClassificationModel(nn.Module):
    def __init__(self, input_size, output_size, hidden_size, dropout_rate=0.2):
        super(BinaryClassificationModel, self).__init__()
        self.lin1 = nn.Linear(input_size, hidden_size)
        self.lin2 = nn.Linear(hidden_size, output_size)
        self.relu = nn.ReLU()
        self.dropout = nn.Dropout(dropout_rate)

    def forward(self, x):
        x = self.lin1(x)
        x = self.relu(x)
        x = self.dropout(x)
        x = self.lin2(x)
        return x
```

Listing 3.6 Binary Classification: Model Class Definition

We create the model instance in Listing 3.7. To be able to apply the code flexibly to other data, we determine the input_size based on the training data. We then create the model by passing the corresponding input_size, hidden_size, and output_size parameters.

```
input_size = X_train.shape[1]

model = BinaryClassificationModel(input_size=input_size, output_size=OUTPUT_
SIZE, hidden_size=HIDDEN_SIZE).to(DEVICE)
```

Listing 3.7 Binary Classification: Model Instance Creation

We use the now familiar Adam as the optimizer (see Listing 3.8), but the loss function is more exciting. As previously explained, we use BCEWithLogitsLoss, which already includes sigmoid activation.

```
# %% optimizer and loss function with weight decay for regularization
optimizer = optim.Adam(model.parameters(), lr=LEARNING_RATE, weight_decay=1e-4)
loss_fn = nn.BCEWithLogitsLoss()
```

Listing 3.8 Binary Classification: Optimizer and Loss Function

We perform the actual model training as shown in Listing 3.9. We follow the usual procedure: we iterate over the number of epochs and perform a forward pass, the loss calculation, a backward pass, and the update of the parameters in each epoch.

```
#%% training loop
train_losses = []
for epoch in range(EPOCHS):
    train_loss = 0
    for X_train_batch, y_train_batch in train_loader:
        # move data to device
        X_train_batch, y_train_batch = X_train_batch.to(DEVICE), y_train_
batch.to(DEVICE)
        # forward pass
        y_train_batch_pred = model(X_train_batch)
        # calculate loss
        loss = loss_fn(y_train_batch_pred, y_train_batch.reshape(-1, 1))
        # backward pass
        loss.backward()
        # update weights
        optimizer.step()
        # reset gradients
        optimizer.zero_grad()
        # update train loss
        train_loss += loss.item()
    # append train loss
    train_losses.append(train_loss)
    print(f"Epoch {epoch+1}/{EPOCHS}, Loss: {train_losses[-1]:.4f}")
```

Listing 3.9 Binary Classification: Model Training

Now, let's look at whether our training was successful. To do this, we use seaborn as an sns alias (as in Listing 3.10) and visualize the training loss and the epochs.

```python
# %% visualize training loss
plt.figure()
sns.lineplot(x=list(range(EPOCHS)), y=train_losses)
plt.xticks(range(EPOCHS))
plt.xlabel('Epoch')
plt.ylabel('Loss')
plt.title('Training Loss Over Time')
```

Listing 3.10 Binary Classification: Loss Visualization

Figure 3.11 shows the progression of training losses over the epochs. The training is exemplary because the losses initially decrease sharply and then hardly any improvements are achieved from around 20 epochs onward.

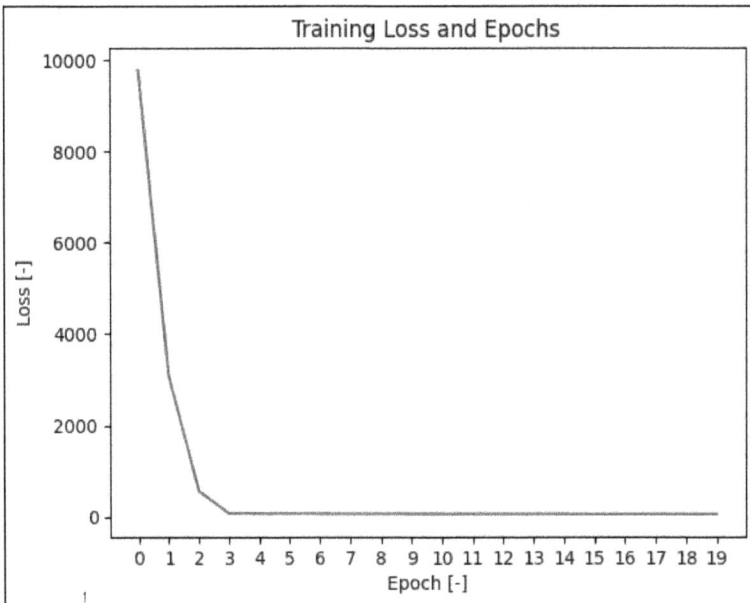

Figure 3.11 Binary Classification: Training Loss and Epochs

3.4.3   Evaluation

Now, let's move on to checking the model by using the test data. Listing 3.11 shows how we let the model predict the y_test_pred_proba probabilities to then determine the y_test_pred_class classes. We use the THRESHOLD limit value for this: whenever the prediction is above the threshold value, it's rounded up to 1, and whenever the prediction is below it, it's rounded down accordingly.

```
y_test_pred_proba, y_test_pred_class, y_test_true = [], [], []
THRESHOLD = 0.5

model.eval()   # Set model to evaluation mode
with torch.no_grad():
    for X_test_batch, y_test_batch in test_loader:
        X_test_batch, y_test_batch = X_test_batch.to(DEVICE),
                                     y_test_batch.to(DEVICE)
        y_test_batch_pred = model(X_test_batch)
        y_test_batch_pred = y_test_batch_pred.cpu().numpy()

        # Store probabilities for ROC curve
        y_test_pred_proba.extend(y_test_batch_pred.flatten().tolist())

        # Store class predictions for confusion matrix
        y_test_batch_pred_class = (y_test_batch_pred > THRESHOLD).astype(int)
        y_test_batch = y_test_batch.cpu().numpy()
        y_test_pred_class.extend(y_test_batch_pred_class.flatten().tolist())
        y_test_true.extend(y_test_batch.flatten().tolist())
```

Listing 3.11 Binary Classification: Model Evaluation

We use these together with the real y_test_true target values to display them in a confusion matrix. The code we use for this is shown in Listing 3.12.

```
#%% create confusion matrix
cm = confusion_matrix(y_true=y_test_true, y_pred=y_test_pred_class)
plt.figure()
sns.heatmap(cm, annot=True, fmt='d', cmap='Blues')
plt.xlabel('Predicted')
plt.ylabel('True')
plt.title('Confusion matrix (binary Classification)')
```

Listing 3.12 Binary Classification: Confusion Matrix Visualization

Figure 3.12 shows the confusion matrix of the test data. The predicted classes are shown on the horizontal axis, and the actual classes are shown on the vertical axis.

We can see here that the model is apparently doing a good job. There were no FPs (as indicated in the top right field), but there were some FNs (as indicated in the bottom left field).

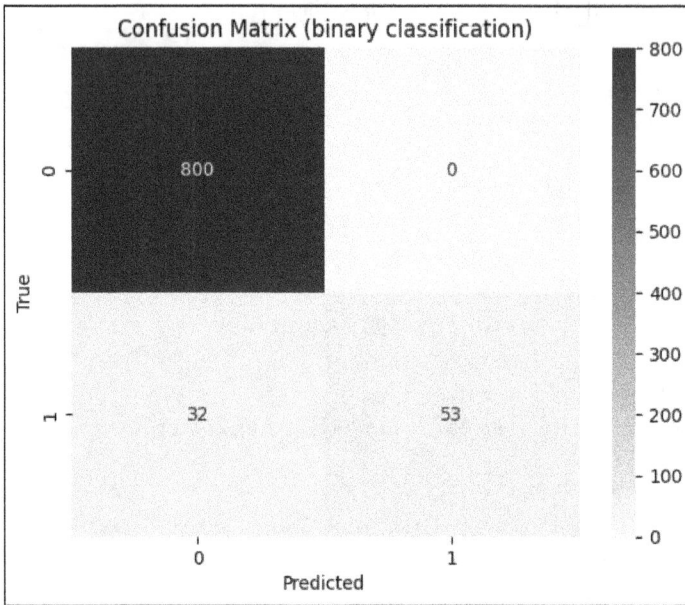

Figure 3.12 Binary Classification: Confusion Matrix

We can also derive the accuracy metric from the confusion matrix. To do this, we can pass the actual and predicted values to the accuracy_score function and obtain the corresponding score, as follows:

```
#%% accuracy score
accuracy_score(y_true=y_test_true, y_pred=y_test_pred_class)
```

0.9683615819209039

The model has an accuracy of 96.8%. That sounds like a very good value, but we must always bear in mind that the target size is very unevenly distributed—there is significantly more normal network traffic (class 0) than malicious network traffic (class 1).

So, to make a definitive statement as to whether the model provides better predictions than pure guessing, we need to train a dummy classifier. This is a class of scikitlearn that helps us compare the quality of the model with a model that is only based on pure guessing.

The dummy classifier always predicts the most frequent class and, in our case, delivers an accuracy of 90.2%, as follows:

```
#%% naive classifier and accuracy score
model_naive = DummyClassifier(strategy='most_frequent').fit(X_train, y_train)
y_test_pred_naive = model_naive.predict(X_test)
```

```
accuracy_score(y_true=y_test_true, y_pred=y_test_pred_naive)
```

0.9016949152542373

We can therefore conclude that our model works much better than pure guesswork.

Finally, we visualize the ROC curve based on the code in Listing 3.13. We use the roc_ curve function, which provides us with the FP rate (fpr) and the TP rate (tpr). We also determine the area under curve (auc) to display it in the graph, and we create the actual diagram with matplotlib.

```
fpr, tpr, thresholds = roc_curve(y_true=y_test_true, y_score=y_test_pred_proba)
roc_auc = auc(fpr, tpr)

plt.figure(figsize=(8, 6))
plt.plot(fpr, tpr, color='blue', lw=2, label=f'AUC = {roc_auc:.2f}')
plt.plot([0, 1], [0, 1], color='red', lw=1, linestyle='--', label='Random Clas-
sifier')
plt.xlim([0.0, 1.0])
plt.ylim([0.0, 1.05])
plt.xlabel('FPR [-]')
plt.ylabel('TPR [-]')
plt.title('ROC Kurve (binäry Classification)')
plt.legend(loc="lower right")
plt.grid(True, alpha=0.3)
```

Listing 3.13 Binary Classification: ROC Curve Creation

Figure 3.13 shows the ROC curve, in which the FPR is plotted horizontally and the TPR is plotted vertically. The dashed red line represents the "random" classifier (dummy classifier), which corresponds exactly to the diagonal.

The ROC curve (the solid blue line) initially runs vertically upward from the lower left-hand corner and ends in the upper right-hand corner. We know that better models show ROC curves that are as close as possible to the upper left-hand corner, and that isn't quite the case here. However, the model still works very well.

Normally, we'd train other models now, and we'd plot their ROC curves in this diagram. I'll leave that for you to do as an exercise.

We have reached the end of the section on binary classification. We've trained a simple model for analyzing network traffic, and we've learned that we must pay particular attention to the activation of the output layer (or its omission) and the loss function used. These two aspects will also play a decisive role in the following section on multiclass classification.

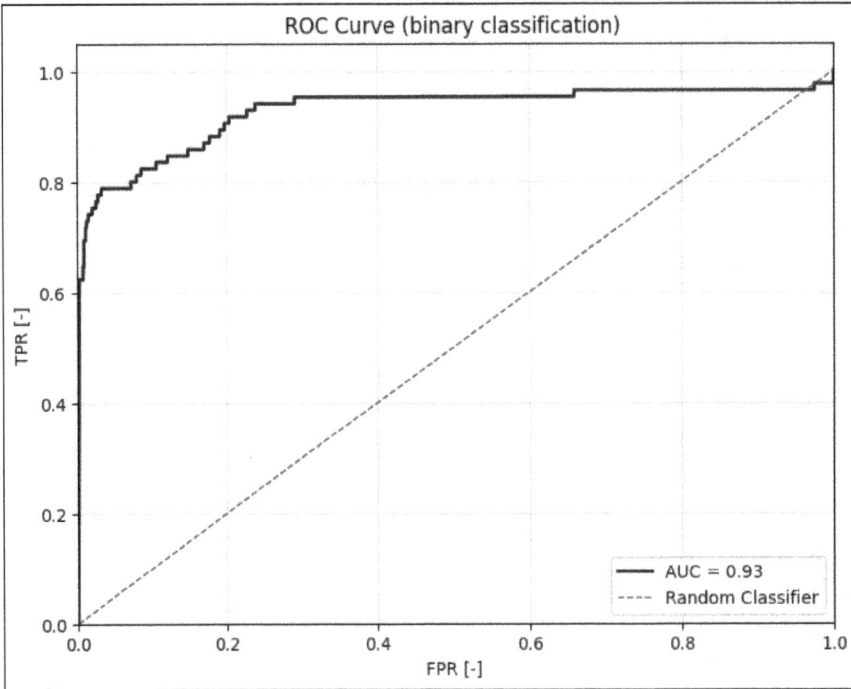

Figure 3.13 Binary Classification: ROC Curve

3.5 Coding: Multiclass Classification

In this example, we'll train a model based on the same dataset as we used in the previous section. What differs is the number of classes of the target variables. Previously, we only had the following two states:

- **0**
  Normal network behavior

- **1**
  Abnormal network behavior

The dataset offers us the opportunity to break down network behavior in more detail, as follows:

- **Normal**
  This class represents normal behavior.

- **BotAttack**
  This is behavior that suggests an attack.

- **PortScan**
  Port scans are performed by both attackers and security personnel to test security. Therefore, this behavior could be an attack, and it isn't normal behavior.

In this section, we'll learn the implications of predicting multiple classes instead of just two.

Our approach is to import and prepare the data (in Section 3.5.1) before we make use of it in the training of the model (in Section 3.5.2). Finally, we'll evaluate the trained model (in Section 3.5.3).

3.5.1 Data Preparation

The entire script for this example can be found in *040_Classification\model_multi-class.py*. As always, we start loading the packages as in Listing 3.14. To help improve your understanding, we've divided the packages into the areas of data import, data processing, modeling, and visualization. There are no new packages on which we need to go into in more detail.

```
# data import
import kagglehub
# data preparation
import os
import pandas as pd
from sklearn.model_selection import train_test_split
# Modeling
import torch
from torch.utils.data import DataLoader, TensorDataset
import torch.nn as nn
import torch.optim as optim
from sklearn.metrics import confusion_matrix, accuracy_score
from sklearn.dummy import DummyClassifier
import numpy as np
# Visualization
import matplotlib.pyplot as plt
import seaborn as sns
```

Listing 3.14 Multiclass Classification: Package Import

In Listing 3.15, the data is loaded directly from Kaggle using the `kagglehub` package and the downloaded CSV file is loaded using `pandas`.

```
#%% data import
path = kagglehub.dataset_download("developerghost/intrusion-detection-logs-nor-mal-bot-scan")
file_path = os.path.join(path, "Network_logs.csv")
df = pd.read_csv(file_path)
```

Listing 3.15 Multiclass Classification: Data Import

Some features are not required, and we can delete them directly. These include the Source_IP and Destination_IP, as well as the target variable from the previous Intrusion section.

```
#%% drop features that are not useful for the analysis
df = df.drop(columns=["Source_IP", "Destination_IP", "Intrusion"])
```

In the next step, we convert *categorical features* into *numerical features* by using *one-hot encoding*, as follows:

```
#%% treat categorical variables
df_cat = pd.get_dummies(df, columns=[ 'Request_Type', 'Protocol', 'User_
Agent','Status'], drop_first=True, dtype=int)
```

Listing 3.16 shows the separation of independent and dependent variables. As previously mentioned, we use Scan_Type as the target variable. We also convert this categorical variable into a numerical format with pd.factorize when we create y. When creating the independent features X, we take into account all features except Scan_Type.

```
#%% separate independent and dependent variables
X = df_cat.drop(columns=['Scan_Type']).astype(float)
y = pd.factorize(df_cat["Scan_Type"])[0].astype(float)

print(f"X shape: {X.shape}, y shape: {y.shape}")
```

X shape: (8846, 16), y shape: (8846)

Listing 3.16 Multiclass Classification: Independent and Dependent Variable Separation

In the next step (in Listing 3.17), we divide the data into training, validation, and test data. If our aim is to split the data into three groups, we must use a two-step procedure because there is no function in sklearn that creates the three groups in one step. Therefore, we begin by splitting the data into temporary and test data so that we can create the training and validation data from the temporary data in a second step.

Here, 10% of the data is reserved for testing, and the remaining data is split up in a second step.

```
#%% split data into training, validation, and testing sets
# First, split off test set
X_temp, X_test, y_temp, y_test = train_test_split(X, y, test_size=0.1, random_
state=42)
# Split remaining data into training and validation
X_train, X_val, y_train, y_val = train_test_split(X_temp, y_temp, test_size=
0.2, random_state=42)
```

```
print(f"X_train shape: {X_train.shape}, X_val shape: {X_val.shape}, X_test
shape: {X_test.shape}")
```

X_train shape: (6368, 16), X_val shape: (1593, 16), X_test shape: (885, 16)

Listing 3.17 Multiclass Classification: Splitting Data into Training, Validation, and Testing

Now, we can pause for a moment and look at the distribution of classes in the training dataset, as shown in Listing 3.18.

```
# Check class distribution
print("Class distribution in training set:")
unique, counts = np.unique(y_train, return_counts=True)
for class_idx, count in zip(unique, counts):
    print(f"Class {class_idx}: {count} samples ({count/len(y_
train)*100:.1f}%)")
```

Number of classes: 3
Class distribution in training set:
Class 0.0: 5759 samples (90.4%)
Class 1.0: 346 samples (5.4%)
Class 2.0: 263 samples (4.1%)

Listing 3.18 Multiclass Classification: Class Distribution Determination

A predominant share of over 90% of the data records are assigned to class 0, and classes 1 and 2 have nearly the same number of data records at just over 4% and 5%, respectively. The data is therefore very unevenly distributed.

When distributing the classes, it's also advisable to ensure that the distribution among the training, validation, and test data is similar. For this purpose, we use the code in Listing 3.19, which determines the percentages and visualizes them in a bar chart.

```
plt.figure()
# Convert counts to percentages for training data
percentages_train = (counts / len(y_train)) * 100

# Get validation data distribution
unique_val, counts_val = np.unique(y_val, return_counts=True)
percentages_val = (counts_val / len(y_val)) * 100

# Get test data distribution
unique_test, counts_test = np.unique(y_test, return_counts=True)
percentages_test = (counts_test / len(y_test)) * 100
```

```
# Plot distributions
width = 0.25
plt.bar(unique - width, percentages_train, width, label='Training')
plt.bar(unique, percentages_val, width, label='Validation')
plt.bar(unique + width, percentages_test, width, label='Test')

plt.xlabel('Class [-]')
plt.ylabel(Percentage [%]')
plt.title('Class distribution in Training-, Validiation- and Testset')
plt.xticks([0, 1, 2])
plt.legend()
```

Listing 3.19 Multiclass Classification: Class Distribution Visualization

The result of the code can be seen in Figure 3.14.

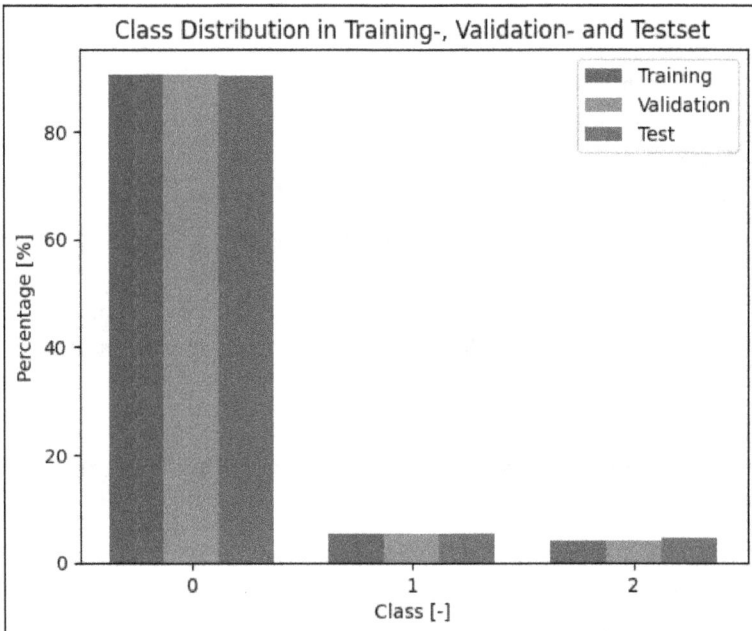

Figure 3.14 Multiclass Classification: Class Distribution in Training, Validation, and Test Datasets

The fact that the classes are very unequal isn't ideal, but it's often unavoidable. However, we need to ensure that the class distribution in the three different datasets is at least similar. In this case, the distributions are very similar, so we can continue with the next step.

Listing 3.20 shows how to create the dataset and DataLoader instances for the training, validation, and test data. It's worth mentioning here that the data is randomly sampled for the `train_dataloader`, whereas this isn't necessary for the `val_dataloader` and `test_dataloader`.

Since most classification loss functions in PyTorch require the class labels as 64-bit integers, we must use a `torch.LongTensor` at this point.

```
train_dataset = TensorDataset(torch.FloatTensor(X_train.values), torch.LongTensor(y_train))
val_dataset = TensorDataset(torch.FloatTensor(X_val.values), torch.LongTensor(y_val))
test_dataset = TensorDataset(torch.FloatTensor(X_test.values), torch.LongTensor(y_test))

# %% create a data loader
train_loader = DataLoader(dataset=train_dataset, batch_size=BATCH_SIZE, shuffle=True)
val_loader = DataLoader(dataset=val_dataset, batch_size=BATCH_SIZE, shuffle=False)
test_loader = DataLoader(dataset=test_dataset, batch_size=BATCH_SIZE, shuffle=False)
```

Listing 3.20 Multiclass Classification: Dataset and DataLoader

At this point, the `DataLoader` instances are available, and we'll use them in the following modeling and evaluation.

3.5.2 Modeling

At the beginning of the modeling, it makes sense for us to define the hyperparameters that influence the training in one place. The following parameters are defined in Listing 3.21 for this purpose:

- **BATCH_SIZE**
  This is the size of the batch.
- **LEARNING_RATE**
  This is the learning rate.
- **EPOCHS**
  This is the number of epochs.
- **HIDDEN_SIZE**
  This is the number of hidden nodes.
- **OUTPUT_SIZE**
  This is the number of output nodes.
- **DEVICE**
  This is the device on which the training will be performed.

```
BATCH_SIZE = 128
LEARNING_RATE = 0.001
EPOCHS = 50
HIDDEN_SIZE = 64
OUTPUT_SIZE = len(df_cat["Scan_Type"].unique())
DEVICE = torch.device("cuda" if torch.cuda.is_available() else "cpu")
print(f"Number of classes: {OUTPUT_SIZE}")
```

Listing 3.21 Multiclass Classification: Hyperparameter Definition

Now, we come to a core element: the structure of the neural network. Figure 3.15 shows the structure of the network that we are implementing. There are three blocks, each of which has a linear layer, batch normalization, an activation using ReLU, and dropout to prevent overfitting.

Here, we can recognize a pattern in which the number of features and thus the number of parameters decreases with the depth of the network. This is very common and useful in many areas of deep learning because it reduces complexity, and the reduction of parameters in the later layers reduces the overall complexity and computational load of the model.

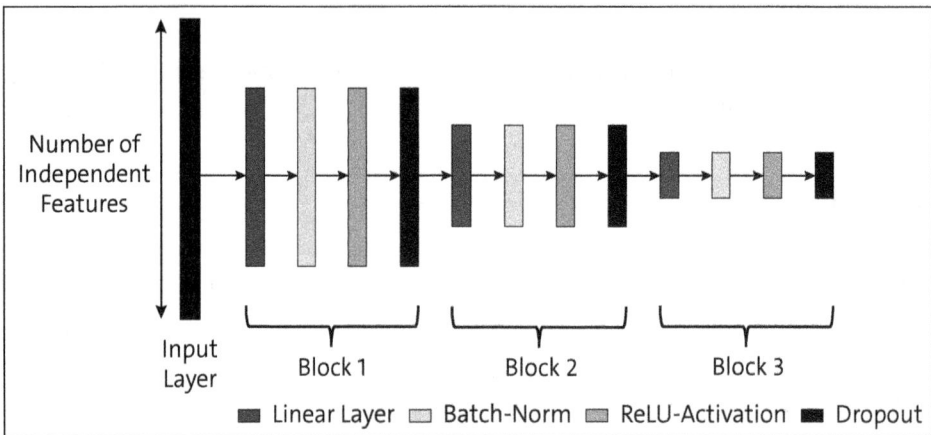

Figure 3.15 Multiclass Classification: Network Structure

The model should be flexibly usable with other datasets. For this reason, the number of input_size independent features, the number of nodes in the hidden_size hidden layer, and the number of output_size output nodes are transferred to the MulticlassModel model class in Listing 3.22. In addition, an optional parameter for the dropout rate (dropout_rate) is defined.

```
class MulticlassModel(nn.Module):
    def __init__(self, input_size, output_size, hidden_size,
        dropout_rate=0.3):
```

```
        super(MulticlassModel, self).__init__()
        self.linear_in = nn.Linear(input_size, hidden_size)
        self.bn1 = nn.BatchNorm1d(hidden_size)
        self.linear_hidden1 = nn.Linear(hidden_size, hidden_size // 2)
        self.bn2 = nn.BatchNorm1d(hidden_size // 2)
        self.linear_hidden2 = nn.Linear(hidden_size // 2, hidden_size // 4)
        self.bn3 = nn.BatchNorm1d(hidden_size // 4)
        self.linear_out = nn.Linear(hidden_size // 4, output_size)
        self.relu = nn.ReLU()
        self.dropout = nn.Dropout(dropout_rate)

    def forward(self, x):
        x = self.linear_in(x)
        x = self.bn1(x)
        x = self.relu(x)
        x = self.dropout(x)

        x = self.linear_hidden1(x)
        x = self.bn2(x)
        x = self.relu(x)
        x = self.dropout(x)

        x = self.linear_hidden2(x)
        x = self.bn3(x)
        x = self.relu(x)
        x = self.dropout(x)

        x = self.linear_out(x)
        return x

input_size = X_train.shape[1]

model = MulticlassModel(input_size=input_size, output_size=OUTPUT_SIZE, hidden_
size=HIDDEN_SIZE).to(DEVICE)
```

Listing 3.22 Multiclass Classification: Model Class and Instance

Adam is used as the optimizer, and CrossEntropyLoss is used as the loss function, as shown in Listing 3.23.

```
optimizer = optim.Adam(model.parameters(), lr=LEARNING_RATE)
loss_fn = nn.CrossEntropyLoss()
```

Listing 3.23 Multiclass Classification: Optimizer and Loss Function

The model training is shown in Listing 3.24. The training loop iterates over the previously defined number of EPOCHS, and in each epoch, the model is improved based on the training data and validated using the validation data. The train_loss and val_loss losses are determined so we can visualize them later.

```
#%% training loop
train_losses = []
val_losses = []
for epoch in range(EPOCHS):
    train_loss = 0
    val_loss = 0
    for X_train_batch, y_train_batch in train_loader:
        # move data to device
        X_train_batch, y_train_batch = X_train_batch.to(DEVICE),
        y_train_batch.to(DEVICE)
        # forward pass
        y_train_batch_pred = model(X_train_batch)
        # calculate loss
        loss = loss_fn(y_train_batch_pred, y_train_batch)
        # backward pass
        loss.backward()
        # update weights
        optimizer.step()
        # reset gradients
        optimizer.zero_grad()
        # update train loss
        train_loss += loss.item()
    # normalize and append train loss
    train_losses.append(train_loss / len(train_loader))
    print(f"Epoch {epoch+1}/{EPOCHS}, Loss: {train_losses[-1]:.4f}")
    for X_val_batch, y_val_batch in val_loader:
        # move data to device
        X_val_batch, y_val_batch = X_val_batch.to(DEVICE),
            y_val_batch.to(DEVICE)
        # forward pass
        y_val_batch_pred = model(X_val_batch)
        # calculate loss
        loss = loss_fn(y_val_batch_pred, y_val_batch)
        # update val loss
        val_loss += loss.item()
    # normalize and append val loss
    val_losses.append(val_loss / len(val_loader))
    print(f"Epoch {epoch+1}/{EPOCHS}, Val Loss: {val_losses[-1]:.4f}")
```

```
Epoch 1/50, Loss: 0.9680
Epoch 1/50, Val Loss: 0.7753
...
Epoch 50/50, Loss: 0.1326
Epoch 50/50, Val Loss: 0.1137
```

Listing 3.24 Multiclass Classification: Model Training

The losses will be output on the console, and they will continue to decrease with the number of epochs. In the next section, we'll take a closer look at the evaluation.

3.5.3 Evaluation

We can display the previously determined losses of the training and validation data in a diagram by using the code in Listing 3.25.

```
# %% visualize training and validationloss
plt.figure()
sns.lineplot(x=list(range(EPOCHS)), y=train_losses, label='Training Loss')
sns.lineplot(x=list(range(EPOCHS)), y=val_losses, label='Validation Loss')
plt.xticks(range(0, EPOCHS, 5))
plt.xlabel('Epoch')
plt.ylabel('Loss')
plt.title('Training and Validation Loss Over Time')
plt.legend()
```

Listing 3.25 Multiclass Classification: Loss Visualization

Figure 3.16 shows the progress of the model training and how the losses for the training and validation data decrease as the number of epochs increases. The training stops after 50 epochs because that's the previously defined total number of epochs. This number of epochs is also a sensible choice because there are hardly any improvements left to be made after that point.

Now, we want to create predictions for the test data as in Listing 3.26 (i.e., data that the model has never seen before) to check whether the model can really generalize well. To do this, we let the trained model create y_test_batch_pred predictions for the X_test_batch independent test data.

The predictions are referred to as *logits*, and they are the direct outputs of the last linear layer of a classification model before a normalizing activation function is applied. Their value range is between $-\infty$ and $+\infty$, and we can determine probabilities from these predictions by applying torch.softmax. These probabilities are normalized outputs of the logits after the application of softmax, and they have a value range from 0 to 1. This gives you the probabilities for the three y_test_batch_pred_proba classes.

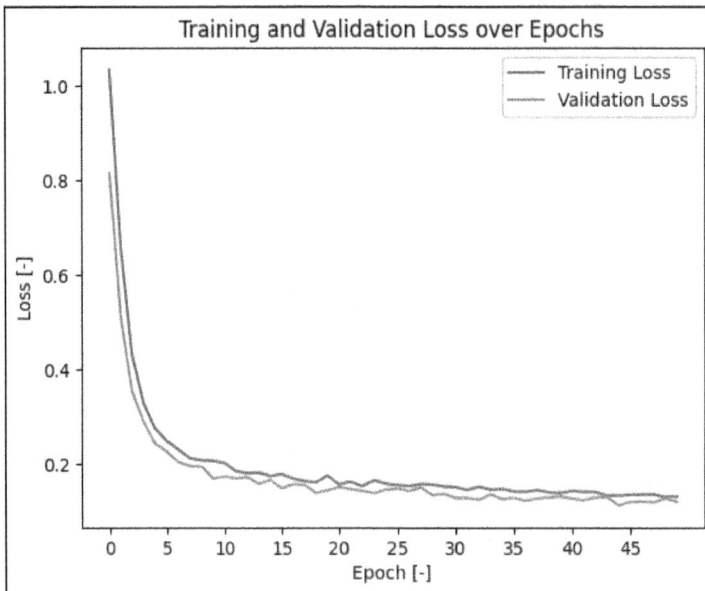

Figure 3.16 Multiclass Classification: Losses

To determine the class with the highest probability, we use torch.argmax. This allows us to determine the index of the corresponding element, and we receive a list with the indices of the predicted classes in the y_test_batch_pred_class object. The real classes are summarized in the y_test_true, and the predicted classes are summarized in the y_test_pred_class object. We'll use the lists in subsequent steps, for example, to determine the confusion matrix.

```
#%% create test predictions for multiclass classification
y_test_pred_proba, y_test_pred_class, y_test_true = [], [], []

model.eval()  # Set model to evaluation mode
with torch.no_grad():
    for X_test_batch, y_test_batch in test_loader:
        X_test_batch, y_test_batch = X_test_batch.to(DEVICE),
                                     y_test_batch.to(DEVICE)
        y_test_batch_pred = model(X_test_batch)

        # Apply softmax to get probabilities
        y_test_batch_pred_proba = torch.softmax(y_test_batch_pred, dim=1)
        y_test_batch_pred_proba = y_test_batch_pred_proba.cpu().numpy()

        # Get predicted classes (argmax)
        y_test_batch_pred_class = torch.argmax(y_test_batch_pred,
        dim=1).cpu().numpy()
        y_test_batch = y_test_batch.cpu().numpy()
```

```
# Store predictions and true labels
y_test_pred_proba.extend(y_test_batch_pred_proba.tolist())
y_test_pred_class.extend(y_test_batch_pred_class.tolist())
y_test_true.extend(y_test_batch.tolist())
```

Listing 3.26 Multiclass Classification: Create Predictions for Test Data

We can use the `confusion_matrix` function to directly determine the confusion matrix by passing the real and predicted values. We can then create a graphic from this table with the code in Listing 3.27.

```
#%% create confusion matrix
cm = confusion_matrix(y_true=y_test_true, y_pred=y_test_pred_class)
plt.figure()
sns.heatmap(cm, annot=True, fmt='d', cmap='Blues')
plt.xlabel('Predicted class)
plt.ylabel('Actual class')
plt.title('Confusion matrix')
```

Listing 3.27 Multiclass Classification: Confusion Matrix Creation

The confusion matrix of the test data for our multiclass classification model is shown in Figure 3.17. The predicted classes are plotted horizontally, and the true classes are plotted vertically. The correct predictions are shown on the main diagonal, and outside the main diagonal are the incorrect predictions.

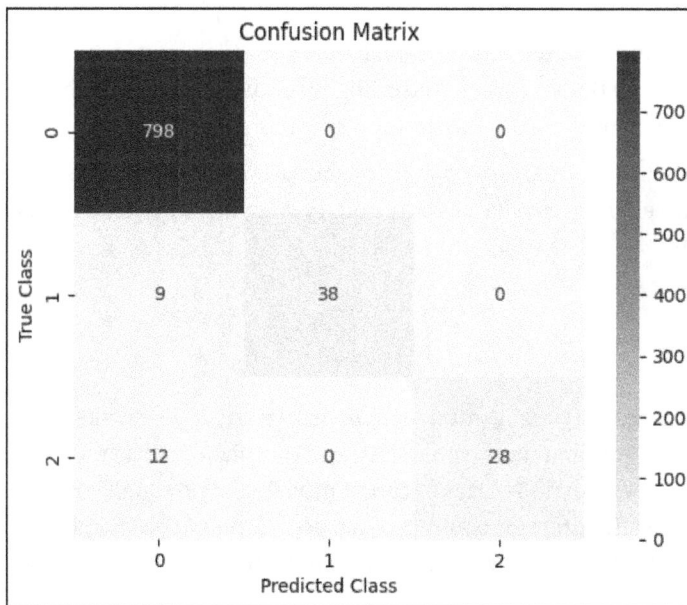

Figure 3.17 Multiclass Classification: Confusion Matrix

Basically, the result looks good because most of the values can be found in the diagonal. However, there are still a few misclassifications. For example, the real class 1 (BotAttack) was often incorrectly predicted as class 0 (Normal), meaning it was an FN. Often, we'll need to make a decision based on a single metric, so at this point, we can determine the accuracy of the model with the code in Listing 3.28.

```
#%% accuracy score
accuracy_score(y_true=y_test_true, y_pred=y_test_pred_class)
```

0.9740112994350283

Listing 3.28 Multiclass Classification: Model Accuracy Score

The model delivers the correct prediction in 97.4% of cases, and we must categorize this result because we know that the data is unbalanced (i.e., there are significantly more elements of one class than of the other classes). For this purpose, we'll determine the dummy classifier accuracy with the code in Listing 3.29, which reflects the results of a model based purely on guesswork.

```
#%% naive classifier and accuracy score
model_naive = DummyClassifier(strategy='most_frequent').fit(X_train, y_train)
y_test_pred_naive = model_naive.predict(X_test)
accuracy_score(y_true=y_test_true, y_pred=y_test_pred_naive)
```

0.9016949152542373

Listing 3.29 Multiclass Classification: Accuracy Score for Dummy Classifier

The dummy classifier has an accuracy of 90.2%, and with such high values, it makes sense to talk about error rather than accuracy. Accordingly, the dummy classifier has an error of just under 10%, whereas the model we trained only has an error of 2.6%.

We have now reached the end of our section on multiclass classifiers. We've learned that the number of output nodes changes relative to the binary classifier and that we must use a different loss model.

3.6   Summary

In this chapter, we've explored the classification models. In Section 3.1, we learned about the different types of classification: binary, multiclass, and multilabel classification. In Section 3.2, we dove into the evaluation of classification models and provided you with insights into how to create and interpret confusion matrices. Building on that, in Section 3.3, we learned about the ROC curve, which is suitable for comparing different models with one another.

After this theoretical preliminary work, in Section 3.4, we ventured into the first model training and trained a binary classification model. In Section 3.5, we extended this concept so that we could predict not only two classes but any number of classes. This chapter on classification models has given you a sound understanding of what to look out for when training such models.

3

Chapter 4

Computer Vision

"The eye sees only what the mind is prepared to comprehend."
—Henri Bergson

This quote can also be applied to deep learning: a deep learning algorithm can only recognize those patterns and objects that it has been designed and trained to recognize. The decisive factor here is that the model is defined by the training data and the model architecture. Therefore, in this chapter, we'll focus on how you can teach a model to see and understand what it sees.

The field of *computer vision* is also known as *machine vision* or *image understanding*. It's about the algorithmic ability to process visual input, such as pixels, in a meaningful way. You can think of it like we humans do. We use our eyes and our brains to see, interpret, and interact with the world around us. Computer vision aims to give computers a similar ability.

We'll focus on the most important areas of computer vision, starting with the simplest tasks and progressing to more complex and elaborate models and skills. Figure 4.1 shows several common computer vision tasks applied to a picture of my dog Kiki on the couch.

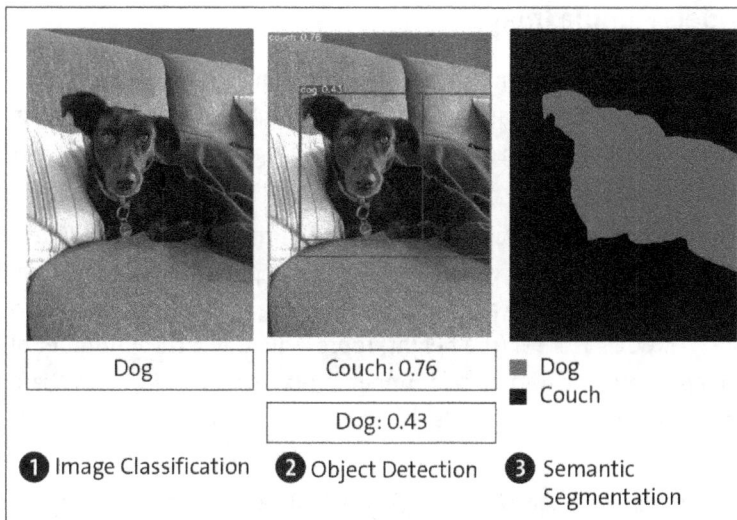

Dog	Couch: 0.76	■ Dog
		■ Couch
	Dog: 0.43	
❶ Image Classification	❷ Object Detection	❸ Semantic Segmentation

Figure 4.1 Tasks in Field of Computer Vision

The simplest type of image recognition is *image classification*, which applies a stamp to an entire image. The technical term for this is stamp is a *label*, and a label refers to the class that the image represents. In our example, image classification occurs in part ❶ of Figure 4.1, where the model predicts as the main class that a dog can be recognized in the image. In part ❷, elements of the image are recognized, and a bounding box is placed around the object. This process is referred to as *object detection*. In part ❸, one of the most complex tasks in the field of image analysis occurs: *semantic segmentation*, which attempts to classify each individual pixel of the image. In this chapter, we'll deal with these three tasks in detail.

Section 4.1 deals with how and to what extent we should handle images differently from other data. In Section 4.2, we'll learn about a new type of network architecture that is particularly well suited to image recognition: convolutional neural networks (CNNs). In Section 4.3, we'll look at image classification, which involves assigning an entire image to a class.

In Section 4.4, we'll learn about object detection, which isn't about assigning an entire image to a class but about assigning image areas to specific classes. In Section 4.5, we'll explore semantic segmentation, which involves further increasing the computational effort and complexity of the model by not only predicting image areas but also assigning each individual pixel to specific classes. Finally, in Section 4.6, we'll explore the style transfer technique, which allows us to transfer the style of a reference image to a target image and thereby create very impressive and atmospheric images.

Now, let's begin with the question of the extent to which images behave differently from other data.

4.1 How Do Models Handle Images?

Images are viewed by computers as a combination of pixels, and each individual pixel is characterized by a color value. Colors are usually coded with the *RGB color model*, in which the primary colors—red (R), green (G), and blue (B)—are added in different ways to create a wide range of other colors.

In computer vision, images are treated as numerical data in the form of *multidimensional tensors*, which is the only form in which computers can process visual information. The most common way to describe a multidimensional tensor is with the following three dimensions: (H, W, C). Here H (the height) represents the vertical number of pixels, W (the width) represents the horizontal number of pixels, and C (the color channels) represents the number of color channels that contain the color information for each pixel.

Figure 4.2 shows how the original image ❶ is transformed into a tensor by applying certain preprocessing steps, which ultimately correspond to a matrix ❷ with the dimensions (1, H, W).

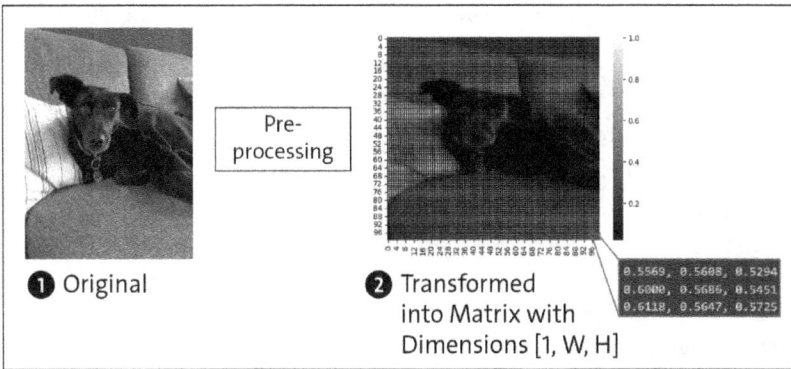

Figure 4.2 From Image to Tensor

In our example, the original image was converted into a tensor with these dimensions: (1, 100, 100). What we understand as colors (or, in this case, as brightness) are numerical values for the computer.

Color Channels

The color channels deserve a little more attention, so let's get into them a little more. There are the following special features regarding the number of color channels:

- **Grayscale images**
  When images are treated as grayscale images, they have only one color channel: $C = 1$. Each pixel represents a *brightness intensity*, which typically ranges from 0 (completely black) to 255 (completely white). The dimensions of the image are (1, H, W).

- **RGB images**
  The most common type of color image is the RGB image, which has three channels ($C = 3$), one each for red, green, and blue. Each pixel in the image has three values that indicate the intensity of the three primary colors. The dimensions of the image are (3, H, W).

- **Images with alpha channel**
  There are also images with four color channels (e.g., RGBA). The additional alpha channel stores information about transparency, and in this case, the dimension is (4, H, W).

Now, let's turn to the most important architecture in the field of computer vision: convolutional neural networks (CNNs) and ViTs.

4.2 Network Architecture

We'll discuss the two common architectures that dominate the field of computer vision. Our later models will be based primarily on CNNs (which we'll get to know in

Section 4.2.1) because they have proven to be extremely powerful while not being very complicated.

The newer ViTs (which we'll get to know in Section 4.2.2) are, as the name suggests, based on the transformer architecture, which dominates the market, particularly in language models. We'll look at how transformers work in Chapter 9, but in this section, we'll focus on the special features of the ViT.

We'll start with CNNs.

4.2.1 Convolutional Neural Networks

Convolutional neural networks (CNNs) are very popular and successful network architectures, especially in computer vision. They differ from classic feedforward networks, which consider each pixel individually. In contrast, CNNs can recognize local patterns and learn different hierarchies of networks through different network layers.

Figure 4.3 illustrates how a CNN trained to classify animals learns simple edges and textures in the early layers and more complex shapes (in this example, body parts such as legs or ears) later. At the end of this network are the model classes that the model is supposed to predict. Probabilities result for the individual classes, and in this example, the model predicts a dog with a very high probability.

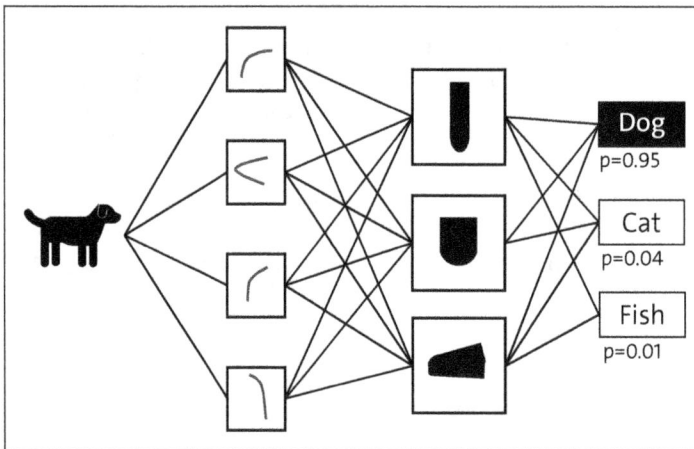

Figure 4.3 CNN Structure

Convolutional layers form the core of a CNN. They can extract relevant features from the input data, and they do this with an operation called *convolution*. What happens with such a fold? Such convolutions are known from image editing programs, and they are used every time edges are extracted or the image is "blurred," for example. Figure 4.4 shows the calculation for such a convolution operation. In this example, the convolution operation (also known as a convolutional filter) corresponds to a 3 × 3 matrix (at the center of the figure).

The CNN systematically moves this over the entire input image (which is depicted in the left-hand matrix in the figure), and at each position where it places the filter, it performs element-by-element multiplication between the values of the filter and the corresponding pixel values of the image area the filter covers. It then adds the results of this multiplication together to produce a single value. The values resulting from all the operations appear in the matrix on the right.

4

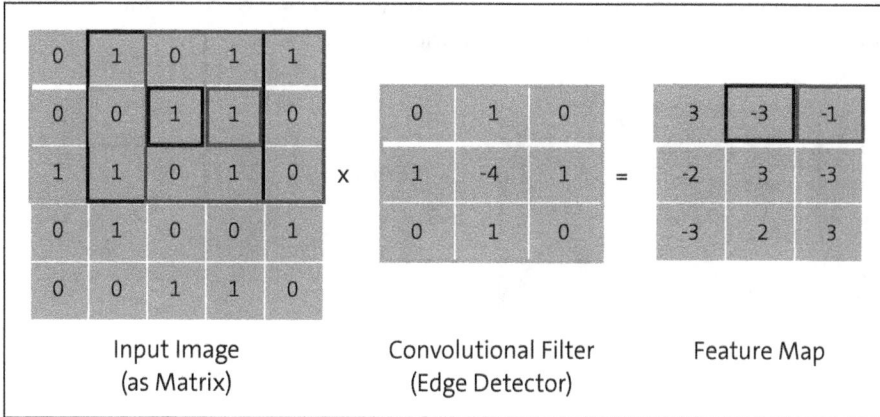

Figure 4.4 CNNs: Convolution Operations

The result of this process is a *matrix*, also known as a *feature map*. It shows where the feature you are looking for is present in the image and how pronounced it is.

Typically, a convolutional layer uses not just one but several such filters, as shown in Figure 4.5. Each filter is trained to recognize a different feature. For example, one filter may emphasize a vertical edge and another may emphasize a horizontal edge.

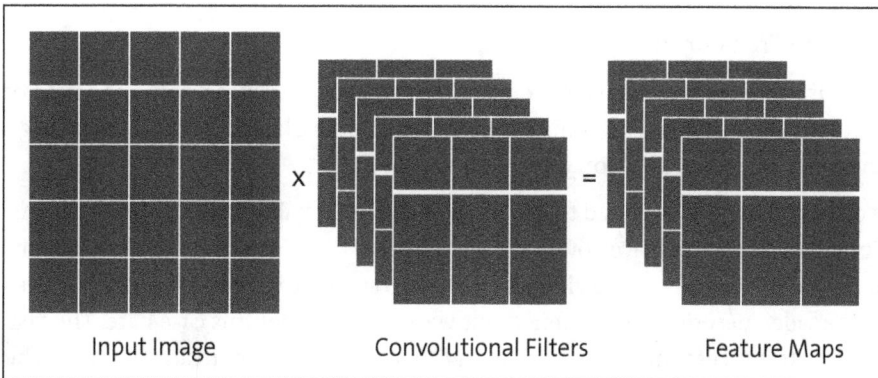

Figure 4.5 CNNs: Feature Maps

We don't set the weights in the convolutional filters manually (as in image processing). Instead, the model learns them during training by adjusting the weights so that it optimally recognizes the features to solve the given task. After the convolution layer, we often apply an activation function such as ReLU to the feature maps to introduce non-linearities that help the model learn more complex relationships.

Another frequently used concept is *pooling*, which works like a filter. Usually, a 2 × 2 matrix is "pushed" over the feature map, and for each currently selected area, the maximum value (with MaxPooling) or the average value (with MeanPooling) is extracted as the output value. Figure 4.6 illustrates the concept. It starts with the top left corner of the feature map and determines the maximum value (in this case, 4) from this 2 × 2 area. The pooling matrix then moves on to determine the next value.

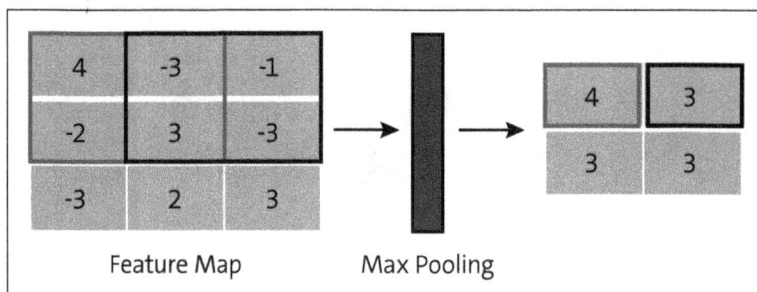

Figure Feature Map   Max Pooling

Figure 4.6 CNNs: Pooling Layer

We mainly use pooling to reduce the dimensions in subsequent layers, which speeds up training. Pooling layers also help the model become less sensitive to shifts in characteristics.

We have now familiarized ourselves with all the relevant layers of a CNN, and we'll see how to combine these in a complete model in the next section. But before that, let's familiarize ourselves with another very popular architecture in computer vision: ViTs.

4.2.2   Vision Transformers

Vision transformers (ViTs) transfer the idea of transformer architecture to the field of computer vision. We'll explain transfers in more detail in Chapter 9, but here, we'll go into the special features of vision transformers.

While transformers are designed to process sequences (usually text), images consist of two-dimensional (2D) matrices of pixels and have no obvious sequences. A sequence would show up in videos, since a video is an ordered sequence of images and the images within the video describe the sequence, but we won't consider this case here. The task of the ViT is to process images in such a way that the transformer can use them. The following three concepts are central to this process:

- **Breaking down the image into smaller patches**
  First, the ViT breaks down a given image into smaller images. These smaller images (also called *patches*) are a fixed number of nonoverlapping, smaller images, and the ViT effectively cuts the overall image into many small squares of equal size.

- **Creating embedding vectors for the patches**
  The ViT combines each of these subimages into a long vector. It also embeds this vector of pixel values, in a way that is very similar to the word or sentence embeddings in natural language processing (which we'll discuss in Chapter 10). The embedding represents the semantic meaning of the words or sentences, and here, the embeddings reflect the image features.

- **Adding positional embeddings**
  In CNNs, spatial information was processed by convolution operations, but transformers don't have any information about the spatial arrangement of the input sequences. Therefore, the visual transformer adds positional embeddings. It adds position information to each embedding of the partial images (in a process called *patch embedding*), and that tells each patch its original position. This allows the transformer to know whether a patch is visible in the upper left of the image, for example.

These concepts are illustrated in Figure 4.7, using an example image.

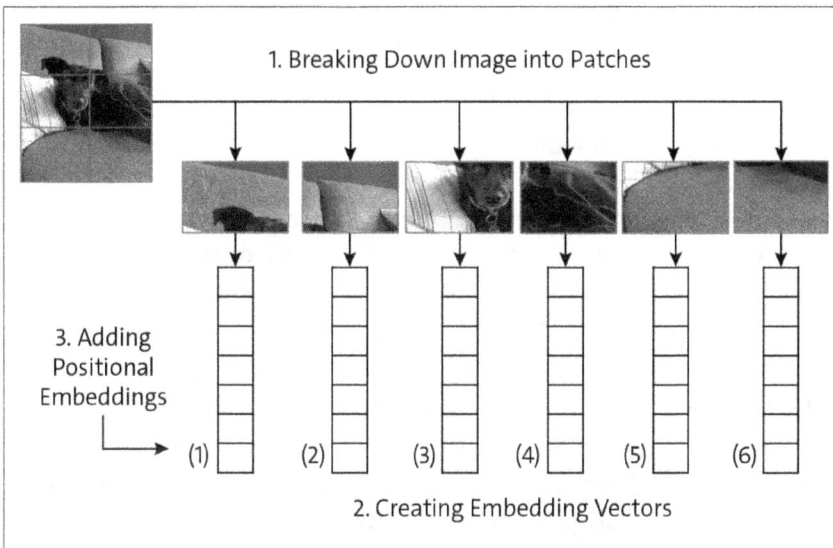

Figure 4.7 ViTs: Core Concepts

After these three steps, which are specific to ViTs, the embedding vectors with their positional embeddings go into the encoder of the transformer, and from there, a standard transformer network follows. Since such transformers form the backbone of language models, we'll discuss them in more detail in Chapter 9.

We won't implement this network architecture from scratch, but rather, we'll fine-tune such a model to our data in Section 4.5.

Next, we go into the simplest task in the field of computer vision: image classification.

4.3   Coding: Image Classification

We can also make a further subdivision via image classification. To illustrate this, we'll use Kiki again as an example. The different types of image classification are illustrated in Figure 4.8.

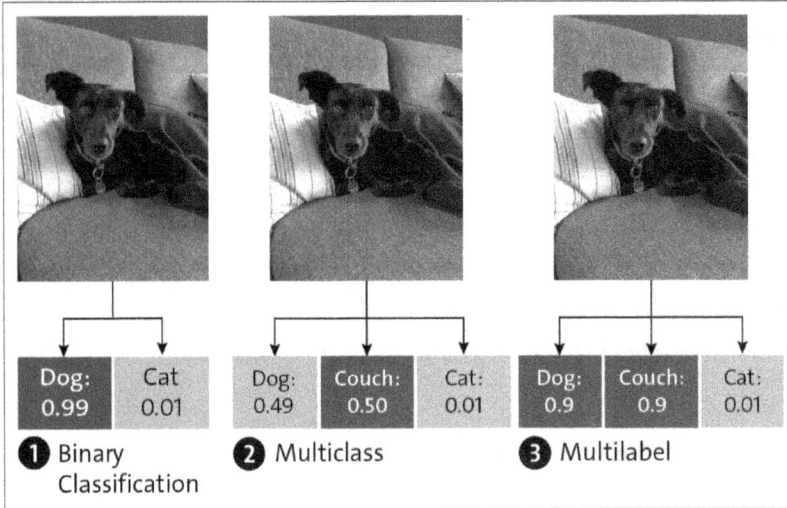

Figure 4.8 Types of Image Classification

The simplest case is *binary classification* ❶, in which there are only two classes. In our example, these are "Dog" and "Cat." The model can therefore recognize only these two different classes, it provides probabilities for these classes, and it generally uses the class with the highest probability as the label for the image.

Of course, the world is more complex than can be described with just two classes, so *multiclass classification* ❷ recognizes three classes. The procedure here is analogous to binary classification: the model provides the probabilities for each class and uses the class with the highest probability. This is generally better because you are not limited to three classes, but there is still a problem: the picture shows my dog on the couch, but the model tries to predict the dominant class and can't predict several classes at the same time.

You can solve this problem with *multilabel classification* ❸, which can simultaneously predict several classes.

We'll look at binary and multiclass classification in the following sections. We don't include a coding section on multilabel classification because the backbone layers remain the same as in the other types of classification. You only have to adapt the final components like the output activations to sigmoid and the loss function to BCE with logits.

4.3.1 Binary Classification

In this example, we'll train a model to assign images to one of two different classes. The model will learn to distinguish muffins from Chihuahuas—a widely underestimated problem that we'll now finally tackle. Figure 4.9 gives an impression of how difficult the task ahead of us will be.

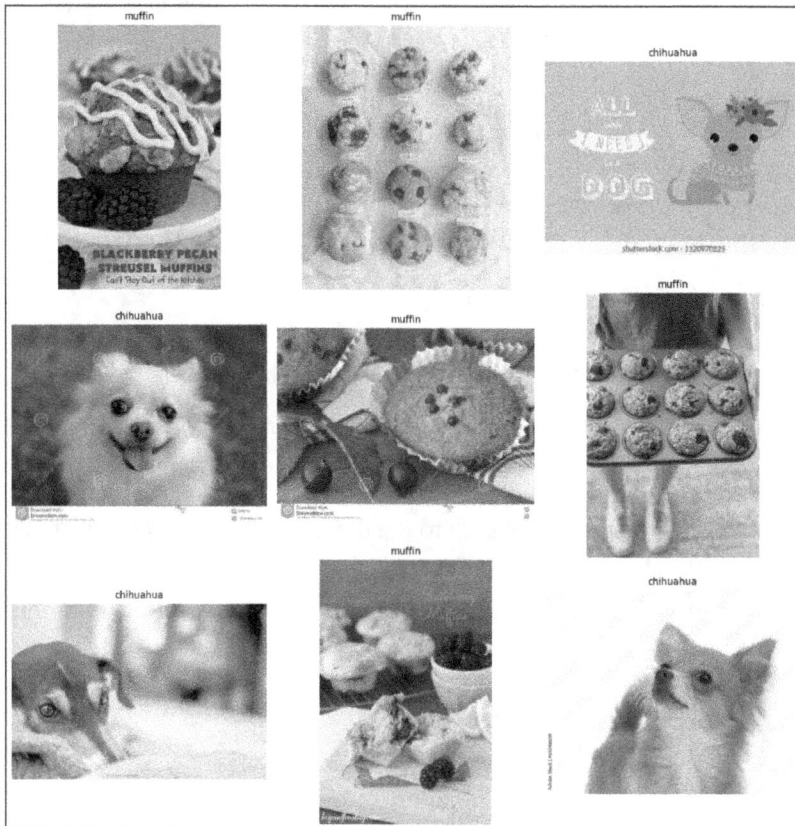

Figure 4.9 Muffin Versus Chihuahua Dataset (Source: https://www.kaggle.com/datasets/samuelcortinhas/muffin-vs-chihuahua-image-classification)

Harder than you thought, isn't it? Now, let's find out how good an algorithm is at distinguishing between muffins and Chihuahuas. First, we'll prepare the data. Then, in computer vision, it's good practice to augment the data, and we'll find out how to do that. Next, we'll use the data to train our model, and finally, we'll evaluate how well our model performs.

Data Preparation

First, we need to download the dataset to our local system by using the code in Listing 4.1. The dataset is hosted on Kaggle, and we can download it with a script.

```
#%% packages
import kagglehub
import json
import os

# Download latest version
path = kagglehub.dataset_download("samuelcortinhas/muffin-vs-chihuahua-image-
classification")

print("Path to dataset files:", path)

#%% path train and test
path_train = os.path.join(path, "train")
path_test = os.path.join(path, "test")

# %% store datapath reference in json file
with open('data_path.json', 'w') as f:
    json.dump({"path_train": path_train, "path_test": path_test}, f)
```

Listing 4.1 Binary Classification: Data Preparation (Source: 050_ComputerVision\010_Binary-ImageClassification\data_prep.py)

After executing the script, we'll save the path to the dataset in the *data_path.json* file. We can load the data from there in the subsequent training.

Another important aspect of training computer vision models is data augmentation, which we'll discuss next.

Data Augmentation

Data augmentation is the artificial enlargement of an existing dataset by creating modified copies of the original images. Instead of capturing new images, which is often very expensive and time-consuming, we can generate new versions of an image with slightly adjusted properties, such as rotation, distortion, or a mirror image. We do this to create greater variety.

The aim of this process is to improve the performance and robustness of models. We train the model with a wider range of variations of the same image, and in this way, the model learns to recognize the essential features of an object and becomes less sensitive to irrelevant changes such as the position or rotation of the objects. The model can thus avoid overfitting.

There are various ways to customize the images, such as rotation (RandomRotation), random upside-down flipping (RandomVerticalFlip), and conversion to grayscale (GrayScale).

An example script illustrates this approach and can be found in *050_ComputerVision\ 00_ImagePreprocessing\ImagePreprocessing.py*.

The most important package in this context is torchvision with the transforms module. This module gives you access to various methods for changing images, and we can load the original image by using Pillow (PIL), as follows:

```
from torchvision import transforms
from PIL import Image
```

We can use the Image.open method to load a sample image and then display it, as follows:

```
# %% import image
img = Image.open('kiki.jpg')
img
```

We can see the result in Figure 4.10.

Figure 4.10 Original Image Before Augmentation

The preprocessing steps are shown in Listing 4.2. The preprocessing steps are transferred as a list to the transforms.Compose class, and they are later processed in the sequence defined here.

```
# %% compose a series of steps
preprocess_steps = transforms.Compose([
    # transforms.Resize(300),  # better (300, 300)
    transforms.RandomRotation(90),
    transforms.CenterCrop((300, 200)),
    transforms.Grayscale(),
    # transforms.RandomVerticalFlip(),
    # transforms.ToTensor(),
    # transforms.Normalize((0.485, 0.456, 0.406), (0.229, 0.224, 0.225)),  # Im-
ageNet values
])
```

Listing 4.2 Data Augmentation (Source: 050_ComputerVision\00_ImagePreprocessing\
ImagePreprocessing.py)

The image is then transferred to the function, and the result is displayed after the steps
have been applied:

```
x = preprocess_steps(img)
```

Some examples of the application of such pre-processing steps are shown in Figure 4.11.
In ❶, the original image can be seen. In ❷, the image has been converted to grayscale
and turned upside down, and in ❸, the image has been randomly rotated with a maximum rotation of 90° (RandomRotation(90)) and then zoomed into by using CenterCrop.

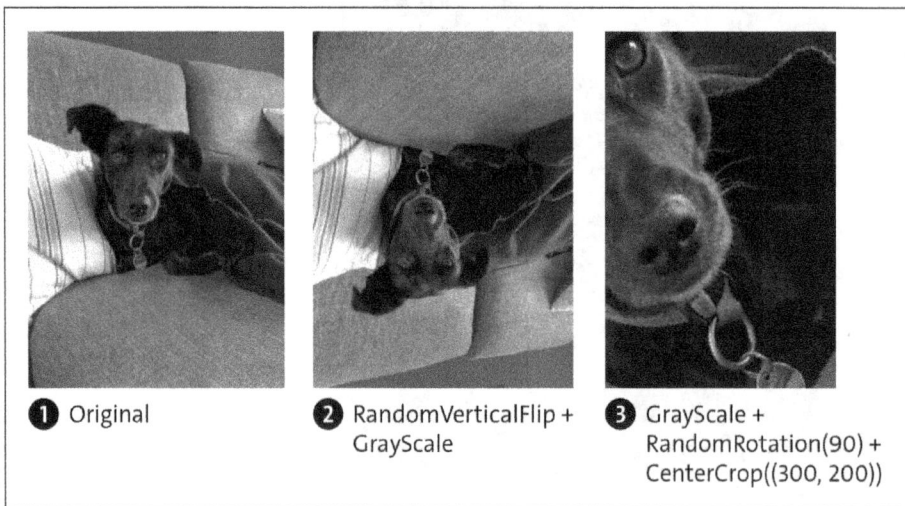

❶ Original   ❷ RandomVerticalFlip +   ❸ GrayScale +
                      GrayScale   RandomRotation(90) +
                                                   CenterCrop((300, 200))

Figure 4.11 Data Augmentation Examples

After we've applied such processing steps to the images, we must convert the bid into a
tensor with transforms.ToTensor(). Often, the images should also be similar in terms of
color ratios, and we can achieve this by normalizing them with transforms.Normalize().

Model Training

The entire script for model training can be found in *050_ComputerVision\010_Binary-ImageClassification\binary_img_classification.py*. We start in Listing 4.3 by loading all the required packages.

```
#%% packages
import json
import torch
import torch.nn as nn
from torch.utils.data import DataLoader, random_split
import torchvision
import torchvision.transforms as transforms
from sklearn.metrics import accuracy_score, confusion_matrix, classification_
report
from sklearn.dummy import DummyClassifier
import seaborn as sns
import matplotlib.pyplot as plt
```

Listing 4.3 Binary Classification: Required Packages (Source: 050_ComputerVision\010_BinaryImageClassification\binary_img_classification.py)

Now, we need the path to the training and test data that we downloaded earlier. We can store this information in the data_path.json file, as follows:

```
#%% load data path
with open('data_path.json', 'r') as f:
    data_path_json = json.load(f)
path_train = data_path_json["path_train"]
path_test = data_path_json["path_test"]
```

Ideally, we should make the code flexible so that it uses GPU capabilities if possible or uses the CPU if no GPU is available. We achieve this by creating a device object that checks whether Cuda is installed, using the following script:

```
#%% check if cuda is available
device = torch.device("cuda" if torch.cuda.is_available() else "cpu")
print(f"Using device: {device}")
```

```
Using device: cuda
```

Cuda is a platform for parallel computing developed by Nvidia, and it can significantly accelerate the computing power of GPUs. For this reason, GPUs have established themselves as the standard for training and inferencing large models in the field of deep learning. In this book, however, we choose examples in which you can run all the scripts even if you don't have a powerful GPU.

I have a GPU at my disposal, but don't worry—the code also runs on a CPU without any problems. The training just takes longer.

We must also modify the images for model training. This is necessary not only to adjust the size of all the images but also to convert them into tensors. We learned about this data augmentation step in the previous section, and we'll now apply this knowledge in practice. An important aspect is that the transformations for model training are more elaborate and complex than the transformations for evaluating and testing the model.

Let's start with the augmentation for training shown in Listing 4.4. We perform the following steps one after the other:

1. To keep the computing effort low, we reduce the images to a size of 32 × 32 pixels.

2. For the same reason, we reduce the images from three color channels (one each for red, green, and blue) to grayscale with only a one-color channel.

3. Using `RandomHorizontalFlip`, we randomly flip the images upside down.

4. We can also use random rotation to help the model during training. The maximum rotation angle is limited to 10° here.

5. At this point, the image is still a `Pillow` object, but what we need for training are tensors. Therefore, we must convert the image into a tensor by using `transforms.ToTensor`.

6. The model should not be confused by different lighting conditions, so may want to normalize the images with appropriate mean values and standard deviations by using `transforms.Normalize`.

```
train_transforms = transforms.Compose([
    transforms.Resize((32, 32)),   # 1.
    transforms.Grayscale(),        # 2.
    transforms.RandomHorizontalFlip(), # 3.
    transforms.RandomRotation(10),     # 4.
    transforms.ToTensor(),   # 5.
    transforms.Normalize(mean=[0.5], std=[0.5])
])
```

Listing 4.4 Binary Classification: Augmentation Steps for Model Training

The augmentation steps for validation and testing in Listing 4.5 are much simpler. Here, we only need to ensure that the images have the correct sizes (i.e., that they match in size [horizontal and vertical pixels] and in the number of color channels [one, since we're using grayscale]).

```
# Transformation for validation and testing (augmentation steps)
test_val_transforms = transforms.Compose([
    transforms.Resize((32, 32)),
    transforms.Grayscale(),
```

```
    transforms.ToTensor(),
    transforms.Normalize(mean=[0.5], std=[0.5])
])
```

Listing 4.5 Binary Classification: Augmentation Steps for Validation and Testing

The hyperparameters for the batch size, number of epochs, and learning rate are defined as follows:

```
# %% Hyperparameter
BATCH_SIZE = 256
EPOCHS = 30
LEARNING_RATE = 0.001
```

The approach to preparing datasets in PyTorch in the context of computer vision is interesting. If we prepare the data as shown in Figure 4.12, it will be very easy to use the images in the context of model training.

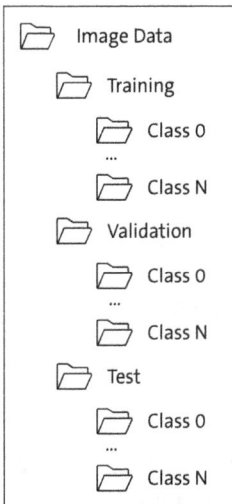

Figure 4.12 Folder Structure for Computer Vision Training

Ideally, the complete dataset will already be divided into training, validation, and test data, and each of these subfolders will have as many subfolders as there are classes to be recognized. In this example, there's a muffin folder and a Chihuahua folder in the training folder. All muffin pictures should be stored in the muffin folder, and all Chihuahua pictures should be stored in the Chihuahua folder. We can check this with the following command:

```
tree C:\Users\BertGollnick\.cache\kagglehub\datasets\samuelcortinhas\muffin-vs-
chihuahua-image-classification\versions\2
```

```
C:.
├───test
│   ├───chihuahua
│   └───muffin
└───train
    ├───chihuahua
    └───muffin
```

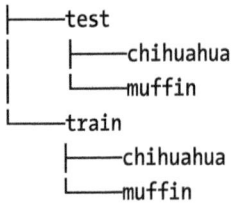

If the data follows this schema, then we can create the data records with a single line of code. The ImageFolder functionality is used for this in Listing 4.6. The folder is passed to this class, along with the transformations to be applied (e.g., train_transformations in this case).

The data is only available in a training and test folder. In this project, we want to check the progress during the model training by using validation data and check the model with test data at the very end. We can prepare the data accordingly, and the first step is to create a training dataset, which we then split into training and validation based on a VALIDATION_SPLIT ratio, as shown in Listing 4.6.

```
train_dataset = torchvision.datasets.ImageFolder(root=path_train,
    transform=train_transforms)
VALIDATION_SPLIT = 0.2
train_size = int((1 - VALIDATION_SPLIT) * len(train_dataset))
val_size = len(train_dataset) - train_size
train_dataset, val_dataset = random_split(train_dataset, [train_size, val_size])
val_dataset.dataset.transform = test_val_transforms
test_dataset = torchvision.datasets.ImageFolder(root=path_test, transform=test_
val_transforms)
```

Listing 4.6 Binary Classification: Datasets for Training, Validation, and Testing

In Listing 4.7, the DataLoader is created based on the generated datasets, the BATCH_SIZE is passed as a parameter, and the randomized data distribution (sampling) is defined with the shuffle=True parameter.

```
train_loader = DataLoader(dataset=train_dataset,
                          batch_size=BATCH_SIZE,
                          shuffle=True)
val_loader = DataLoader(dataset=val_dataset,
                        batch_size=BATCH_SIZE,
                        shuffle=True)
test_loader = DataLoader(dataset=test_dataset,
                         batch_size=BATCH_SIZE,
                         shuffle=True)
```

Listing 4.7 Binary Classification: DataLoader for Training, Validation, and Testing

We then check the size of the data records to assess whether the data breakdown meets our expectations by using the following code:

```
#%% Check sizes of datasets
print(f"Train dataset size: {len(train_dataset)}")
print(f"Validation dataset size: {len(val_dataset)}")
print(f"Test dataset size: {len(test_dataset)}")
```

```
Train dataset size: 3786
Validation dataset size: 947
Test dataset size: 1184
```

The training dataset is four times as large as the validation dataset, and the validation dataset is also large enough to obtain statements about the quality of the model. In addition, the size of the test dataset is similar to that of the validation dataset. Therefore, we can continue to work with these values.

Next, we need to develop the model. We've already learned about the general structure of a convolutional neural network, so now, we'll apply this practically to our dataset. Figure 4.13 shows the schematic structure of our network.

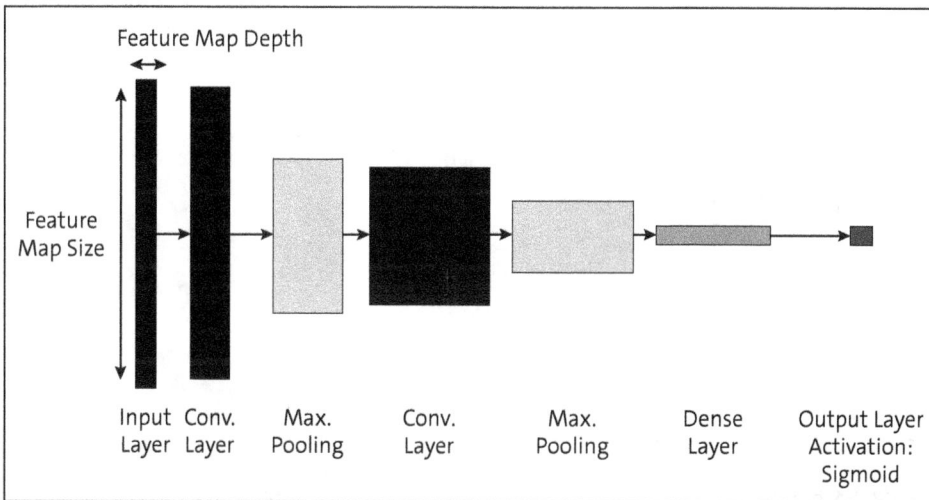

Figure 4.13 Structure of Binary Classification Model

The input layer of the network is determined by the dataset or the transformed images. We'll then reduce the images to 32 × 32 pixels and only use them as grayscales, which means that they'll be available as a tensor with the dimensions (1, 32, 32). We then pass the images through several convolutional and pooling layers.

Next, we must convert the dimensions of the tensor into a format that can be processed by subsequent fully connected layers. This is because the output of the convolutional and pooling layers is a three-dimensional (3D) tensor and the fully connected layers expect a one-dimensional (1D) tensor. Therefore, we must reduce the dimensions, and we can do this by using nn.Flatten.

In Listing 4.8, we now turn to the model structure. Again, we create the custom model based on nn.Module:

```python
class ImageClassificationModel(nn.Module):
    def __init__(self):
        super().__init__()
        self.conv1 = nn.Conv2d(in_channels=1, out_channels=6, kernel_size=3)
        self.conv2 = nn.Conv2d(in_channels=6, out_channels=16, kernel_size=3)
        self.pool = nn.MaxPool2d(kernel_size=2)
        self.relu = nn.ReLU()
        self.fc1 = nn.Linear(16*6*6, 64)
        self.fc2 = nn.Linear(64, 1)
        self.flatten = nn.Flatten()
        self.sigmoid = nn.Sigmoid()

    def forward(self, x):
        x = self.conv1(x)   # [BS, 6, 30, 30]
        x = self.relu(x)
        x = self.pool(x)   # [BS, 6, 15, 15]
        x = self.conv2(x)   #  [BS, 16, 13, 13]
        x = self.relu(x)
        x = self.pool(x)   # [BS, 16, 6, 6]
        x = self.flatten(x) # [BS, 16*6*6]
        x = self.fc1(x) # [BS, 64]
        x = self.relu(x)
        x = self.fc2(x) # x   # output [BS, 1]
        return x

model = ImageClassificationModel().to(device)
# dummy_input = torch.randn(1, 1, 32, 32)   # (BS, C, H, W)
# model(dummy_input).shape
```

Listing 4.8 Binary Classification: Model Structure

In Listing 4.9, we rely on the well-known Adam optimizer. The loss function must fit the problem, and therefore, the binary cross entropy nn.BCEWithLogitsLoss loss function is best suited for this purpose.

BCELoss or BCEWithLogitsLoss

There are basically two suitable loss functions: nn.BCELoss and nn.BCEWithLogitsLoss. It's generally recommended to use nn.BCEWithLogitsLoss. This loss function is more numerically stable and has the advantage that sigmoid activation of the output layer isn't necessary, as sigmoid activation is already integrated into the loss function. We could also use nn.BCELoss, but in this case, it's mandatory to use nn.Sigmoid as the activation of the output layer.

```
# %% optimizer and loss function
optimizer = torch.optim.Adam(model.parameters(),
                             lr = LEARNING_RATE)

loss_fun = nn.BCEWithLogitsLoss()
```

Listing 4.9 Muffin Versus Chihuahua: Optimizer and Loss Function

The training loop from Listing 4.10 follows the familiar pattern. Within the batch loop, we train the model based on the training data and evaluate it using the validation data.

```
train_losses, val_losses = [], []
best_val_loss = float('inf')
for epoch in range(EPOCHS):
    model.train()
    running_train_loss = 0.0
    running_train_loss = 0
    for i, (X_train_batch, y_train_batch) in enumerate(train_loader):
        # move data to device
        X_train_batch = X_train_batch.to(device)
        y_train_batch = y_train_batch.to(device)

        # zero gradients
        optimizer.zero_grad()

        # forward pass
        y_train_pred = model(X_train_batch)

        # loss calc
        loss = loss_fun(y_train_pred, y_train_batch.reshape(-1, 1).float())

        # backward pass
        loss.backward()

        # update weights
        optimizer.step()

        # extract losses
        running_train_loss += loss.item()
    avg_train_loss = running_train_loss / len(train_loader)
```

```
train_losses.append(avg_train_loss)
print(f"Epoch {epoch}: Train Loss: {avg_train_loss}")

# Validation
model.eval()
running_val_loss = 0.0
with torch.no_grad():
    for X_val_batch, y_val_batch in val_loader:
        X_val_batch = X_val_batch.to(device)
        y_val_batch = y_val_batch.to(device)
        y_val_pred = model(X_val_batch)
        val_loss = loss_fun(y_val_pred, y_val_batch.reshape(-1, 1).float())
        running_val_loss += val_loss.item()
avg_val_loss = running_val_loss / len(val_loader)
val_losses.append(avg_val_loss)

# store best model
if avg_val_loss < best_val_loss:
    best_val_loss = avg_val_loss
    torch.save(model.state_dict(), 'best_model.pth')

print(f"Epoch {epoch}: Validation Loss: {avg_val_loss}")
```

Listing 4.10 Binary Classification: Model Training

Model Evaluation

Next, let's look at how the losses developed during the model training. We've calculated them for the training data as well as the validation data, and we can visualize the losses over the epochs. The code required to create the mapping is shown in Listing 4.11.

```
# %%
plt.figure(figsize=(10,6))
sns.lineplot(x=range(EPOCHS), y=train_losses, label='Train Loss [-]')
sns.lineplot(x=range(EPOCHS), y=val_losses, label='Validation Loss [-]')
plt.xlabel('Epoch [-]')
plt.ylabel('Verlust [-]')
plt.title('Train and Validation Loss')
plt.xticks(range(0, EPOCHS, 5))
plt.legend()
```

Listing 4.11 Binary Classification: Model Evaluation

The corresponding Figure 4.14 shows the training and validation loss of the model during training.

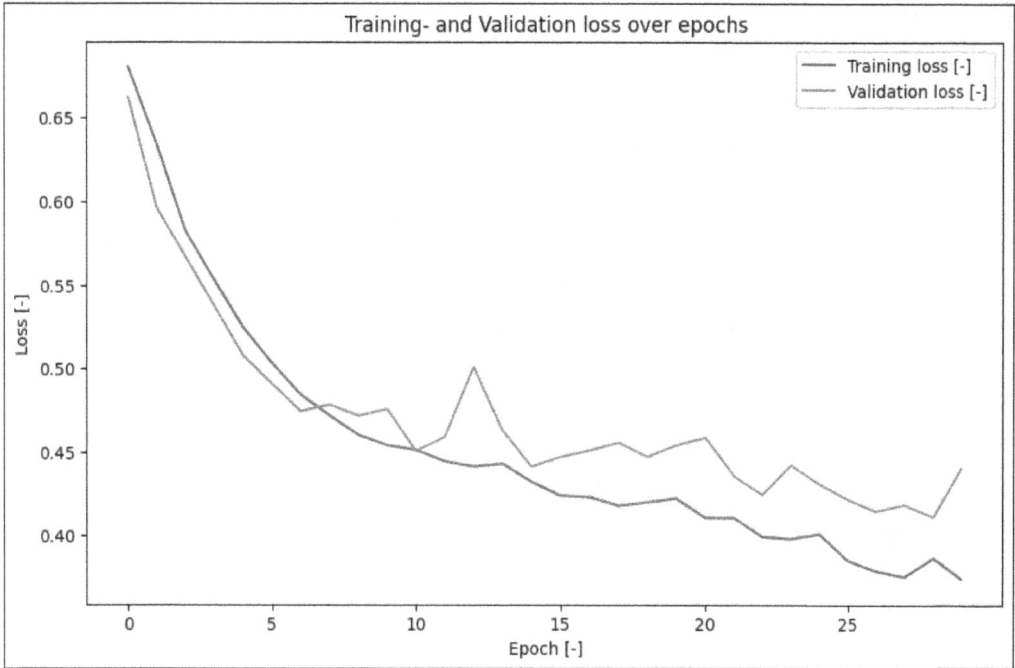

Figure 4.14 Muffin Versus Chihuahua: Training and Validation Process Loss

Over time, or with an increasing number of epochs, the training loss continues to decrease. If we continued training, it would be highly likely that the training loss would decrease even further. But that isn't necessarily the point. What's more important is that the model works well on the validation data, and here, we can observe that there is hardly any improvement in the model (and possibly even a slight increase in training loss) from around 20 epochs onward.

We can assume that the model tends more towards overfitting from about 20 epochs onward and learns training data "by heart," instead of internalizing general patterns. You can load the best model, as follows:

```
#%% load best model
model.load_state_dict(torch.load('best_model.pth'))
```

With this best model, we can create predictions for the test data. This data has never been seen by the model before and is therefore ideal for evaluating the final model's quality.

Listing 4.12 shows how to create predictions. The procedure is like in previous trainings: we create the predictions for each batch by using a forward pass. A special feature here is that the code should work on both CPU and GPUs, so we must load the data onto the device where the model is located. We do this by using the .to(device) method.

```
#%% test loop
y_test_true, y_test_pred = [], []
for i, (X_test_batch, y_test_batch) in enumerate(test_loader):
    # Move input to same device as model
    X_test_batch = X_test_batch.to(device)
    with torch.no_grad():
        y_test_pred_batch = model(X_test_batch)
        y_test_true.extend(y_test_batch.cpu().detach().numpy().tolist())
        y_test_pred.extend(y_test_pred_batch.cpu().detach().numpy().tolist())
```

Listing 4.12 Binary Classification: Predictions for Test Data

Now, we must check which class is assigned to which number. Using the .classes property, we can determine the original classes and find out that "Chihuahua" is at the top of the list and therefore has index 0 while "Muffin" has index 1, as follows:

```
# check the classes of the test set
test_dataset.classes
```

```
['chihuahua', 'muffin']
```

At this point, the predictions are still probabilities (i.e., values between 0 and 1), and we need the associated classes so that we can compare the predictions with the "real" values. The normal approach in binary classification is to assign values below a threshold value to class 0 and values above the threshold value to class 1. If no further information is available, we can set the threshold value to 0.5 for the sake of simplicity, as follows:

```
threshold = 0.5
y_test_pred_class = ['chihuahua' if float(i[0]) > threshold else 'muffin' for i
in y_test_pred]
y_test_true_labels = ['chihuahua' if i == 1 else 'muffin' for i in y_test_true]
```

Now, we've arrived at the goal of it all and can create the confusion matrix, as in Listing 4.13. We can calculate the values by calling the confusion_matrix function and then visualize them as a heat map.

```
cm = confusion_matrix(y_pred=y_test_pred_class, y_true=y_test_true_labels)
plt.figure(figsize=(8,6))
sns.heatmap(cm, annot=True, fmt="d",
            xticklabels=['true_muffin', 'true_chihuahua'],
            yticklabels=['pred_muffin', 'pred_chihuahua'],
            cbar=False)
plt.title('Confusion Matrix')
```

Listing 4.13 Binary Classification: Confusion Matrix Creation

Figure 4.15 shows the confusion matrix. The correct predictions are on the main diagonal, and the incorrect predictions are on the secondary diagonal.

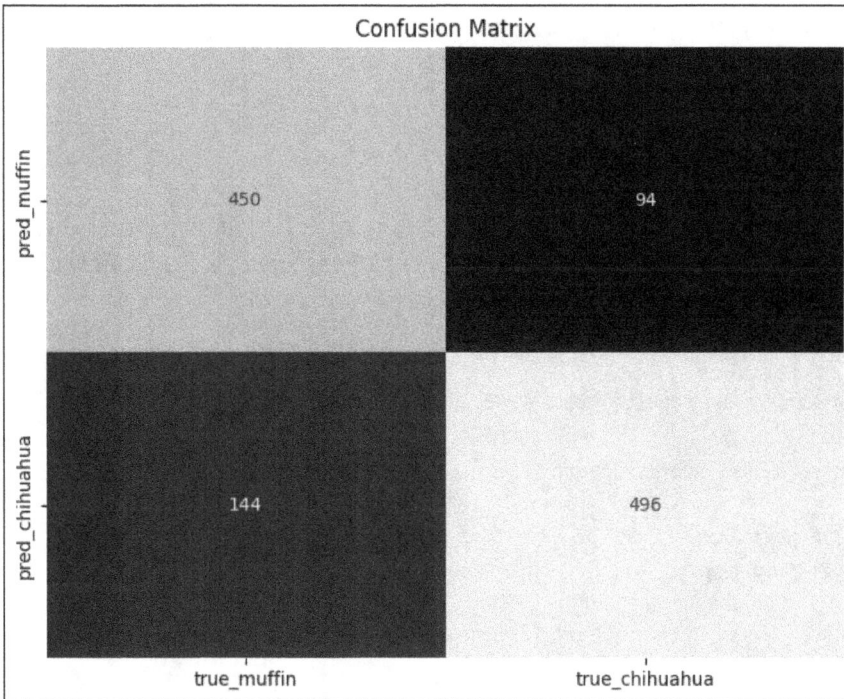

Figure 4.15 Muffin Versus Chihuahua: Confusion Matrix

There are some wrong predictions, but does that make our model useless? We don't know, but we can find out by comparing the performance of our model with that of a model that's based purely on guesswork.

Accuracy is our performance metric. It shows how many predictions were made correctly, and we can determine it by using the accuracy_score function, as follows:

```
accuracy = accuracy_score(y_test_pred_class, y_test_true)
print(f"Model Accuracy: {accuracy*100:.2f}%")
```

Model Accuracy: 79.90%

The model we trained has an accuracy of just under 80%, so let's look at how pure guessing performs for comparison purposes. With the same number of images of both classes, the accuracy should be 50%.

For this purpose, the sklearn package contains the DummyClassifier, which looks at the frequency of the classes in the test data and makes it easy by always predicting the most frequent class. We can also determine the accuracy by using the following:

```
y_test_pred_class = [1 if float(i[0]) > threshold else 0 for i in y_test_pred]
dummy_clf = DummyClassifier(strategy="most_frequent")
dummy_clf.fit(y_test_true, y_test_pred_class)
dummy_clf.score(y_test_true, y_test_pred_class)
print(f"Dummy Classifier Accuracy: {dummy_clf.score(y_test_true, y_test_pred_
class)*100:.2f}%")
```

Dummy Classifier Accuracy: 50.17%

We can see that the DummyClassifier is roughly 50% accurate (i.e., the two classes are distributed almost evenly). Our model, with just under 80% accuracy, is significantly more accurate and thus shows a satisfactory result.

We can calculate even more metrics with classification_report, as follows:

```
print(classification_report(y_test_true, y_test_pred_class))
```

precision recall f1-score support

0 0.83 0.79 0.81 640
1 0.77 0.81 0.79 544

accuracy 0.80 1184
macro avg 0.80 0.80 0.80 1184
weighted avg 0.80 0.80 0.80 1184

The report summarizes the performance of the model. We have the two classes: 0 and 1, and of all the instances when the model predicted class 0, 83% were class 0 (i.e., if the model said, "It is class 0," then it was correct 83% of the time). This is what is meant by *precision*.

Recall is about completeness or *sensitivity*. We can interpret it as follows: of all actual instances of class 0, the model has correctly identified 79% as class 0, so it has found 79% of the images that can actually be assigned to class 0.

In our model, performance is relatively balanced between the two classes, so there is no indication that the model classifies one class significantly better than the other. Both measures, precision and recall, can be summarized in the *F1 score*, which represents the harmonic mean of precision and recall. You'll frequently use the F1 score because it provides a good balance between the two metrics.

The *support* refers to the data basis. A support of 640 for class 0 means that there are 640 instances of the class in the dataset. We've determined the accuracy beforehand: it's 80%.

The *macro-average* calculates the unweighted average of the precision, recall, and F1 score metrics across all classes. You'll use it when each class should have the same significance for the overall quality of the model, regardless of how many samples it contains. Thus, the macro-average gives a clear picture of the performance of the model on the minority class, and it's therefore an important indicator of robustness in the case of imbalanced datasets (which is generally referred to as *imbalance*). There's also the *weighted average*, which uses the number of samples per class for weighting.

This concludes our introduction to binary image classification, and we now turn our attention to the classification of multiple classes.

4.3.2   Multiclass Classification

Now, we come to the prediction of several classes (a number of classes greater than or equal to three). Basically, nothing changes in the training, and we can fall back on convolutional layers again. What does change is the number of output nodes and the loss function. Figure 4.16 illustrates the key differences between binary and multiclass classification.

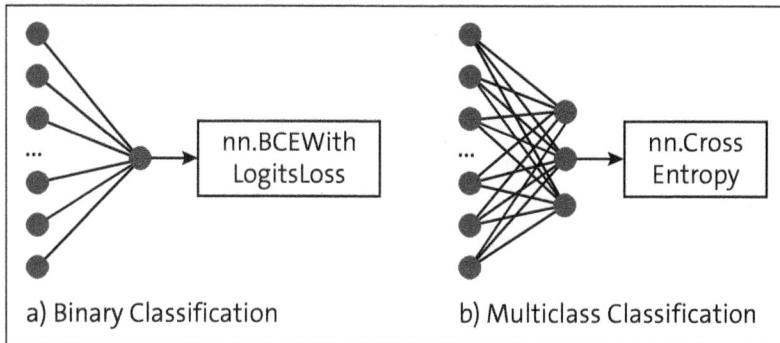

Figure 4.16 Differences in Model and Loss Function for Binary and Multiclass Classification

Whereas with binary classification, there is only one output node whose predicted value corresponds to class 0 or 1, with multiclass classification, there are as many output nodes as there are classes to be predicted.

In addition, there are differences regarding the loss function. We've seen that nn.BCE-WithLogitsLoss is the most sensible choice for binary classification, but with multiclass classification, it's nn.CrossEntropy. With this knowledge, we can now move on to programming.

Data Preparation

The dataset is the "Fingers" dataset provided by Pavel Koryakin on Kaggle. It contains a total of 21,600 images of hands showing different numbers of fingers. Each image is then assigned to the class that corresponds to the number of fingers, so for example, if three fingers can be seen, the picture is assigned to class 3.

A few examples can be seen in Figure 4.17. The images range from no fingers to five fingers, so we already know that there are six different classes.

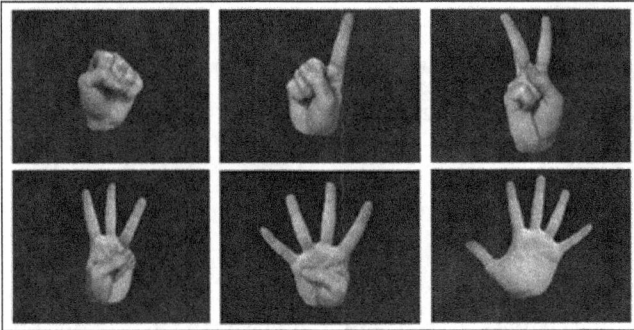

Figure 4.17 Example Images from Fingers Dataset

The complete script is stored at *050_ComputerVision\010_BinaryImageClassification\ binary_img_classification.py.*

First, we load the packages shown in Listing 4.14. We've already familiarized ourselves with all packages at earlier stages.

```
#%% packages
import os
from PIL import Image
import kagglehub
import pandas as pd
import numpy as np
import torch
import torch.nn as nn
from sklearn.metrics import confusion_matrix, accuracy_score
from sklearn.dummy import DummyClassifier
from torch.utils.data import Dataset, DataLoader
import torchvision.transforms as transforms
import seaborn as sns
import matplotlib.pyplot as plt
```

Listing 4.14 Multiclass Classification: Required Packages

Now, we can import the data and follow the steps shown in Listing 4.15. First, we load the dataset using kagglehub to ensure the download to the local system. The data is divided into training and testing data, so we save the corresponding paths in the image_path_ train and image_path_test objects.

```
#%% data import
path = kagglehub.dataset_download("koryakinp/fingers")
```

```
print("Path to dataset files:", path)

image_path_train = os.path.join(path, "train")
image_path_test = os.path.join(path, "test")
```

Listing 4.15 Multiclass Classification: Data Import

It's good practice to define all hyperparameters that influence model training in a central location, so we define them in the following code:

```
#%% Hyperparameters
BATCH_SIZE = 32
EPOCHS = 5
IMG_SIZE = 32
LEARNING_RATE = 0.001
```

The preparation of the dataset is significantly more complex here than in Section 4.3.1 of the binary classification. The images are not stored in subfolders of corresponding classes but are all in the same train or test folder.

Figure 4.18 illustrates the existing folder structure, in which all images are stored next to each other with equal rights. The target size is coded as part of the file name, and the numbers in bold indicate the number of fingers shown. The dataset class we want to create has the task of processing all images and returning the respective image tensor X and the target size y using its __getitem__ method.

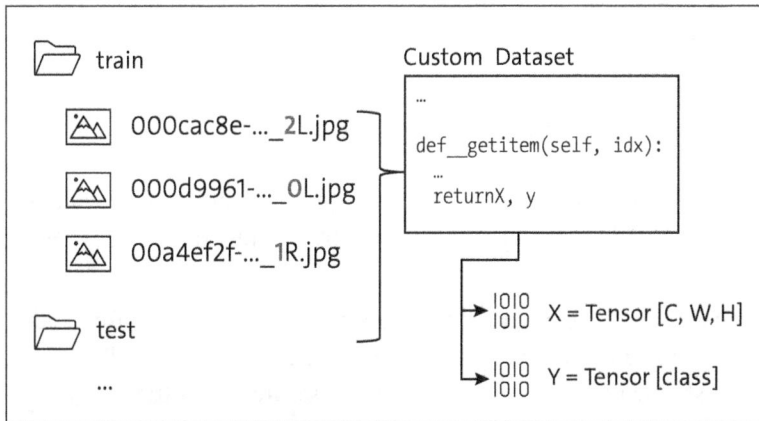

Figure 4.18 Multiclass Classification: Folder Structure and Dataset

Listing 4.16 shows the dataset class, which is initialized with the path to the images and the transformations to be applied. The __init__ method ensures that all image_files images and their target size label are captured. In the __getitem__ method, each individual image is then processed, the transformations are applied, and the image is returned as a tensor and the corresponding label.

```
# %% Dataset
class FingersDataset(Dataset):
    def __init__(self, image_path, transform=None):
        self.transform = transform
        self.image_files = [f for f in os.listdir(image_path)
            if f.endswith('.jpg') or f.endswith('.png')]

        # Load all images during initialization
        self.images = []
        self.labels = []
        for img_name in self.image_files:
            image = Image.open(os.path.join(image_path, img_name))
            if image.mode != 'RGB':
                image = image.convert('RGB')
            self.images.append(image)
            self.labels.append(int(img_name[-6]))

    def __len__(self):
        return len(self.images)

    def __getitem__(self, idx):
        image = self.images[idx]
        y = self.labels[idx]

        # Apply transforms
        X = self.transform(image)
        return X, y
```

Listing 4.16 Multiclass Classification: Dataset Class

We need to augment the images for training to achieve a better training result, and the steps we use for this are listed in Listing 4.17. First, we must adjust the image sizes by using Resize.

We'll perform various steps or the augmentation: RandomHorizontalFlip, ColorJitter, and RandomAffine). RandomHorizontalFlip randomly flips the image horizontally; Color-Jitter randomly adjusts the brightness, contrast, hue or saturation of the image; and RandomAffine performs random rotations and shifts of the image.

We can then convert the image into a tensor with ToTensor, normalize the color values with Normalize, and transform the image into grayscale with Grayscale. When testing, we only need to transform the images into tensors of the corresponding size but not augment them. Therefore, transform_test has significantly fewer steps than transform_train.

```
transform_train = transforms.Compose([
    transforms.Resize((IMG_SIZE, IMG_SIZE)),
    transforms.RandomHorizontalFlip(p=0.5),
    transforms.ColorJitter(brightness=0.2, contrast=0.2, saturation=0.2),
    transforms.RandomAffine(degrees=0, translate=(0.1, 0.1)),
    transforms.ToTensor(),
    transforms.Normalize(mean=[0.485], std=[0.229]),
    transforms.Grayscale(num_output_channels=1),
])

transform_test = transforms.Compose([
    transforms.Resize((IMG_SIZE, IMG_SIZE)),
    transforms.ToTensor(),
    transforms.Normalize(mean=[0.485], std=[0.229]),
    transforms.Grayscale(num_output_channels=1),
])
```

Listing 4.17 Multiclass Classification: Image Transformations

We use the transformations to create the instances of the `FingersDataset` class—`train_dataset` and `val_test_dataset`—as follows:

```
train_dataset = FingersDataset(image_path_train, transform=transform_train)
val_test_dataset = FingersDataset(image_path_test, transform=transform_test)
```

Next, we split the `val_test_dataset` instance to create a `val_dataset` validation dataset and a `test_dataset` test dataset. Then, we check the size of the datasets, as shown in Listing 4.18.

```
# Split validation dataset into validation and test
val_size = len(val_test_dataset)
val_split = val_size // 2
test_split = val_size - val_split

val_dataset, test_dataset = torch.utils.data.random_split(val_test_dataset, [
val_split, test_split])

# Test the splits
print(f"Training set size: {len(train_dataset)}")
print(f"Validation set size: {len(val_dataset)}")
print(f"Test set size: {len(test_dataset)}")

Training set size: 18000
Validation set size: 1800
Test set size: 1800
```

Listing 4.18 Multiclass Classification: Data Splitting

In this case, we have more than enough data for all three classes and can create Data-Loader instances, as follows:

```
train_dataloader = DataLoader(train_dataset, batch_size=BATCH_SIZE, shuffle=
True)
val_dataloader = DataLoader(val_dataset, batch_size=BATCH_SIZE, shuffle=True)
test_dataloader = DataLoader(test_dataset, batch_size=BATCH_SIZE, shuffle=
False)
```

We have now completed the data preparation and can start training the model.

Model Training

The model will be based on the already known convolutional layer and pooling layer. In addition, we'll make use of two further network layers: *batch normalization* and *dropout*.

Before we create the model, we'll familiarize ourselves with batch normalization.

Batch Normalization Layer

Batch normalization is a network layer used to make training more stable, faster, and more effective.

Imagine that a deep neural network with multiple layers is being trained. During training, the model weights are constantly adjusted. The data is analyzed in batches, and if the data in the batches is extremely different, the weights of the layers must constantly adapt to a changing input distribution. It's like someone constantly moving the carpet under your feet as you try to balance, and the technical term for this problem is *internal covariate shift*. This can make training become unstable and make the optimizer experience convergence problems. You can counter this problem by reducing the learning rate, but that slows down training.

Batch normalization normalizes the activations of the intermediate layers. You can do this per batch during training, and the procedure is similar to normalization in classical machine learning:

- You calculate the means and variances of the current batch.
- You normalize the activations.
- You perform scaling and shifting for batch normalization, and the scaling and shifting factors you use are themselves learnable parameters of the batch norm layer.

PyTorch provides readymade functions such as nn.BatchNorm1d and nn.BatchNorm2d, depending on whether 1D or 2D data is being processed.

The second new network layer that we'll be using is the dropout layer. We'll also get to know this layer before creating the model.

Dropout Layer

Dropout layers serve to reduce overfitting in networks. Essentially, certain neurons and their connections are randomly "deactivated" during training, and the probability p of deactivation is a hyperparameter that the developer can specify. The deactivated neurons don't contribute to the forward pass and don't participate in backpropagation.

Here is a brief analogy: Imagine working in a large project team. Every day, a random decision is made as to which team members have the day off, and the remaining team members have to do all the work. This allows you to determine which key people are particularly important for the project.

During model inference, all neurons in the network are used, but the fact that only part of the network was active during training is scaled with the corresponding probabilities.

Frameworks, including PyTorch, often take the opposite approach, known as *inverted dropout*:

- During training, the output of a neuron that remains active is divided by $(1 - p)$. When a neuron is deactivated, its output is 0.

- During inference, all neurons are active and no scaling is necessary, as this has already been done during training.

In the network class, we'll now use convolutional layers, max pooling layers, batch normalization, and dropout in addition to convolutional layers.

Figure 4.19 shows the structure of the network. There are two convolutional blocks (i.e., groups of layers that start with a convolutional layer), followed by `BatchNorm`, `MaxPooling`, and `Dropout`. For better readability, networks are often subdivided into individual blocks. After the two convolutional blocks, there are fully connected layers, and the output layer is at the end.

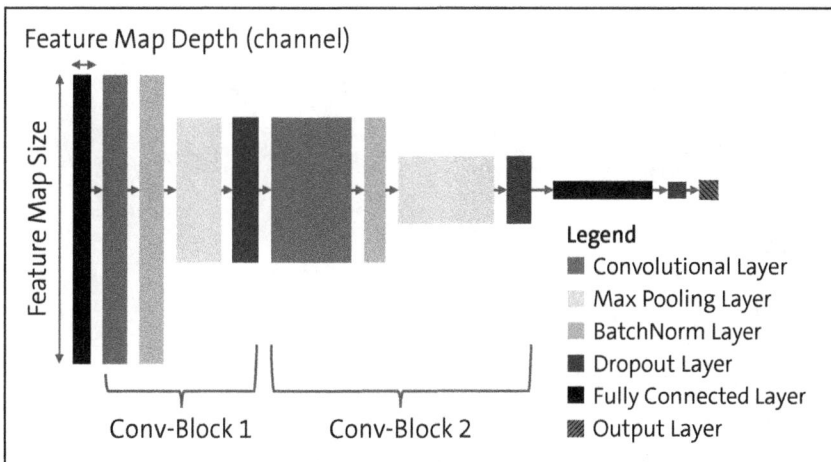

Figure 4.19 Multiclass Classification: Network Structure

The implementation of this network structure is shown in Listing 4.19.

```python
class FingersModel(nn.Module):
    def __init__(self, num_classes):
        super(FingersModel, self).__init__()
        # First conv block
        self.conv1 = nn.Conv2d(1, 32, kernel_size=3, stride=1, padding=1)
        self.bn1 = nn.BatchNorm2d(32)

        # Second conv block
        self.conv2 = nn.Conv2d(32, 64, kernel_size=3, stride=1, padding=1)
        self.bn2 = nn.BatchNorm2d(64)

        self.pool = nn.MaxPool2d(2, 2)
        self.flatten = nn.Flatten()

        # Calculate input features for first FC layer
        self._to_linear = 64 * 8 * 8  # After 2 pooling layers: 32->16->8

        # Fully connected layers
        self.fc1 = nn.Linear(self._to_linear, 256)
        self.fc2 = nn.Linear(256, num_classes)

        # Activations and regularization
        self.relu = nn.ReLU()
        self.dropout = nn.Dropout(0.5)
        self.dropout2d = nn.Dropout2d(0.25)

    def forward(self, x):
        # Block 1
        x = self.conv1(x)
        x = self.bn1(x)
        x = self.relu(x)
        x = self.pool(x)
        x = self.dropout2d(x)

        # Block 2
        x = self.conv2(x)
        x = self.bn2(x)
        x = self.relu(x)
        x = self.pool(x)
        x = self.dropout2d(x)
```

```
# FC Layers
x = self.flatten(x)
x = self.fc1(x)
x = self.relu(x)
x = self.dropout(x)

x = self.fc2(x)
return x
```

Listing 4.19 Multiclass Classification: Network Structure

An important aspect is the capitalization of expenses, as this is linked to the loss function used. There are the following two options for implementing this:

- Activation using nn.LogSoftmax() and use of the loss function nn.NLLLoss().
- Direct use of raw outputs (logits) without further activation and use of nn.CrossEntropyLoss(). This approach is generally recommended as it's numerically more stable and easier to implement.

After creating the class, we create a model instance and define the loss function and the optimizer. It's important here for the loss function to match the problem, and since we're dealing with a multiclass classification problem here, CrossEntropyLoss is a sensible choice, as follows:

```
model = FingersModel(num_classes=6)
criterion = nn.CrossEntropyLoss()
optimizer = torch.optim.Adam(model.parameters(), lr=LEARNING_RATE)
```

In Listing 4.20, we carry out the actual training. It's important to know here that the loss function can process the data in different formats. The y_pred predictions come in the form of [BATCH_SIZE, N_CLASS] (i.e., the logits [the raw predictions] are created for each image prediction). These are compared with the real y_train values.

```
#%% training loop
losses_train, losses_val = [], []
for epoch in range(EPOCHS):
    model.train()  # Set model to training mode
    epoch_loss = 0
    for X_train, y_train in train_dataloader:
        # Forward pass
        y_pred = model(X_train)
        # calculate loss
        loss_train = criterion(y_pred, y_train)
        # zero gradients
        optimizer.zero_grad()
```

```
        # backward pass
        loss_train.backward()
        # update weights
        optimizer.step()
        # update epoch loss
        epoch_loss += loss_train.item()
    losses_train.append(epoch_loss / len(train_dataloader))


    # Validation loop
    model.eval()  # Set model to evaluation mode
    with torch.no_grad():
        epoch_loss = 0
        for X_val, y_val in val_dataloader:
            # forward pass
            y_pred = model(X_val)
            # calculate loss
            loss_val = criterion(y_pred, y_val)  # CrossEntropyLoss expects raw
logits and target as class indices
            # update epoch loss
            epoch_loss += loss_val.item()
        losses_val.append(epoch_loss / len(val_dataloader))
    print(f"Epoch {epoch+1}/{EPOCHS}, Loss: {losses_train[-1]:.4f}, Val Loss:
{losses_val[-1]:.4f}")
```

Listing 4.20 Multiclass Classification: Training Loop

Once we've successfully trained the model, we can use the code in Listing 4.21 to visualize the losses over the training and validation data to check whether the model has converged.

```
sns.lineplot(x=range(len(losses_train)), y=losses_train, label='Training')
sns.lineplot(x=range(len(losses_val)), y=losses_val, label='Validierung')
plt.xlabel('Epoch [-]')
plt.ylabel('Verlust [-]')
plt.title('Verlustkurve für Trainings- und Validierungsdaten')
plt.legend()
plt.xticks(range(len(losses_train)))  # Set integer ticks
plt.show()
```

Listing 4.21 Multiclass Classification: Training and Validation Losses

The training losses decrease, and the validation losses don't seem to improve with longer model training, as shown in Figure 4.20.

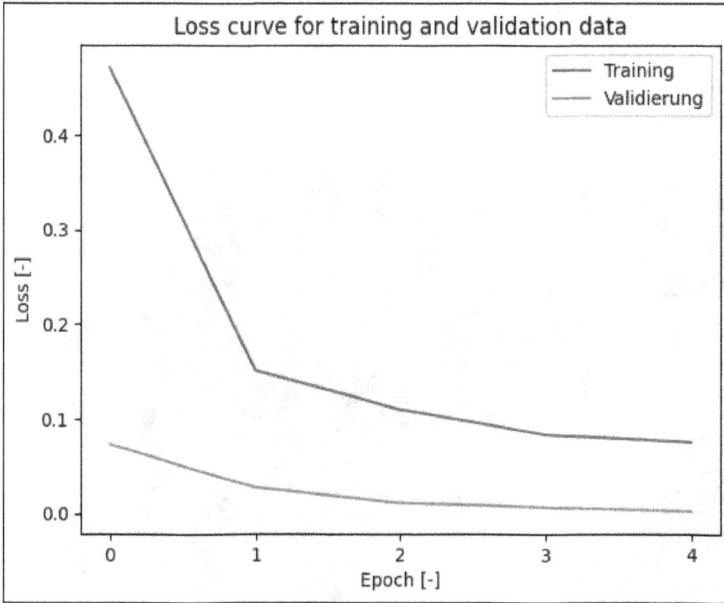

Figure 4.20 Multiclass Classification: Loss Curve

Model Evaluation

Now, we'll use the confusion matrix to check how good the predictions for the test data are. The corresponding code is listed in Listing 4.22.

```
# %% evaluate test data
y_true = []
y_pred = []
model.eval() # Set model to evaluation mode
with torch.no_grad():
for X_test, y_test in test_dataloader:
outputs = model(X_test)
_, predicted = torch.max(outputs.data, 1)
y_pred.extend(predicted.cpu().numpy().tolist())
y_true.extend(y_test.cpu().numpy().tolist())

cm = confusion_matrix(y_true=y_true, y_pred=y_pred)
plt.figure(figsize=(10,8))
sns.heatmap(cm, annot=True, fmt='.0f')
plt.xlabel('Predicted class')
plt.ylabel('Actual class')
plt.title('Finger model:Confusion matrix')
plt.show()
```

Listing 4.22 Multiclass Classification: Confusion Matrix Visualization

Figure 4.21 shows the confusion matrix with the predicted and actual classes. The model was able to predict the number of fingers very well, and there are hardly any misclassifications.

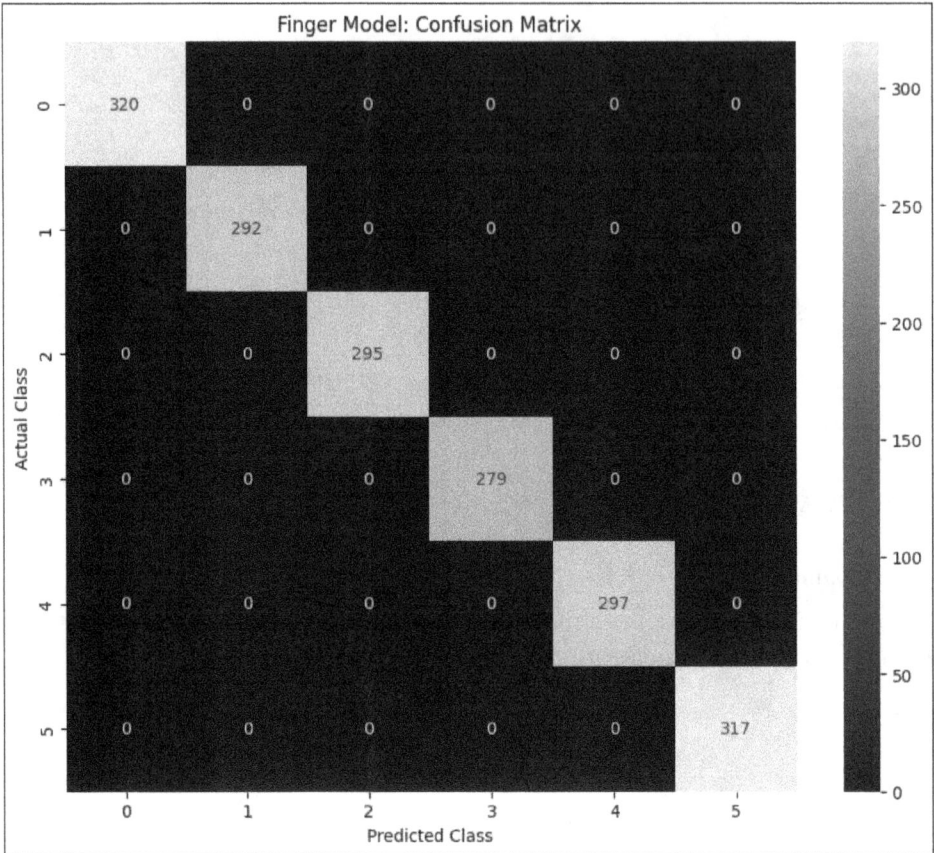

Figure 4.21 Multiclass Classification: Confusion Matrix

We can also condense the very good results in the confusion matrix into a single metric: accuracy, which we can output as follows:

```
accuracy_score(y_true, y_pred)
```

0.9994444444444445

Our model returns the correct class in 99.94% of cases, which is an outstanding result. To be on the safe side, let's look at the results of pure guessing as a reference. For this purpose, we can create the dummy classifier, which always predicts the most frequent class, as follows:

```
#%% dummy classifier
y_pred_dummy = np.zeros(len(y_true))
accuracy_score(y_true, y_pred_dummy)
```

0.16666666666666666

Since the data is perfectly balanced and there are the same number of images of each class, the percent of correct predictions by guessing is 16.67%. This gives us the final proof that our model delivers a very good result.

4.4   Object Detection

So far, we've predicted classes at the level of an entire image. Now, we'll go beyond that with object detection and create predictions for image areas in which we can see the classes. The image areas are represented as rectangular boxes (called *bounding boxes*). This can involve many different boxes of the same class or different classes.

An example is shown in Figure 4.22. A test image ❶ is shown on the left-hand side, and the test image with superimposed predictions ❷ is shown on the right-hand side. In this example, we'll analyze drone images showing sheep and then predict their positions afterwards.

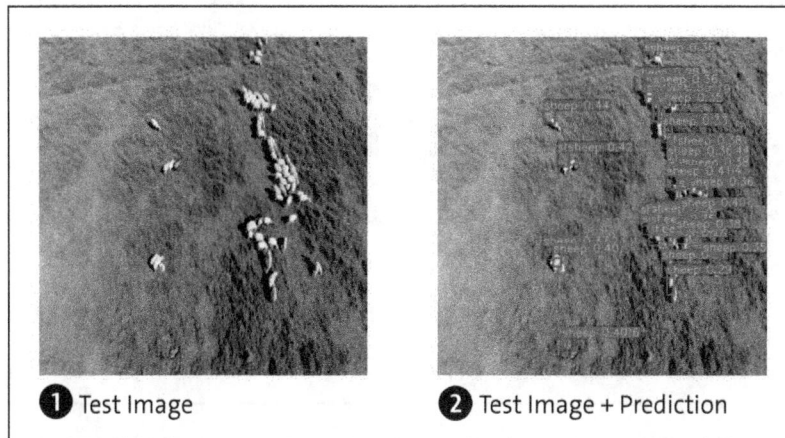

❶ Test Image   ❷ Test Image + Prediction

Figure 4.22 Object Detection: Model Inference

First, we should gain a rough understanding of how object detection works, and then, we'll perform our typical process, which consists of data preparation, model training, model evaluation, and model inference.

4.4.1   How Does Object Detection Work?

Object detection is based on convolutional neural networks, and it's designed to automatically learn features and thus simultaneously localize and classify objects. A basic distinction is made between two-stage and single-stage models. Initially, researchers developed two-stage models that first suggest regions in the image where objects could be located and then analyze and classify those suggested regions. This approach is very precise but also slower.

Since object detection is often used in real-time applications, researchers have developed one-step models that can make predictions faster. With single-stage models, localization and classification take place in a single run. The best-known representative of this model type is *you only look once* (YOLO), and it's also the model we'll look at in more detail. The general procedure for YOLO is as follows:

1. YOLO places a grid over the image, and each cell in this image is used to make predictions about objects whose center lies within that cell. A neural network is used for this purpose, and its structure is shown in Figure 4.23, where we can see the familiar convolutional layers and pooling layers.

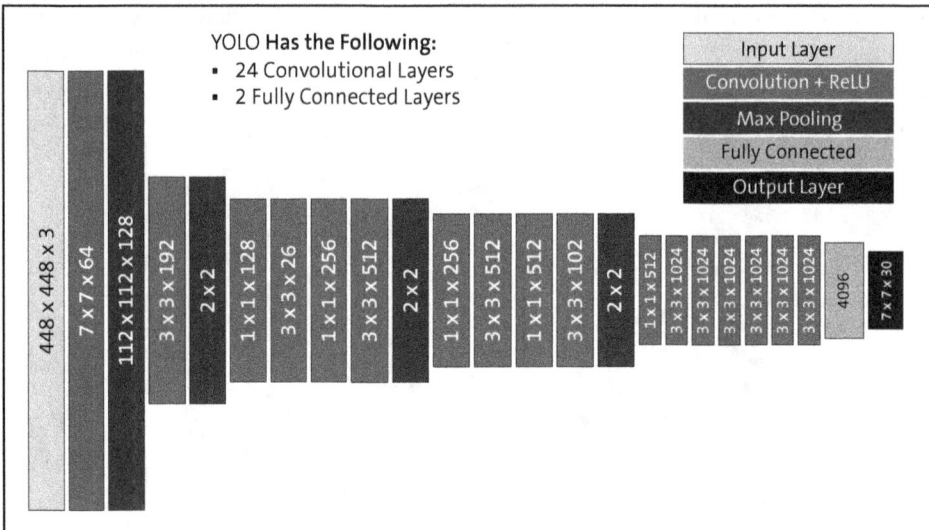

Figure 4.23 YOLO Architecture

2. We now have many overlapping boxes as a result. Each box has a confidence value that the corresponding object can be detected in it, and boxes with confidence values below a threshold are filtered out.

3. There may still be several slightly different boxes for the same object. This is where an algorithm called nonmax suppression comes into play. It identifies the box with the highest confidence and filters out all other overlapping boxes that could represent the same object. This leaves only the best and most reliable prediction for each object.

Fortunately, we don't have to define this already relatively complex model ourselves. Instead, we can rely on a readymade framework, and you can see exactly how this works in the following practical implementation.

4.4.2   Data Preparation

For this example, we'll use the HuggingFace "Aerial Sheep Dataset". It shows sheep from a bird's eye view and contains boxes around each sheep. We'll prepare a script for data preparation, and you can find it at *050_ComputerVision\040_ObjectDetection\10_data_prep_sheep.py*.

Listing 4.23 shows the required packages and functions. Worth mentioning here is the draw_bounding_boxes function, which we can use later to overlay the boxes in the images.

```
#%% packages
from datasets import load_dataset
from torchvision.utils import draw_bounding_boxes
import torch
import numpy as np
import matplotlib.pyplot as plt
import yaml
import os
from PIL import Image
```

Listing 4.23 Object Detection: Import Required Packages

Now, we can download the dataset with load_dataset:

```
ds = load_dataset("keremberke/aerial-sheep-object-detection")
```

The dataset is already set up well, so we can easily create the training dataset (ds_train), validation dataset (ds_val), and test dataset (ds_test), as follows:

```
#%% create datasets
ds_train = ds['train']
ds_val = ds['validation']
ds_test = ds['test']
```

It's also good to look at a concrete example image img and the corresponding boxes, if only to check access to the data:

```
#%% show dataset with true labels (bounding boxes)
img = ds_train['image'][0]
boxes = ds_train['objects'][0]['bbox']
```

Next, we must convert the image and the boxes into tensors, as follows:

```
# Convert PIL Image to tensor and ensure correct format
img_tensor = torch.from_numpy(np.array(img)).permute(2, 0, 1)
boxes_tensor = torch.tensor(boxes)
```

We can define the boxes in a variety of ways. In this case, they are in *COCO* format, but we must convert them to YOLO format.

Box Label Formats in Object Detection

At the beginning of this chapter, we talked about different model approaches. The various models were defined by different development teams with the corresponding label formats, and the best-known formats are as follows:

- Pattern analysis, statistical modeling, and computational learning visual object classes (Pascal VOC)
- Common objects in context (COCO)
- You only look once (YOLO)

Pascal VOC was developed by the University of Oxford, the label format is XML-coded, and there is ultimately one file per image. *COCO* was trained by Microsoft with 2.5 million images, and the labels are all combined in a single JSON file. YOLO has one label file per image, and the boxes are described line by line.

Another key difference among the three formats is how the boxes are defined. Figure 4.24 illustrates how the different algorithms describe the box coordinates.

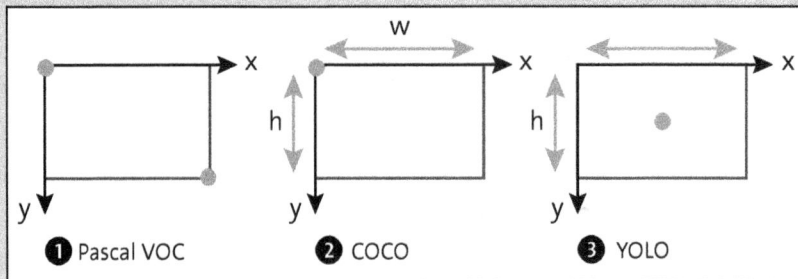

Figure 4.24 Object Detection: Box Label Formats

In Pascal VOC ❶, the upper left and lower right corners are used as references for the box. In COCO ❷, the reference is located at the upper left corner of the box and the horizontal and vertical extensions are based on this reference point. With YOLO ❸, the reference point is exactly in the middle of the box and the horizontal and vertical extensions are also required.

If the data used is in a format that doesn't match the algorithm used, we must convert it. This is the case in our example, and we can do it for the example image by using the code in Listing 4.24.

```
# Convert boxes from [x_center, y_center, width, height] to [x_min, y_min, x_
max, y_max] format
boxes_xyxy = torch.zeros_like(boxes_tensor)
boxes_xyxy[:, 0] = boxes_tensor[:, 0] - boxes_tensor[:, 2]/2 + 0 # x_min
boxes_xyxy[:, 1] = boxes_tensor[:, 1] - boxes_tensor[:, 3]/2 + 10  # y_min
boxes_xyxy[:, 2] = boxes_tensor[:, 0] + boxes_tensor[:, 2]/2 + 10 # x_max
boxes_xyxy[:, 3] = boxes_tensor[:, 1] + boxes_tensor[:, 3]/2 + 10  # y_max
```

Listing 4.24 Object Detection: Data Preparation; Extraction of Boxes

At this point, we have completed the main work and can now use `draw_bounding_boxes` and `matplotlib` for the visualization, as shown in Listing 4.25.

```
img_with_boxes = draw_bounding_boxes(img_tensor, boxes_xyxy, colors='red',
width=2)

#%% show dataset with bounding boxes
plt.figure(figsize=(10,10))
plt.imshow(img_with_boxes.permute(1,2,0))  # Convert from CHW to HWC format for
matplotlib
plt.axis('off')
plt.show()
```

Listing 4.25 Object Detection: Data Preparation; Visualization of Sample Image Including Boxes

The output was shown in Figure 4.22, at the beginning of this section.

Now, we can determine and output the number of images available to us in the various buckets, as follows:

```
# %% number of images in train, val, test
print(f"Number of images in train: {len(ds_train)}")
print(f"Number of images in val: {len(ds_val)}")
print(f"Number of images in test: {len(ds_test)}")

Number of images in train: 3609
Number of images in val: 350
Number of images in test: 174
```

There are usually a lot of sheep in each of these pictures, and that brings us to the question of how many pictures or boxes you need for good model training. If you want to train your own dataset, the question arises as to how much data you need to annotate beforehand to achieve a good result. This depends on many different factors:

- One important influencing factor is the quality of the images. As a rule, the images should be sharp and the boxes must be placed precisely. Incorrect annotation misleads the model and prolongs the training process.

- The images should also be diverse. As is so often the case in life, quality is more important than quantity. The complexity of the images that the model will process later should be reflected in the training data, so it's best to include a wide range of variations, such as different angles, different sizes of objects, and diverse lighting conditions. Partial occlusions (i.e., in our case, one sheep partially obscuring another) can also be helpful.

- We should almost always use a pretrained network for object detection. The model has already "seen" millions of images and now only needs to learn to detect the specific objects. If we use pretraining, a few dozen to 100 images per class may be sufficient.

- The more different object classes the model has to distinguish, the more data is required.

- The complexity of the objects also has a major influence. Objects that hardly differ (such as Lego bricks, QR codes, and license plates) certainly require fewer training images than objects that vary greatly (such as people in different poses or wearing different clothes).

- If only a small amount of data is available, applying data augmentation approaches can be helpful. This involves artificially increasing the amount of training data by varying the images through rotation, mirroring, or scaling.

Essentially, there is no fixed number you should shoot for, but there should be at least a few hundred well-annotated images per object class.

We need to prepare the images and labels in such a way that they are stored in the corresponding folders, like in the following setup:

```
images
└──train
└──fall
└──test
labels
└──train
└──fall
└──test
```

To do this, we first create the folders as shown in Listing 4.26.

```
# %% Define paths for saving images and labels
data_dir = '.'
img_train_dir = os.path.join(data_dir, 'images', 'train')
labels_train_dir = os.path.join(data_dir, 'labels', 'train')
```

```
img_val_dir = os.path.join(data_dir, 'images', 'val')
labels_val_dir = os.path.join(data_dir, 'labels', 'val')
img_test_dir = os.path.join(data_dir, 'images', 'test')
labels_test_dir = os.path.join(data_dir, 'labels', 'test')

# Create directories if they don't exist
os.makedirs(img_train_dir, exist_ok=True)
os.makedirs(labels_train_dir, exist_ok=True)
os.makedirs(img_val_dir, exist_ok=True)
os.makedirs(labels_val_dir, exist_ok=True)
os.makedirs(img_test_dir, exist_ok=True)
os.makedirs(labels_test_dir, exist_ok=True)
```

Listing 4.26 Object Detection: Data Preparation; Folder Creation

We must now populate these folders with the images and label files. To do this, we write the process_dataset_split function in Listing 4.27, which is passed to the splitting object. The function is then called for training, validation, and test data.

```
def process_dataset_split(dataset_split, img_dir, label_dir):
    for i, item in enumerate(dataset_split):
        img = item['image']
        boxes = item['objects']['bbox']

        img_filename = f"{i:05d}.jpg" # Padding required
        label_filename = f"{i:05d}.txt"

        # Save image
        img_path = os.path.join(img_dir, img_filename)
        img.save(img_path)

        # Process and save labels
        label_path = os.path.join(label_dir, label_filename)
        with open(label_path, 'w') as f:
            for bbox in boxes:
                # Get image dimensions
                img_width, img_height = img.size

                x_center, y_center, width, height = bbox

                # Normalize bounding box coordinates
                # Ensure float conversion for division
                x_center_norm = x_center / img_width
                y_center_norm = y_center / img_height
                width_norm = width / img_width
```

```
                     height_norm = height / img_height
                     f.write(f"0 {x_center_norm} {y_center_norm} {width_norm}
{height_norm}\n")


print("Processing training data...")
process_dataset_split(ds_train, img_train_dir, labels_train_dir)
print("Processing validation data...")
process_dataset_split(ds_val, img_val_dir, labels_val_dir)
print("Processing test data...")
process_dataset_split(ds_test, img_test_dir, labels_test_dir)
```

Listing 4.27 Object Detection: Data Preparation; Image and Label File Creation

We use the ultralytics framework, which requires a YAML configuration file, for model training. Using such a framework is helpful because we don't have to reinvent the wheel—it's much easier if we use an existing system such as ultralytics. Listing 4.28 shows how we prepare this.

In the YAML file, we must specify the base path to the dataset (path) and the folder containing the training, validation, and test images. The nc parameter describes how many different classes are to be detected, and the corresponding class names are described in names. Once we've created this dictionary, we can save it to a JSON file by using yaml.dump.

```
# %% create yaml file for yolo training
yaml_file_content = {
    'path': f'./{data_dir}', # Base path to your dataset
    'train': 'images/train',
    'val': 'images/val',
    'test': 'images/test', # Add test set for evaluation after training
    'nc': 1,
    'names': ['sheep']
}


# Save the YAML file
yaml_path = os.path.join(data_dir, 'sheep_yolo.yaml')
with open(yaml_path, 'w') as file:
    yaml.dump(yaml_file_content, file, default_flow_style=False)


print(f"\nYAML configuration saved to: {yaml_path}")
```

Listing 4.28 Object Detection: Data Preparation; YAML Configuration File

The YAML file can be found at *050_ComputerVision\040_ObjectDetection\data.yaml*, and its content can be seen in Listing 4.29.

```
names:
- sheep
nc: 1
train: train
val: val
```

Listing 4.29 Object Detection: Data Preparation: YAML Configuration File

At this point, we have prepared the data and can direct our attention to model training.

4.4.3   Model Training

We carry out the training via the file found at *050_ComputerVision\040_ObjectDetection\20_train_sheep_detect.py*.

We initialize the model by passing the name of an existing model, as shown in the following code. The models can be found in the package descriptions.

```
#%% packages
from ultralytics import YOLO
import os
#%% base-model loading
model = YOLO('yolov8n.pt')
```

Before we can start the training, we need to define the model tracking, and we use MLflow for that. MLflow is generally used to manage, track, and deploy machine learning experiments, while PyTorch is used to create and train models. We should save the training results in a local folder, and we should therefore adjust the MLFLOW_TRACKING_ URI environment variable accordingly, as follows:

```
#%% train
# Set MLflow tracking URI to local file system
os.environ["MLFLOW_TRACKING_URI"] = "file:./mlruns"
```

Now, we can really get started. With the code in Listing 4.30, we'll pass the prepared *sheep_yolo.yaml* configuration file to the model.train method, alongside model-specific parameters such as the number of training epochs, the image size (imgsz), or the number of batches (batch).

```
# Train the model
yaml_path = 'sheep_yolo.yaml'
print("\nStarting YOLOv8 training...")
```

```
results = model.train(data=yaml_path, epochs=10, imgsz=320, batch=16) # Adjust
epochs and batch as needed

print("\nTraining complete!")
```

Listing 4.30 Object Detection: Model Training

Training takes a few minutes or more, depending on the hardware equipment.

4.4.4   Model Evaluation

The model training metrics provide a good overview of the model's progress, which is shown in Figure 4.25. This can be found in the *mlruns* folder under the *results.png* name. You can see various metrics of model training over 10 epochs.

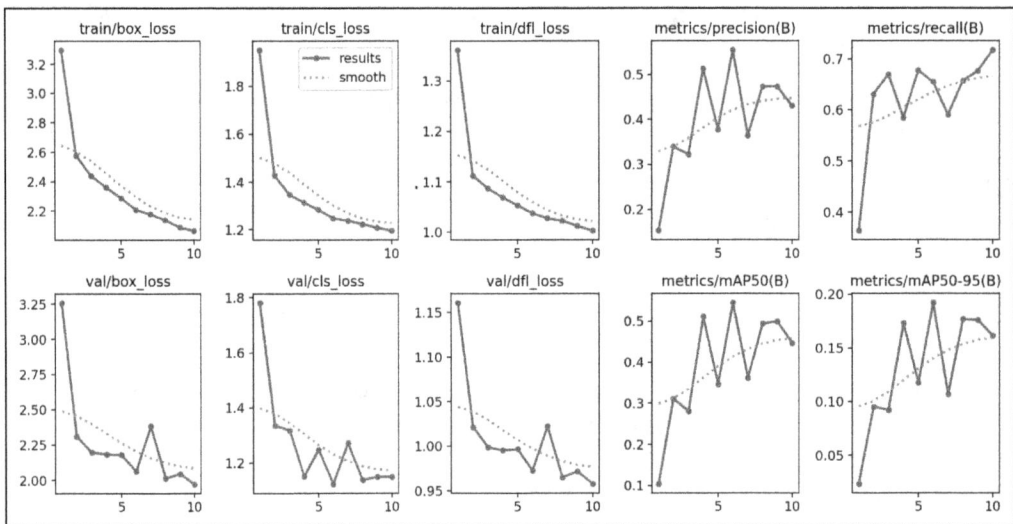

Figure 4.25 Object Detection: Training Metrics

The first three diagrams in each row represent the loss function, the first row shows the training results, and the second row shows the validation results. The diagrams track the following kinds of losses:

- **Box losses**
  Theses include `train/box_loss` and `val/box_loss`, and they measure how well the model predicts the bounding boxes of the objects.

- **Classification losses**
  These include `train/cls_loss` and `val/cls_loss`, and they evaluate the quality of the classification of the objects within the bounding boxes.

- **Distributed focal losses (DFLs)**
  These include `train/dfl_loss`, and they represent special, newer losses that are used to improve localization accuracy.

Before we move on to the other columns in the graph, let's look at some metrics used in object detection. To evaluate the quality of an object detection model, we need some additional metrics that are specifically tailored to this problem. We'll derive these using an example. Imagine that an object detection algorithm has been trained to recognize ships. Figure 4.26 shows a test image with the actual objects (light blue boxes) and the model's predictions (dark blue boxes).

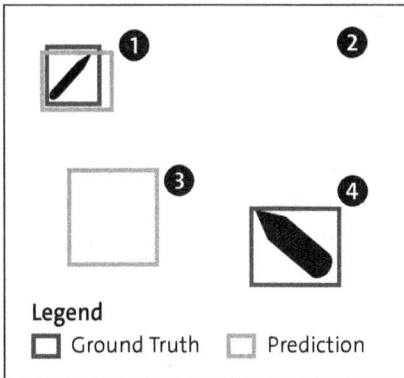

Figure 4.26 Object Detection: Ground Truth and Predictions

This results in the following combinations:

- **True positive ❶**
  The model predicts a ship, and there is a ship at the location.
- **True negative ❷**
  There is no ship, and none was predicted.
- **False positive ❸**
  An object was predicted, and there is no object at the location.
- **False negative ❹**
  An object is at the location, but it wasn't predicted.

We can derive corresponding metrics, such as precision, from these basic metrics.

Figure 4.27 illustrates how to determine precision. We can create a confusion matrix in which we represent the assignment of all boxes in true positive (TP), true negative (TN), false positive (FP), and false negative (FN) in a matrix.

Precision is the ratio of all correctly predicted (TP) boxes to all predicted boxes.

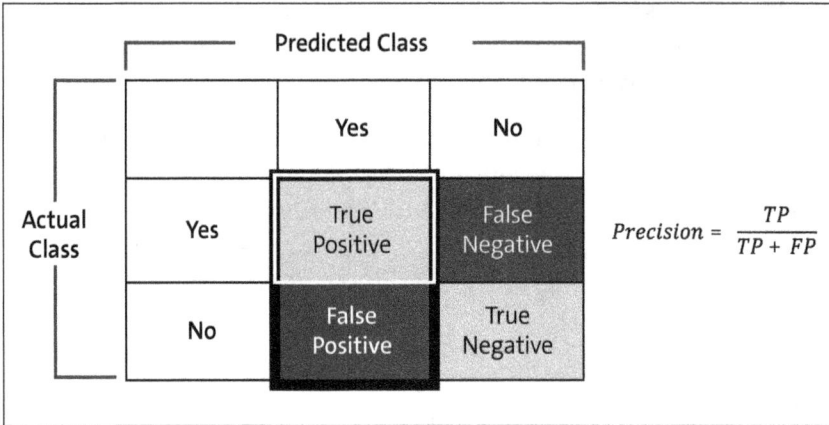

Figure 4.27 Object Detection: Precision Metric

As a rule, there is no perfect match between the boxes, and we need a measure that quantifies the prediction of a box that is only partially correct. This measure is called *intersection over union* (IoU). Figure 4.28 shows the example situation from ❶ in Figure 4.26. A box was predicted reasonably correctly, and we want to quantify the meaning of "reasonably." To do this, we determine the intersection (i.e., the area shared by both boxes), and it represents the numerator of the IoU value we wish to determine. The denominator is the total area of both boxes (which we call the *union*).

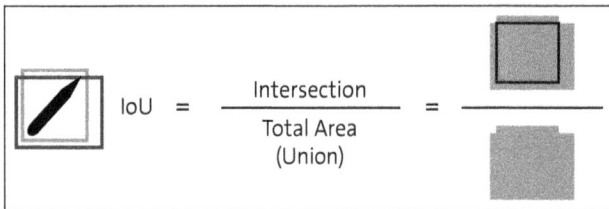

Figure 4.28 Object Detection: Intersection over Union

The prediction can also vary in quality, and we can quantify this quality by using IoU. The IoU value ranges between 0 and 1.

Figure 4.29 shows the box of an actual object, as well as various predictions.

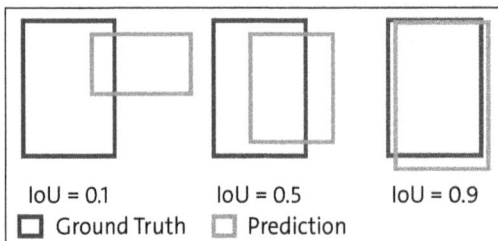

IoU = 0.1   IoU = 0.5   IoU = 0.9
☐ Ground Truth   ☐ Prediction

Figure 4.29 Object Detection: IoU Examples

If the prediction matches the actual box only slightly, then the IoU value will be very small (as in the example on the left). If the prediction matches the actual box relatively well, the IoU value will be around 0.5 (as in the example in the middle). And if the prediction matches the actual box almost perfectly, the IoU value will be nearly 1 (as in the example on the right). We also use the IoU value as a filter for some metrics, so that the model only considers predictions above an IoU threshold. This is the case with the *Mean Average Precision @ IoU=0.5* (mAP50), which represents the average precision (AP) for all classes, using a threshold of 0.5 for the IoU. This metric therefore evaluates the accuracy of the model in object detection by considering the quality of the bounding boxes based on an IoU threshold.

Returning to Figure 4.25, the two columns on the right show important performance metrics determined based on the validation data.

- **Precision**
  This reflects the proportion of correctly recognized objects, and high precision means that there are few FPs.

- **Recall**
  This measures the proportion of correctly recognized objects in relation to all objects that are actually present.

- **mAP50**
  We explained this in detail in the previous info box.

- **mAP50-95**
  This metric is even more comprehensive than map50, and it calculates the average of the mAP values over thresholds from 0.5 to 0.95 in increments of 0.05. It provides a very robust picture of model performance across different localization accuracies.

In our example from Figure 4.25, we see a positive training curve across the epochs. The loss functions decrease continuously and the performance metrics—such as precision, recall, and mAP on the validation dataset—increase. This suggests that the model is successfully learning to recognize and localize the sheep, so we can be satisfied with the training and can turn now to inference in the next section.

4.4.5   Model Inference

We'll start by loading the trained model and the training data, selecting a specific image from the training dataset, and creating the prediction for it. Let's get started. The finished script can be found at *050_ComputerVision\040_ObjectDetection\30_sheep_inference.py*.

We need YOLO to access the model and load_datasets to access the dataset. We'll use matplotlib to display the image and the superimposed boxes, as follows:

```
#%% packages
import matplotlib.pyplot as plt
from ultralytics import YOLO
from datasets import load_dataset
```

We'll save the trained model in the runs folder and the relative path to the *best.pt* model file in the variable best_model, as follows:

```
#%% inference model
try:
    best_model = "../../../runs/detect/train1/weights/best.pt"
    inference_model = YOLO(best_model)
except:
    print("Error loading best model")
```

Now, we load the test data using load_dataset, as we did during data preparation:

```
# %% load test dataset
ds = load_dataset("keremberke/aerial-sheep-object-detection")
ds_test = ds['test']
```

We then select a specific image from the large number of test images. We've selected the image at index position 1 in our code, but you are welcome to select any other image.

```
#%% select a specific image
test_image = ds_test['image'][1]
```

We can output the image with imshow, as follows:

```
#%% show test image
plt.imshow(test_image)
plt.axis('off')
plt.show()
```

Figure 4.30 shows an example test image.

There are no predictions of sheep in it yet, so we'll get to that now. The test_image image is passed to the loaded inference_model model, and the predictions are saved under results:

```
results = inference_model(test_image)
```

These predictions can now be superimposed on the image, and we can see that the sheep have been excellently recognized:

```
plt.imshow(results[0].plot())
plt.axis('off')
plt.show()
```

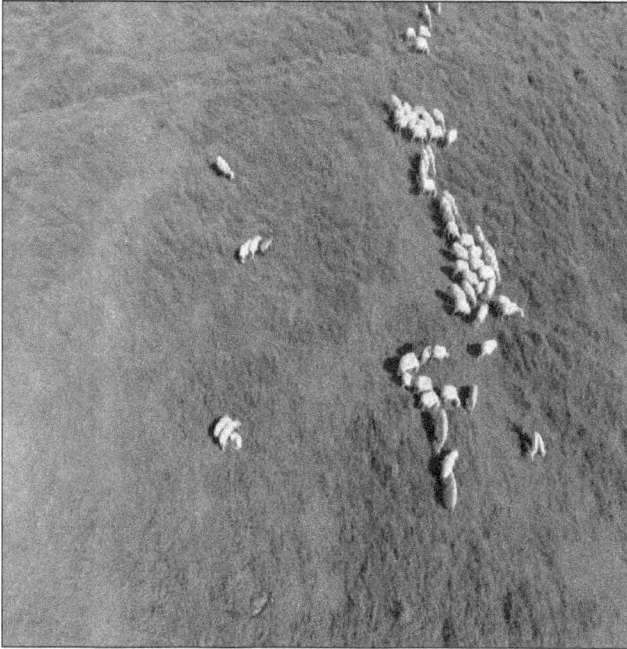

Figure 4.30 Object Detection: Test Image (without Predictions)

Figure 4.31 shows the image of the test image, including the overlay with the predictions.

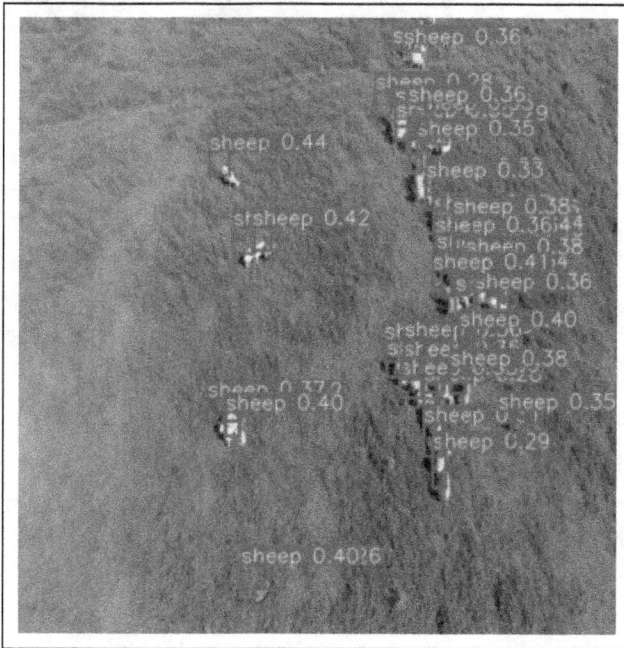

Figure 4.31 Object Detection: Test Image with Predictions

We see that the model has learned to predict sheep quite well.

In this section, we've dealt with the topic of object detection and successfully trained an object detection model. In doing so, we've become acquainted with an extraordinarily capable model architecture that can only be improved by predicting every single pixel of the image instead of image areas of objects. This technique is called semantic segmentation and will be the subject of the following section.

4.5   Semantic Segmentation

In this section, we deal with a model that is more complex in the sense that the associated class is now predicted for each individual pixel.

Figure 4.32 shows an example of what this is all about. On the left, we have an initial image. The segmentation shown on the right was created for this image, and it's an image of the same size, but each pixel represents a class rather than a color value.

Figure 4.32 Example of Semantic Segmentation: Left = Original Image, Right = Segmentation Mask

The images come from the "Flood Area Segmentation" dataset, which contains 290 images of areas affected by flooding as well as a segmentation mask representing the "no flooding," and "flooding" classes.

We'll use this dataset to train a model to identify flooded areas, but before we get to the implementation, let's try to understand how semantic segmentation works.

We'll start by getting an understanding of how semantic segmentation works in Section 4.5.1, and then, we'll cover how computers understand segmentations in Section 4.5.2. After that, we'll go through our usual sequence of data preparation in Section 4.5.3, model training in Section 4.5.4, and model evaluation in Section 4.5.5.

4.5.1   How Does Semantic Segmentation Work?

In this network type, the target doesn't consist of a numerical value (unlike in classification, where the numerical value represents the class). In object detection, a bounding box with its four coordinates is predicted.

During segmentation, our prediction has the pixel size of the original image. If our image has a pixel size of M horizontal, N vertical pixels, then the segmentation algorithm returns dimensions of (M, N, C). What is C? It's a parameter that represents the number of different classes that the model will learn and predict.

Most segmentation models use an encoder-decoder architecture. Due to its appearance, the model architecture is often called U-Net because it has a U-shaped structure. A down-sampling path (the *encoder*) for capturing context information is connected to an up-sampling path (the *decoder*) that is used for precise localization and detail recovery.

The encoder part of the network takes the input image and extracts its features, and we use pretrained CNN or transformer-based models for this. At the same time, the encoder reduces the spatial resolution of the image, learning abstract features (such as edges or textures) and more complex features (such as shapes).

The decoder part of the network records the features extracted by the encoder and attempts to restore the original spatial resolution of the image. For this purpose, it uses the learned features to make a class prediction for each pixel, and the result is a translation of the abstract features back into a pixel-accurate segmentation mask. The decoder then compares the predicted masks with the *ground truth masks* (the masks that best describe the object location) and quantifies the loss by using the loss function. The network structure is usually like the one shown in Figure 4.33.

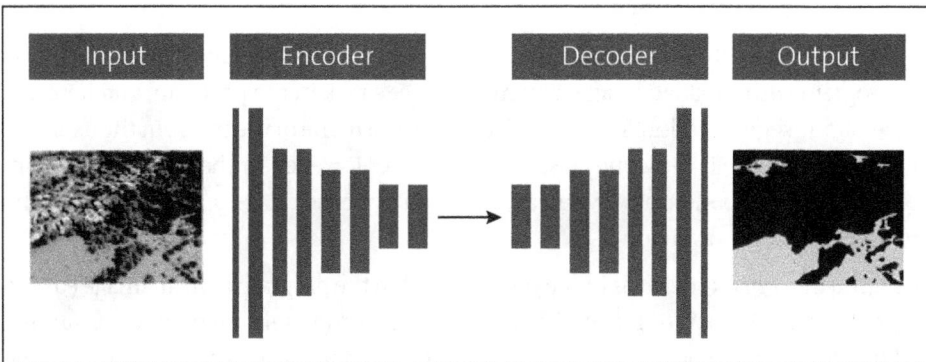

Figure 4.33 Semantic Segmentation: Network Structure

An image passes through the encoder and then the decoder, and the result is an image with a segmentation mask. Where does the segmentation mask that the model learns come from? We'll study that in the next section.

4.5.2   Segmentation Masks

The process of creating the segmentation mask is illustrated in Figure 4.34.

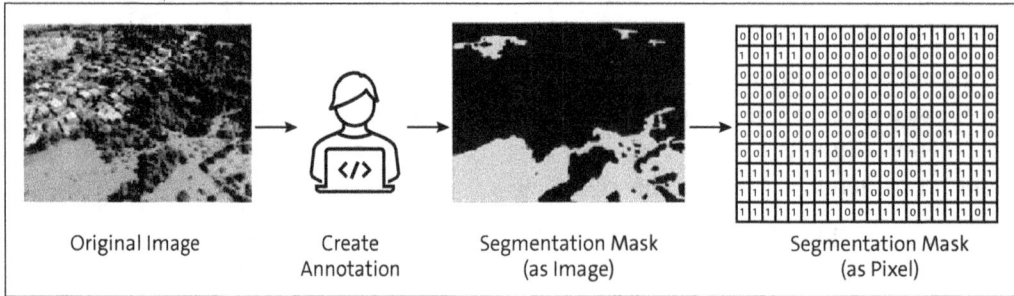

Figure 4.34 Semantic Segmentation: Creating the Mask

In the real world, you'll have to manually annotate each of the source images, which means you'll have to mark the image areas that should correspond to various classes and then assign them to those classes. But don't worry—in this chapter, we're working with a dataset for which the image masks have already been created. Nevertheless, in the rest of this section, we'll describe the process as you'd need to perform it in a real-world project.

There are various providers for this image labeling process. One free service is Label Studio, and you can install the Python package for it by using `pip`, as follows:

```
pip install label-studio
```

You must then start it via the terminal with the following command:

```
label-studio start
```

This starts a local web service that you can use to label the images, and Figure 4.35 shows the program with a loaded image. You must select a task in the program, and here, it's the semantic segmentation. You must also define the name of the mask. In the example in Figure 4.35, there is only one class: `flooded_area`. However, in the real world, you're not limited to just one class—you can define many other classes, depending on the problem you want to solve.

Then, you must mark the areas of the selected class in the program as in an image editing program. This is like painting by numbers, which is unfortunately a very time-consuming and intellectually unchallenging process. But you need to do it to create good training data for the model training.

In this book, we'll focus on model development and therefore skip this step by choosing a dataset that provides both source images and the corresponding masks.

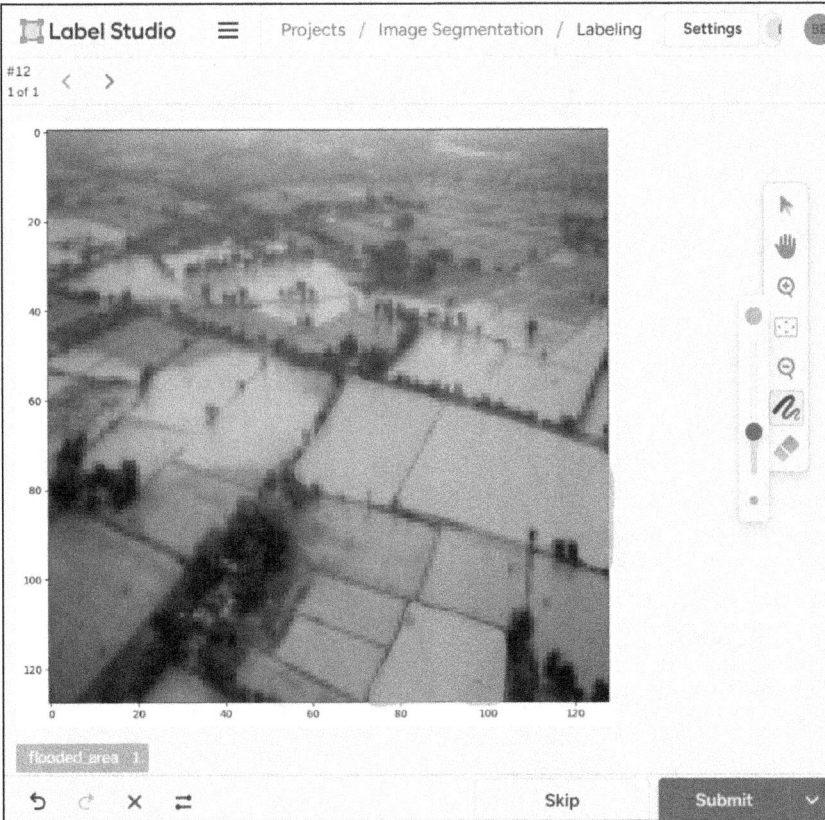

Figure 4.35 Semantic Segmentation: Creating Image Mask with Label Studio

4.5.3   Data Preparation

Listing 4.31 lists all the required packages. The AutoModelForSemanticSegmentation class is required for segmentation, and it allows us to load a pretrained model from Hugging Face. This functionality comes from the transformers package, which also provides the trainer that greatly simplifies model training. We're familiar with the other packages from previous training sections.

```
#%% packages
import kagglehub
import os
import matplotlib.pyplot as plt
import numpy as np
import pandas as pd
from PIL import Image
from torch.utils.data import Dataset, DataLoader
from torchvision import transforms, datasets
import torch
```

```
import torch.nn as nn
from transformers import AutoModelForSemanticSegmentation, TrainingArguments,
Trainer, AutoImageProcessor
from evaluate import load
```

Listing 4.31 Semantic Segmentation: Required Packages

We can download the data to our local computer by using `kagglehub.dataset_download`, as follows:

```
dataset_name = "faizalkarim/flood-area-segmentation"
path = kagglehub.dataset_download(dataset_name)

print("Path to dataset files:", path)
```

Path to dataset files: *C:\Users\BertGollnick\.cache\kagglehub\datasets\ faizalkarim\flood-area-segmentation\versions\1*

The data contains a file called *metadata.csv*, which creates a link between the image and the corresponding mask:

```
metadata = pd.read_csv(os.path.join(path, "metadata.csv"))
metadata.head(2)
```

	Image	Mask
0	0.jpg	0.png
1	1.jpg	1.png

A look in the folder shows that the images are stored in corresponding folders:

Data folder
└──Image
└──Mask

We'll need to know this structure when we create the dataset class in the next section.

4.5.4   Model Training

The most difficult part of the entire model training is the preparation of the dataset class. Listing 4.32 shows the creation of the `FloodSegmentationDataset` class, which has the following methods and functions within it:

▶ `__init__`
This method is where we pass the metadata data frame and the path to the original `img_dir` images and the `mask_dir`. We also define the necessary transformations to ensure that all images have the same dimensions using `transforms.Resize`.

▶ **__len__**

As usual, this function is short and concise. We derive the number of images from the number of rows in the metadata.

▶ **__getitem__**

This method is where the magic happens. First, we use the idx index to access a specific image and its mask, extracting their img_name and mask_name data names. Then, we get these images, load the image and mask, and apply the previously defined transformations so that they are available as tensors. The mask tensor represents the classes as grayscale values, and since we're dealing with a binary classification problem here, we can use (mask > 0.5).long() to reduce all numerical values to 0 or 1.

```python
class FloodSegmentationDataset(Dataset):
    def __init__(self, metadata_df, img_dir, mask_dir):
        self.metadata = metadata_df
        self.img_dir = img_dir
        self.mask_dir = mask_dir
        self.transform = transforms.Compose([
            transforms.Resize((128, 128)),
            transforms.ToTensor()
            ])

    def __len__(self):
        return len(self.metadata)

    def __getitem__(self, idx):
        # Get image and mask paths
        img_name = self.metadata.iloc[idx]["Image"]
        mask_name = self.metadata.iloc[idx]["Mask"]

        # Load image and mask
        img_path = os.path.join(self.img_dir, img_name)
        mask_path = os.path.join(self.mask_dir, mask_name)

        # import image and mask
        image = Image.open(img_path).convert('RGB')
        mask = Image.open(mask_path).convert('L')

        # Apply transformations
        image = self.transform(image)
        mask = self.transform(mask)  # Convert mask to tensor
        mask = mask.squeeze(0)  # Remove channel dimension to get [H,W]
        mask = (mask > 0.5).long()  # Convert to binary mask with long dtype
        return image, mask
```

Listing 4.32 Semantic Segmentation: Dataset Class

After we've created the class, we create an instance of dataset, as follows:

```
dataset = FloodSegmentationDataset(
    metadata_df=metadata,
    img_dir=path_images,
    mask_dir=path_masks
)
```

To avoid overfitting, we'll later use the validation data to check the improvement of the training during the model training. Therefore, we now need to split the data into a training dataset and a validation dataset (train_dataset and val_dataset). At this point, we'll specify that 80% of the data is used for training and 20% for testing.

```
# Split into training and validation sets
train_size = int(0.8 * len(dataset))
val_size = len(dataset) - train_size
train_dataset, val_dataset = torch.utils.data.random_split(dataset, [train_
size, val_size])
```

We want to define all hyperparameters—such as BATCH_SIZE, LEARNING_RATE, the number of EPOCHS, and the hardware DEVICE we want to use—centrally in one place. If we can use a graphics card for model training, we'll select the DEVICE accordingly, or alternatively, we can train it on the CPU, as follows:

```
#%% Hyperparameters
BATCH_SIZE = 8
LEARNING_RATE = 0.001
EPOCHS = 10
DEVICE = torch.device("cuda" if torch.cuda.is_available() else "cpu")
```

Now, we can create the DataLoader train_loader and val_loader, as follows:

```
# Create data loaders
train_loader = DataLoader(train_dataset, batch_size=BATCH_SIZE, shuffle=True)
val_loader = DataLoader(val_dataset, batch_size=BATCH_SIZE, shuffle=False)
```

Let's take a quick look at the number of images available for training and validation:

```
print(f"Training samples: {len(train_dataset)}")
print(f"Validation samples: {len(val_dataset)}")
```

```
Training samples: 232
Validation samples: 58
```

The number of images appears to be relatively small, but we must not forget at this point that 128 × 128 pixels (i.e., 16,384 target values) are available per image.

It's always good to visualize the dimensions of the individual screens and masks. To do this, we iterate over the `train_loader` and terminate with a break at the end of the first iteration, as follows:

```
# %% Extract one image and one mask
for i, (image, mask) in enumerate(train_loader):
    print(f"Image shape: {image.shape}")
    print(f"Mask shape: {mask.shape}")
    break
```

Image shape: torch.Size([8, 3, 128, 128])
Mask shape: torch.Size([8, 128, 128])

We transfer the original images in batches of 8, which is our previously defined batch size. We therefore use them as color images with three color channels, and they have 128 horizontal and vertical pixels. The masks only have dimensions of the batch size and the horizontal and vertical pixels.

We'll transfer the mapping from the IDs (0 or 1) to the labels with the `id2label` object and the perform reverse operation with `label2id` to the model. To do this, we create the following objects:

```
#%% id2label and label2id
id2label = {0: "background", 1: "flood"}
label2id = {v: k for k, v in id2label.items()}
```

At this point, the model is loaded with `AutoModelForSemanticSegmentation`. This is a pretrained model, in which the model architecture and the weights are loaded from the checkpoint, as follows:

```
checkpoint = "nvidia/mit-b0"
model = AutoModelForSemanticSegmentation.from_pretrained(checkpoint, id2label=
id2label, label2id=label2id)
```

During model training, we must ensure that the images and masks are the same sizes. We define `collate_fn` for this purpose, as you can see in the following code, and we pass this to the trainer later as a parameter.

```
# Define data collator function
def collate_fn(batch):
    return {
        "pixel_values": torch.stack([i[0] for i in batch]),
        "labels": torch.stack([i[1] for i in batch])
    }
```

We define the training via various hyperparameters, which we transfer with `Training-Arguments`, as follows:

```
training_args = TrainingArguments(
    output_dir="flood-segmentation-128",
    learning_rate=LEARNING_RATE,
    num_train_epochs=EPOCHS,
    per_device_train_batch_size=BATCH_SIZE,
    per_device_eval_batch_size=BATCH_SIZE,
    save_total_limit=3,
    logging_dir=None
)
```

Now, we come to the actual model training. This is handled by the trainer class, to which we pass the previously created TrainingArguments, the datasets, the data_collator, and the model:

```
trainer = Trainer(
    model=model,
    args=training_args,
    train_dataset=train_dataset,
    eval_dataset=val_dataset,
    data_collator=collate_fn
)
```

We start model training via trainer.train, as follows:

```
trainer.train()
```

Depending on the hardware available, training can take anything from a few minutes to hours. We'll look at the results in the following section.

4.5.5   Model Evaluation

We carry out the evaluation by using an example image. For an image with a sample_idx index, we extract the original sample_image and the "ground truth" sample_mask from the validation dataset, as follows:

```
# Get a sample image from validation dataset
sample_idx = 1
sample_image, sample_mask = val_dataset[sample_idx]
```

Since we want to display the image and the mask graphically, we must convert the objects into NumPy arrays and adjust the dimensions with permute to get from the tensor format [C, H, W] to the image format [H, W, C], as follows:

```
sample_image = sample_image.permute(1, 2, 0).numpy()
sample_mask = sample_mask.numpy()
```

Now, we can create the predictions for the mask_prediction validation data and the predicted mask for the selected sample_mask_pred index, as follows:

```
mask_prediction = trainer.predict(val_dataset)
sample_mask_pred = mask_prediction[0][1][sample_idx]
```

The predictions of class membership are expressed as probabilities, and to obtain the actual binary classes in the sample_mask_pred_binary object, we can set all values >0 to 1 and all values <=0 to 0, as follows:

```
# adapt mask to have value 1 if >0, 0 otherwise
sample_mask_pred_binary = (sample_mask_pred > 0)
```

At this point, we have the actual image, the actual mask, and the mask predicted by the model, and we can display all three in one image. Listing 4.33 shows the code created based on matplotlib to visualize all three images side by side.

```
# Create figure with subplots
fig, (ax1, ax2, ax3) = plt.subplots(1, 3, figsize=(12, 6))

# Plot original image
ax1.imshow(sample_image)
ax1.set_title('Beispiel Validierungsbild')
ax1.axis('off')

# Plot ground truth mask
ax2.imshow(sample_mask)
ax2.set_title('Ground Truth Maske')
ax2.axis('off')

# Plot prediction mask
ax3.imshow(sample_mask_pred_binary)
ax3.set_title('Vorhergesagte Maske')
ax3.axis('off')

plt.tight_layout()
plt.show()
```

Listing 4.33 Semantic Segmentation: Model Evaluation

Figure 4.36 shows that our model already provides fairly good predictions that almost match those of the actual mask.

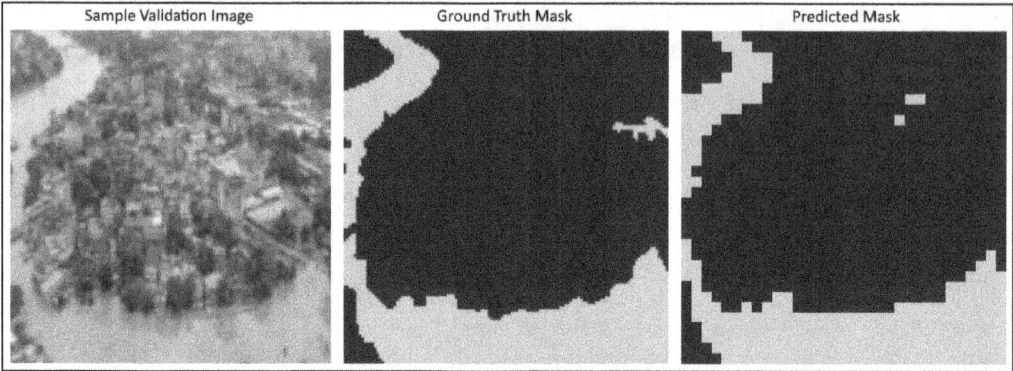

Figure 4.36 Semantic Segmentation: Model Evaluation

4.6   Style Transfer

Style transfer is a technique in which the style of an artistic image is transferred to a reference image. An example is shown in Figure 4.37. A style transfer model processes the reference image together with a style image, and the result is the artistic image shown on the right, which contains features of both original images.

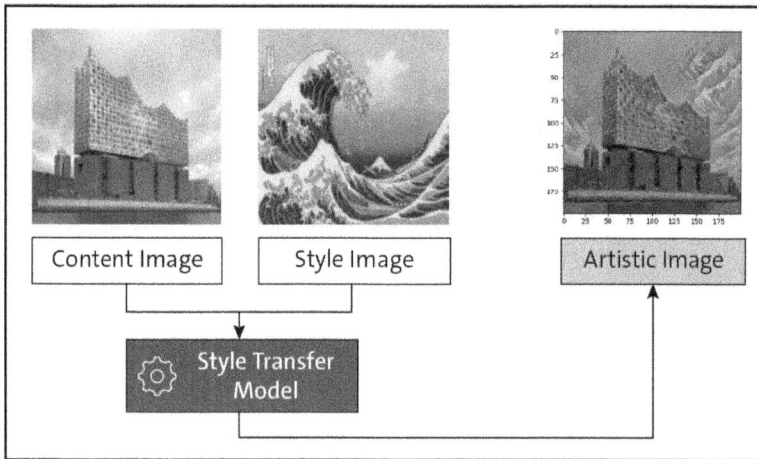

Figure 4.37 Style Transfer: General Process

We'll develop a style transfer model, but first, we'll need to get an understanding of the functionality of style transfer (in Section 4.6.1). Then, we'll prepare the data (in Section 4.6.2), train the model (in Section 4.6.3), and evaluate the training result (in Section 4.6.4).

4.6.1 How Does Style Transfer Work?

How exactly does this approach work? In fact, style transfer isn't model training in the classic sense because no model weights are adjusted. This becomes clear when looking at Figure 4.38. In conventional model training, the input and target value are predefined and the model adjusts its model weights during training. But with style transfer, the weights are frozen and the target value is continuously adjusted.

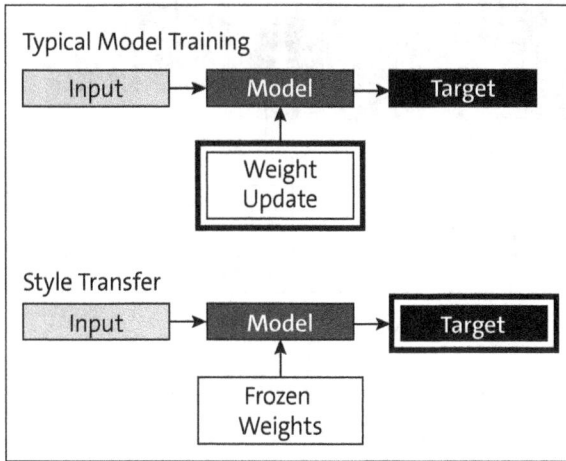

Figure 4.38 Style Transfer: General Approach Compared with Typical Model Training

In principle, you can use any pretrained network to implement the style transfer, but in our example, we use *VGG19*. This model was originally developed for the image classification portion of the ImageNet competition. ImageNet is a large image database that is considered the most important resource for research in computer vision and deep learning, and the annual competition was instrumental in the deep learning revolution.

The model has 19 layers: 16 convolutional layers, 5 pooling layers, and 3 fully connected layers. This means it actually has 24 layers, but only 19 layers have learnable parameters, as pooling layers have no trainable parameters. This CNN-based network remains unchanged during training.

We know that the first layers at the beginning of the network are responsible for recognizing simple features such as pixels, lines, and edges. The deeper you go into the network, the higher the abstraction level of the features, and larger components or objects are then recognized. We take advantage of this behavior by extracting features at various points in the network.

Figure 4.39 shows the structure of the VGG19 network, from the input layer to the output layer. We extract the features for the style from various points in the network, and we extract the style for the content from another point.

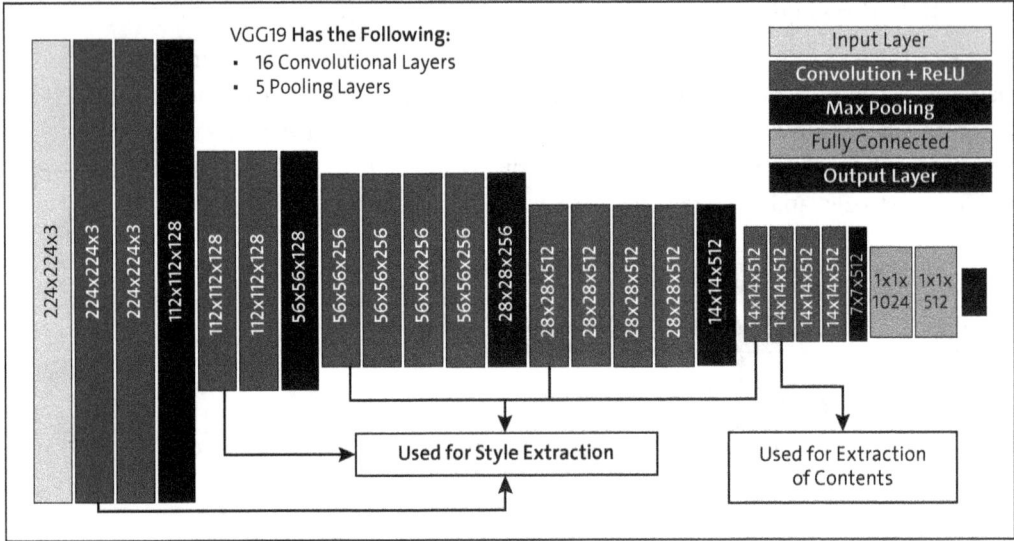

Figure 4.39 Extraction of Style and Content Information from Network

Thus, we extract the content from the original image and the style from the artistic image, creating a new image that looks as if the artist had created it.

4.6.2   Data Preparation

The complete script for this example can be found at *050_ComputerVision\060_Style-Transfer\style_transfer.py*. We start in Listing 4.34 by importing the required packages. Only old acquaintances can be found here.

```
#%% packages
import numpy as np
from PIL import Image
import torch
from torch.optim import Adam
from torch.nn.functional import mse_loss
from torchvision import transforms, models
import matplotlib.pyplot as plt
```

Listing 4.34 Style Transfer: Required Packages

Now, we can define the hyperparameters via the code in Listing 4.35. There is no batch size here, as we're only processing a single image. Likewise, no learning rate is required in this case. It's sufficient to define the number of epochs. Instead, we define the scaling factors for the CONTENT_WEIGHT and STYLE_WEIGHT losses. We need them because the losses work in very different orders of magnitude, so it's helpful to apply a large factor to the style loss to obtain losses of a comparable order of magnitude.

The VGG19 algorithm is based on ImageNet data. If we analyze all the data, we'll obtain corresponding VGG_MEAN mean-values and VGG_STD standard deviations.

```
EPOCHS = 500
CONTENT_WEIGHT = 1 # Or whatever factor you deem appropriate
STYLE_WEIGHT = 2000000
vgg_mean = [0.485, 0.456, 0.406]
VGG_STD = [0.229, 0.224, 0.225]
DEVICE = torch.device('cuda' if torch.cuda.is_available() else 'cpu')
```

Listing 4.35 Style Transfer: Hyperparameters

We selected the VGG19 model, and since we want to use it together with its weights, we'll pass the parameter pretrained. We're only interested in the features of the entire model, so to ensure that all weights are frozen, we'll set the requires_grad parameters accordingly and bring the model into the evaluation model with vgg.eval(), as follows:

```
vgg = models.vgg19(pretrained=True).features.to(DEVICE)
for param in vgg.parameters():
param.requires_grad = False
vgg.eval()
```

Now, let's move on to preparing the images in Listing 4.36. We set up the transformation steps with transforms.Compose, and here, we only need to resize the image and then convert it into a tensor. We also use the previously defined mean values and standard deviations of the training data of the VGG algorithm to normalize our images.

We must also ensure that the image has the correct dimensions. The model is expected to have dimensions of [1, C, H, W], so we must add the first dimension with torch.unsqueeze.

```
preprocess_steps = transforms.Compose([
    transforms.Resize((224, 224)),
    transforms.ToTensor(),
    transforms.Normalize(mean=VGG_MEAN, std=VGG_STD)
])
content_img = Image.open('hamburg.jpg').convert('RGB')
content_img = preprocess_steps(content_img)
content_img = torch.unsqueeze(content_img, 0).to(DEVICE)
print(content_img.shape)
```

torch.Size([1, 3, 224, 224])

Listing 4.36 Style Transfer: Preprocessing Steps

We apply the same steps to the styling image, as follows:

```
style_img = Image.open('The_Great_Wave_off_Kanagawa.jpg').convert('RGB')
style_img = preprocess_steps(style_img)
style_img = torch.unsqueeze(style_img, 0).to(DEVICE)
print(style_img.shape)
```

torch.Size([1, 3, 224, 224])

Now, let's take another look at the images side by side by using the code from Listing 4.37. We'll need to use the Unnormalize class to undo the previously performed normalization, and then, we'll define the imshow function to turn the tensor back into a real image.

```
class UnNormalize(object):
    def __init__(self, mean, std):
        self.mean = mean
        self.std = std

    def __call__(self, tensor):
        """
        Args:
            tensor (Tensor): Tensor image of size (C, H, W) to be un-normalized.
        Returns:
            Tensor: UnNormalized image.
        """
        for t, m, s in zip(tensor, self.mean, self.std):
            t.mul_(s).add_(m)
        return tensor

unnormalize_transform = UnNormalize(mean=VGG_MEAN, std=VGG_STD)

def imshow(tensor, title=None):
    image = tensor.cpu().clone()
    image = image.squeeze(0)
    image = unnormalize_transform(image)
    image.clamp_(0, 1)
    # Convert from (C,H,W) to (H,W,C) for matplotlib
    image = image.permute(1, 2, 0).numpy()
    return image
# Create a figure with two subplots side by side
fig, (ax1, ax2) = plt.subplots(1, 2, figsize=(10, 5))

# Show content image in first subplot
content_image = imshow(content_img)
ax1.imshow(content_image)
```

```
ax1.set_title('Reference Image')
ax1.axis('off')

# Show style image in second subplot
style_image = imshow(style_img)
ax2.imshow(style_image)
ax2.set_title('Style Image')
ax2.axis('off')

plt.show()
```

Listing 4.37 Style Transfer: Visualization of Content Image and Style Image

Figure 4.40 shows the content image and the style image.

Figure 4.40 Style Transfer: Content Image and Style Image

Now, we come to the actual model training.

4.6.3 Model Training

As described earlier, we want to extract the content and style from certain layers. We define which layers should be in Listing 4.38 in a LOSS_LAYERS dictionary.

```
STYLE_LOSS_LAYERS = { '0': 'conv1_1',
'5': 'conv2_1',
'10': 'conv3_1',
'19': 'conv4_1',
'21': 'conv4_2',
'28': 'conv5_1'}
CONTENT_LOSS_LAYER = 'conv4_2'
```

Listing 4.38 Style Transfer: Layers for Determination of Losses

We use the extract_features function shown in Listing 4.39 to extract the tensors of the corresponding layers from the network. We create a features dictionary, and we extract the weights of the previously defined network layers. We apply the function directly to the content image and the style image to represent the features in the content_img_features and style_img_features objects, and these features are the ground truth that we use for comparison purposes during training in the loss function.

```
def extract_features(x, model):
    features = {}
    for name, layer in model._modules.items():
        x = layer(x)

        if name in LOSS_LAYERS:
            features[LOSS_LAYERS[name]] = x

    return features

content_img_features = extract_features(content_img, vgg)
style_img_features   = extract_features(style_img, vgg)
```

Listing 4.39 Style Transfer: Function for Feature Extraction

We can carry out the calculation of the loss of image content classically—more on this later in this section. However, we need to use a different approach to calculate the loss of style. Every image has its own unique "look and feel," meaning its own style. Artistically, we can recognize this in things such as color patterns, brush strokes, and textures. Mathematically, we can describe the style of an image by using the *Gram matrix*, which determines which features occur together and how often. This is clearly shown in Figure 4.41, where we can see recurring elements in different areas of the image. For the algorithm, these are correlated features.

Figure 4.41 Style Transfer: Correlated Features

The Gram matrix represents precisely these correlations between different feature filters of a network layer, and thus, it captures the visual style of an image. The Gram matrix is independent of the exact spatial arrangement of the features, and we calculate it in Listing 4.40 as follows.

1. We determine the number of channels (or filters) C, as well as the number of horizontal (W) and vertical (H) elements.

2. We determine and transform feature maps in the layers. C represents the number of individual channels, which is equivalent to an image of a recognized feature.

3. Now we come to the core of the calculation. We multiply these tensors by their transposed matrices (tensor.t()) by using torch.mm. This results in a square matrix. Each entry in this matrix is the scalar product of two feature vectors, and if this scalar product is high, it means that the features occur together often and strongly. By analyzing the correlation between all pairs of features in a layer, we get an abstract measure of style.

4. Finally, normalization takes place because the feature maps are of different sizes. This is important to keep the losses stable.

```
def calc_gram_matrix(tensor):
    _, C, H, W = tensor.size()
    tensor = tensor.view(C, H * W)
    gram_matrix = torch.mm(tensor, tensor.t())
    gram_matrix = gram_matrix.div(C * H * W)

style_features_gram_matrix = {layer: calc_gram_matrix(style_img_features[layer]) for layer in style_img_features}
```

Listing 4.40 Style Transfer: Gram Matrix Calculation

Now, we can weight the different layers and their style information differently, and we use the following mapping for this purpose:

```
weights = {'conv1_1': 1.0, 'conv2_1': 0.8, 'conv3_1': 0.6,
           'conv4_1': 0.4, 'conv5_1': 0.2}
```

We generate a start screen and create an optimizer—Adam, as usual—as follows:

```
generated_image = content_img.clone().requires_grad_(True).to(DEVICE)

optimizer = Adam([generated_image], lr=0.003)
```

Now, the actual training of the model begins as shown in Listing 4.41. The total losses total_loss result from the style_loss and content_loss losses. We determine the two separately. In each epoch, we extract features from the current generated_image. The content_loss measures how much the content of the generated image differs from the original content image, while the style_loss evaluates the stylistic similarity by comparing Gram matrices. The sum of these two total_loss losses is weighted by STYLE_WEIGHT and CONTENT_WEIGHT, and every 100 epochs, we save the current progress as an image to track the development of the style transfer.

```
for epoch in range(1, EPOCHS):

    target_features = extract_features(generated_image, vgg)

    content_loss = mse_loss(target_features[CONTENT_LOSS_LAYER], content_img_
features[CONTENT_LOSS_LAYER])

    style_loss = 0
    for layer in weights:

        target_feature = target_features[layer]
        target_gram_matrix = calc_gram_matrix(target_feature)
        style_gram_matrix = style_features_gram_matrix[layer]

        layer_loss = mse_loss (target_gram_matrix, style_gram_matrix) *
                              weights[layer]
        style_loss += layer_loss

    total_loss = STYLE_WEIGHT * style_loss +
                 CONTENT_WEIGHT * content_loss
    if epoch % 100 == 0:
        output_image = imshow(generated_image.detach())
        plt.imsave(f"output_epoch_{epoch}.jpg", output_image)

    optimizer.zero_grad()

    total_loss.backward()

    optimizer.step()
```

Listing 4.41 Style Transfer: Model Training

4.6.4   Model Evaluation

We saved the images during the model training in specified epochs. Figure 4.42 shows the progress of the model training after several epochs, revealing how the model uses the sky more and more as an image surface to represent the waves of the stylistic image.

This brings us to the end of this section on style transfer and also to the end of the chapter.

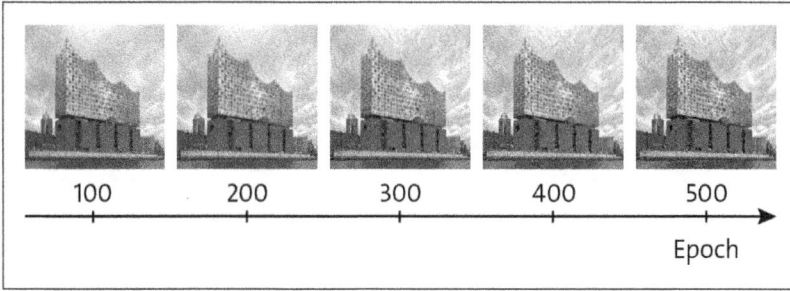

Figure 4.42 Style Transfer: Results of Different Training Steps

4.7 Summary

The chapter has dealt with various aspects of computer vision, a discipline that deals with the analysis of images to classify entire images or parts of images.

First, it covered the basics, and it clarified how images are handled in models and which network architectures you can use for modeling images. It discussed convolutional neural networks and ViTs, and it dealt with several important tasks on this basis. These tasks included the following:

- **Image classification**
  This task assigns entire images to a class.

- **Object detection**
  This task doesn't predict labels for entire images but for individual image areas. The predictions are bounding boxes for the different classes, whereby there can be several bounding boxes per class and image.

- **Semantic segmentation**
  The segmentation task takes the principle of image classification to the extreme by predicting the class for each individual pixel.

- **Style transfer**
  This task modifies a source image with the help of an artistic reference image in such a way that the source image is still recognizable but has unmistakably taken on the style of the artistic image.

Chapter 5
Recommendation Systems

"Information is a selection from alternatives. This presupposes that one distinguishes a realm of possibilities against others."
—*Niklas Luhmann*

In today's world, we're flooded with content and data. The question is not so much about whether we have access to information, but rather, which information is relevant for us. This is exactly where recommendation systems come in. They help us to filter out the information that could be relevant to us personally from the sheer volume of possible options. We encounter such systems everywhere in our everyday lives: when streaming movies and TV series in the evening, when shopping online, and on social networks.

In this chapter, we'll look at how we can implement such recommendation systems technically. With the help of PyTorch, we'll develop a recommendation system step by step, one that not only processes data but can also make intelligent predictions as to which content or products may be relevant for the user.

We'll get to know the basic concepts and gain an understanding of the underlying models in Section 5.1, and then, we'll set up an initial functional system ourselves in Section 5.2.

5.1 Theoretical Foundations

We'll learn basic concepts of recommendation systems in Section 5.1.1 and gain an understanding of the mathematical concept of matrix factorization in Section 5.1.2.

5.1.1 Basic Concepts

The basic concepts of recommendation systems can be roughly divided into three categories:

- Content-based filtering
- Collaborative filtering
- Hybrid models

Figure 5.1 shows the two main systems of content-based filtering and collaborative filtering. The whole thing is illustrated using a streaming example, in which the next movie is to be suggested to a user.

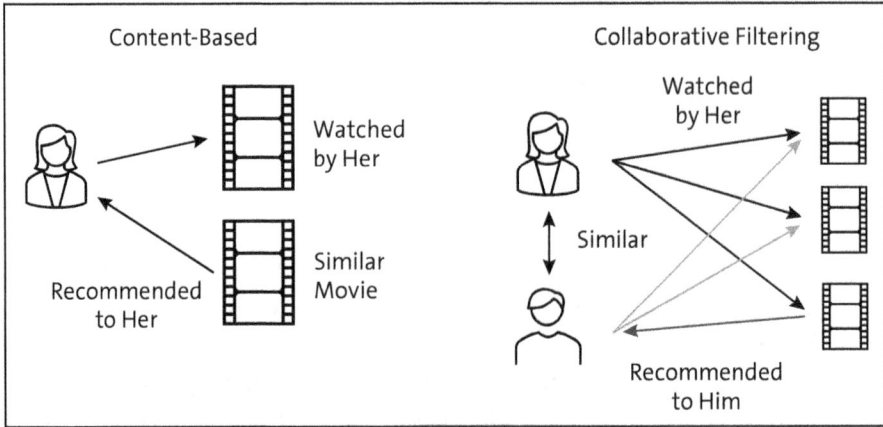

Figure 5.1 Categories of Recommendation Systems

With *content-based filtering*, the system analyzes the user's behavior and makes a recommendation based on the similarities among films the user has watched. For example, if the user has watched the movie *Titanic* all the way through, the system will interpret the fact that she watched all three-hours of it as meaning that she liked it. It will then search for similar movies, which could be films of the same genre, set in the same time period, or with the same actors. It will them recommend the movie that is the closest match.

With *collaborative filtering*, the system makes a recommendation based on the preferences of similar users. This means that it recommends films that other users with preferences that are similar to the user's have liked. The core idea is that if person A and person B liked the same movies X, Y, and Z, and if person B also liked movie Q, then the probability is very high that person A will also like movie Q.

The algorithm ignores the specific characteristics of the films. Instead, it concentrates on user behavior by using one of two approaches:

- **User-based**
  This algorithm finds users who have preferences that are similar to those of the user it is currently analyzing, and it recommends movies that these "similar" users also liked and that the user it is currently analyzing hasn't watched yet.

- **Item-based**
  The algorithm finds films that were liked by similar user groups and recommends to the current user films that other users who also liked the same film liked.

Common algorithms for collaborative filtering have the problem that they scale poorly. So, to stay with the example of our streaming provider, if the store is buzzing and the number of users is exploding, the computing effort and memory requirements of our algorithm will suddenly increase exponentially.

5.1.2   Matrix Factorization

A special implementation that addresses this problem is *matrix factorization*. This is a common technique that is classified as collaborative filtering, and it solves the problem of scalability and complexity.

We can imagine this matrix as shown in Figure 5.2. The rows correspond to the users, and the columns correspond to the films. The values in this matrix represent the ratings or interactions (such as clicks and screen time) of the users with these movies.

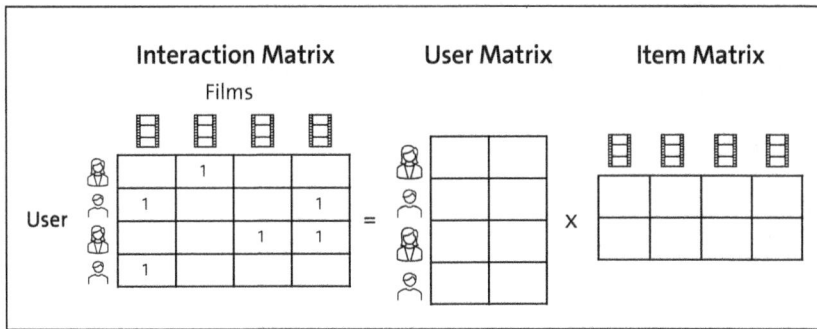

Figure 5.2 Matrix Factorization

As most users only watch a small number of the streaming service's large offerings, this matrix is very sparsely populated, so it has many empty fields.

Matrix factorization breaks down this large, sparse matrix into two smaller, dense matrices: a user feature matrix and an element feature matrix.

- **User feature matrix**
  In this matrix, each row represents a user and contains a series of values (known as latent factors) that describe the user's preferences.
- **Item feature matrix**
  In this matrix, each column represents the films offered by the service.

By multiplying these two smaller matrices, the algorithm can approximate the original matrix, predict the missing values, and present those missing values to the user as recommendations.

However, problems arise when this matrix becomes larger, and that's where PyTorch comes to our aid. We can train a neural network to learn the complex patterns in user behaviors, and in the following section, we'll put this concept into practice.

5.2 Coding: Recommendation Systems

The recommendation system we're developing will recommend games to gamers. The complete script can be found at *060_RecommendationSystems\recommender_system.py*. As usual, we'll prepare the data, create and train the model, and then evaluate it.

5.2.1 Data Preparation

We start by loading the required packages in Listing 5.1. These packages are divided into subpackages, as follows:

- **Data preparation**
  For this, we use kagglehub, pandas, numpy, and os.

- **Modeling**
  For this, we use torch (with multiple submodules) and sklearn.

- **Visualization**
  For this, we use matplotlib and seaborn.

```
# data handling
import kagglehub
import pandas as pd
import numpy as np
import os
from sklearn.model_selection import train_test_split
# modeling
import torch
import torch.nn as nn
from torch.utils.data import Dataset, DataLoader
# visualizations
import matplotlib.pyplot as plt
import seaborn as sns
```

Listing 5.1 Recommendation Systems: Required Packages

Listing 5.2 demonstrates how to download our dataset from kagglehub and load it directly into a pandas DataFrame. By using dataset_download, we can download the dataset and then read it in with pd.read_csv. As the file doesn't contain any column names, we must define the column names ourselves. The dataset contains a column with the user ID (user_id), the title of the game (item_title), the information as to whether the game was purchased (purchase) and the behavior. This column represents the game usage in hours by the player.

```
#%% Data Download and Load
path = kagglehub.dataset_download("tamber/steam-video-games")
print("Path to dataset files:", path)
```

```
file_name = "steam-200k.csv"
df = pd.read_csv(os.path.join(path, file_name), header=None, names=["user_id",
"item_title","purchase","behavior", "value"])
```

Listing 5.2 Recommendation Systems: Data Loading

The value column doesn't contain any values, and we can delete it directly, as shown in Listing 5.3.

```
df = df.drop(columns=["value"])
df.head(2)
```

Listing 5.3 Recommendation Systems: Data Preparation; Column Deletion

It's always good to familiarize yourself with the data, especially in the behavior column, which should reflect the behavior that will be predicted by the algorithm, as follows:

```
#%% visualize the distribution of "behavior"
plt.hist(df["behavior"], bins=100, range=(0,100))
plt.show()
```

The histogram shown in Figure 5.3 illustrates that some games were played for up to a hundred hours but that the majority were only played for a few hours.

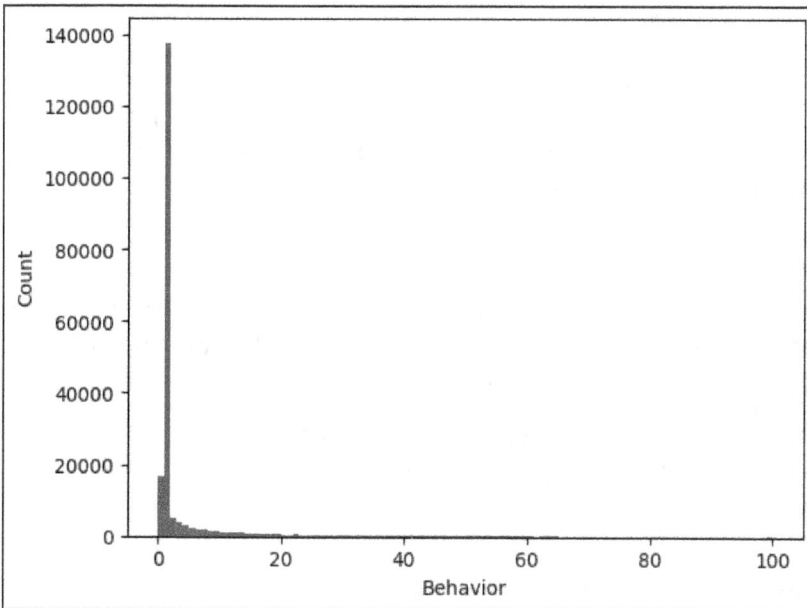

Figure 5.3 Feature Behavior Histogram

We need to create an algorithm that makes a recommendation with values between 1 and 5, where 1 stands for low interest (low usage time) and 5 for extremely high interest (very frequent and long play). To do this, in Listing 5.4, we create a new rating column that divides the playing time in hours in the behavior column into 5 groups. Here, we bring in limits, which were set arbitrarily to obtain enough values in the individual five groups.

```
#%% create rating column on "behavior" with range 1-5
df["rating"] = df["behavior"].apply(lambda x: 1 if x < 10 else 2 if x < 20 else
3 if x < 40 else 4 if x < 60 else 5)
```

Listing 5.4 Recommendation Systems: Data Preparation; Rating Feature

Let's take a closer look at this new rating column and the number of values per group, as follows:

```
#%% visualize the group count of "rating"
df["rating"].value_counts()
```

```
rating
1    174590
5      8357
2      8227
3      6078
4      2748
Name: count, dtype: int64
```

The vast majority can be found in group 1. This is hardly surprising, as most players have clear preferences for certain games in a genre, meaning that the majority of what is on offer is not relevant. But the other groups also have enough elements from which the model can learn behavior.

We know that PyTorch can only work with tensors. Currently, our user_ids are not consecutive numbers, and the game titles (item_title) are only available as strings. For modeling, we need to create two new columns for user_id and item_id, which are zero-based integer values. We can use the pd.factorize function for this, as shown in Listing 5.5.

```
#%% implement label encoder for "user_id" and "item_id"
# Convert user_id and item_id to 0-based indices
df["user_id"] = pd.factorize(df["user_id"])[0]
df["item_id"] = pd.factorize(df["item_title"])[0]
```

Listing 5.5 Recommendation Systems: Label Encoder

At this point, we have the three columns—user_id, item_id, and rating—which contain the ratings given for each user and which we need for modeling, as follows:

```
df[["user_id", "item_id", "rating"]].head(2)
```

	user_id	item_id	rating
0	0	0	1
1	0	0	5

This brings us to the creation of the PyTorch dataset class in Listing 5.6. This class returns the user_id, item_id, and rating for a given user via its __getitem__ method. In the __init__ method, we must include a check for the data type, because if the df object passed is already a Torch dataset, it must first be converted back into a DataFrame.

```python
class RecommenderSystemDataset(Dataset):
    def __init__(self, df):
        # Convert RandomSplit subset back to DataFrame if needed
        if isinstance(df, torch.utils.data.dataset.Subset):
            self.df = pd.DataFrame(df.dataset.iloc[df.indices])
        else:
            self.df = df
        self.user_ids = self.df["user_id"].unique()
        self.item_ids = self.df["item_id"].unique()
        self.ratings = self.df["rating"].values

    def __len__(self):
        return len(self.df)

    def __getitem__(self, idx):
        user_id = self.df.iloc[idx]["user_id"]
        item_id = self.df.iloc[idx]["item_id"]
        rating = self.df.iloc[idx]["rating"]
        return user_id, item_id, rating
```

Listing 5.6 Recommendation Systems: Dataset Class

We bundle all hyperparameters that are relevant for training into one block, as shown in Listing 5.7. In addition to the DEVICE, we define the number of EPOCHS and the BATCH_SIZE here.

```python
DEVICE = torch.device('cuda' if torch.cuda.is_available() else 'cpu')
EPOCHS = 8
BATCH_SIZE = 128
```

Listing 5.7 Recommendation Systems: Hyperparameters

Based on this dataset class, we can now create the training dataset (train_df) and validation dataset (val_df). Here, we reserve 80% of the data for training and 20% for validation. Then, we can create the Dataloader instances (train_dataloader and val_dataloader) in Listing 5.8.

```
# Split data into training and validation sets
train_size = int(0.8 * len(df))
val_size = len(df) - train_size
train_df, val_df = train_test_split(df, test_size=val_size/len(df), random_
state=42)

#%% Create datasets
train_dataset = RecommenderSystemDataset(train_df)
val_dataset = RecommenderSystemDataset(val_df)

# Create dataloaders
train_dataloader = DataLoader(train_dataset, batch_size=BATCH_SIZE, shuffle=
True)
val_dataloader = DataLoader(val_dataset, batch_size= BATCH_SIZE, shuffle=
False)
```

Listing 5.8 Recommendation Systems: Dataset and DataLoader instances

At this point, we have completed the preparation of the data and can start modeling the model.

5.2.2   Modeling

Figure 5.4 shows the model architecture that we'll implement. The user and item inputs are embedded. Embedding is a technique for transforming discrete objects (which are the properties of the user and the items in this case) into dense vectors in a low-dimensional space. This allows the content-related (semantic) relationships to be represented by the proximity of these vectors. The topic of embedding is fundamentally important, and we'll discuss it in detail in this context.

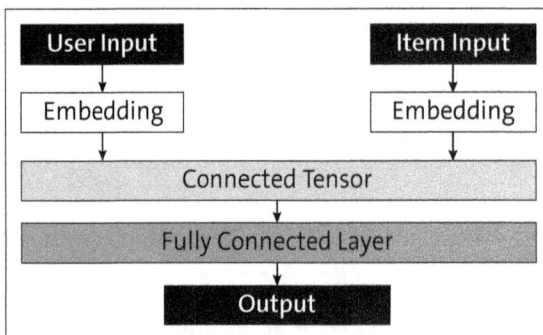

Figure 5.4 Model Architecture

The embedding tensor of both embeddings is then connected, and this embedding layer is passed through a linear layer before the result of the model is output.

Listing 5.9 shows the RecommenderSystemModel model class. It implements a simple recommendation system that uses the player embedding (user_embedding) and game embedding (item_embedding). The embedding layers and the linear layer for processing the embeddings are created in the __init__ method.

The embeddings enable the model to capture latent characteristics and relationships between users and items. Two players who have similar preferences receive embedding vectors that are close to each other in the vector space because of the training, and the same applies to the items: games of similar genre or popularity also have embedding vectors that are close to each other.

These vectors are dense and compact, and they therefore represent an efficient representation of the information. Another advantage is the model's ability to generalize. If two players have only a few games in common, the model can still make good recommendations because it compares the preferences of the two players in the embedding space and can thus find commonalities. This is a key point of collaborative filtering, as the system generates recommendations based on the behavior of other, similar users.

```
class RecommenderSystemModel(nn.Module):
    def __init__(self, num_users, num_items):
        super(RecommenderSystemModel, self).__init__()
        self.user_embedding = nn.Embedding(num_users, 32)
        self.item_embedding = nn.Embedding(num_items, 32)
        self.fc = nn.Linear(64, 1)

    def forward(self, user_id, item_id):
        user_embedding = self.user_embedding(user_id)
        item_embedding = self.item_embedding(item_id)
        x = torch.cat([user_embedding, item_embedding], dim=1)
        x = self.fc(x)
        return x
```

Listing 5.9 Recommendation Systems: Model Class

Now, let's look at Listing 5.10 to see how many games and matches are in our data. Just over twelve thousand players and five thousand games are represented in the dataset.

```
#%% get numbers of users and items
num_users = df["user_id"].nunique()
num_items = df["item_id"].nunique()
print(f"Number of users: {num_users}")
print(f"Number of items: {num_items}")
```

```
Number of users: 12393
Number of items: 5155
```

Listing 5.10 Recommendation Systems: Numbers of Users and Games (Items)

As these numbers are relevant for the size of the network, they are transferred to the model as parameters when the instance is created in Listing 5.11.

```
#%% model instance
model = RecommenderSystemModel(num_users, num_items)
model.to(DEVICE)
```

Listing 5.11 Recommendation Systems: Model Instance

In Listing 5.12, the instances of the loss function (loss_fn) and the optimizer (optimizer) are created. The mean-squared error loss and Adam are used as optimizers.

```
#%% loss function and optimizer
loss_fn = nn.MSELoss()
optimizer = torch.optim.Adam(model.parameters(), lr=0.001)
```

Listing 5.12 Recommendation Systems: Loss Function and Optimizer

Listing 5.13 shows the actual training of the model. The user_id and item_id are used in each batch to create predictions with the model, which are then compared with the actual values by the loss function (loss_fn).

```
# %% training loop
for epoch in range(EPOCHS):
    model.train()
    for user_id, item_id, rating in train_dataloader:
        user_id = user_id.to(DEVICE)
        item_id = item_id.to(DEVICE)
        rating = rating.float().to(DEVICE)
        optimizer.zero_grad()
        rating_pred = model(user_id, item_id)
        loss = loss_fn(rating_pred, rating.unsqueeze(1))
        loss.backward()
        optimizer.step()
    print(f"Epoch {epoch+1}/{EPOCHS} - Loss: {loss.item():.4f}")

Epoch 1/8 - Loss: 0.6219
...
Epoch 8/8 - Loss: 0.5057
```

Listing 5.13 Recommendation Systems: Training Loop

The model training has no surprises in store. We'll therefore now turn our attention to model evaluation.

5.2.3   Model Evaluation

Listing 5.14 describes how to create the rating_pred validation predictions.

```
model.eval()
rating_true, rating_pred = [], []
with torch.no_grad():
    for user_id, item_id, rating in val_dataloader:
        user_id = user_id.to(DEVICE)
        item_id = item_id.to(DEVICE)
        rating_pred_batch = model(user_id, item_id)
        rating_pred.extend(rating_pred_batch.tolist())
        rating_true.extend(rating.tolist())
```

Listing 5.14 Recommendation Systems: Creating Predictions for Validation Data

We'll first look at the correlation between actual and predicted values. The code for creating Figure 5.5 is shown in Listing 5.15.

```
#%% visualize the rating_true and rating_pred
sns.regplot(x=rating_true, y=rating_pred)
plt.xlabel('Actual Ratings')
plt.ylabel('Predicted Ratings')
plt.title('Actual vs. Predicted Ratings')
```

Listing 5.15 Recommendation Systems: Actual and Predicted Ratings

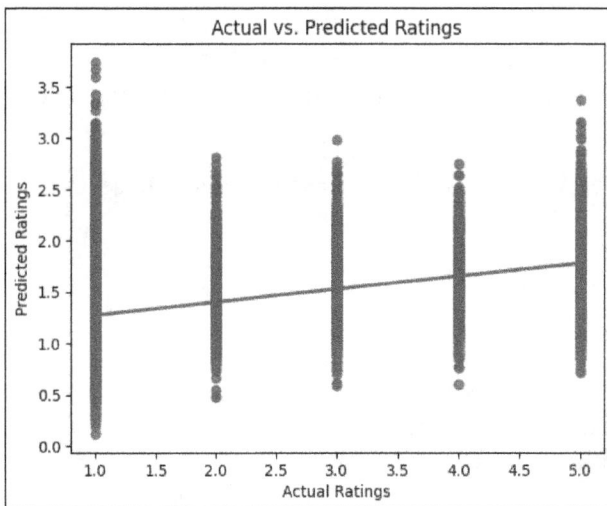

Figure 5.5 Recommendation Systems: Actual and Predicted Ratings

There is a positive correlation, as indicated by the ascending blue line. More common than the correlation plot, however, is the determination of metrics such as *Recall@k* and *Precision@k*. Before we implement these, let's get an understanding of how they work.

Let's start with Recall@k. Imagine you have a recommendation system that suggests ten movies to you. The number 10 corresponds to the k = 10 parameter. At the same time, you have a list of movies that you like—the relevant items. Recall@k calculates the percentage of relevant items that are included in your list of top-*k* recommendations. The formula is as follows:

$$Recall@k = \frac{Count\ of\ relevant\ items\ in\ Top - k\ recommendations}{Total\ count\ of\ relevant\ items}$$

Let's look at an example. You like movies A, B, C, D, and E, and the algorithm recommends movies A, B, F, G, H, I, J, K, L, and M. The relevant movies in the top ten recommendations are A and B, and these two movies represent the numerator in the equation. The denominator describes the total number of relevant movies, and in our example, that's 5 (A, B, C, D, and E).

In this case, the Recall@k is 2/5, or 40%. This means that the algorithm found 40% of the movies you like in its top ten recommendations. The Recall@k metric should be as high as possible, as it indicates that the system is more effective at finding relevant movies. It focuses on finding as many relevant movies as possible without considering the order or quality of the other movies.

Listing 5.16 shows the implementation of a Recall@k function. This function receives the trained model (model) and the validation data loader (val_dataloader) as parameters. In addition, we can adjust the k parameter and define a threshold that determines when an item is classified as relevant.

We want to determine the recall for each user and then average it across all values, and we use the rating threshold to determine the recall value. In addition, we want to filter out users with too few interactions because it's difficult to make a meaningful prediction for such users.

```
def calculate_recall_at_k(model, val_dataloader, k=10, threshold=4.0, device=
'cpu'):
    """
    Calculate Recall@k for the validation set.

    Args:
        model: Trained recommender system model
        val_dataloader: Validation data loader
        k: Number of top recommendations to consider
        threshold: Rating threshold to consider an item as relevant
        device: Device to run computations on
```

```
    Returns:
        float: Average recall@k across all users
    """
    model.eval()
    user_recalls = []

    # Group validation data by user
    user_items = {}
    user_ratings = {}

    with torch.no_grad():
        for user_id, item_id, rating in val_dataloader:
            for i in range(len(user_id)):
                u = user_id[i].item()
                item = item_id[i].item()
                r = rating[i].item()

                if u not in user_items:
                    user_items[u] = []
                    user_ratings[u] = []

                user_items[u].append(item)
                user_ratings[u].append(r)

    # Calculate recall@k for each user
    for user in user_items.keys():
        if len(user_items[user]) < 2:  # Skip users with too few interactions
            continue

        # Get relevant items for this user (items with rating >= threshold)
        relevant_items = set([item for item, rating in zip(user_items[user],
user_ratings[user])
                              if rating >= threshold])

        if len(relevant_items) == 0:
            continue

        # Get predictions for all items for this user
        user_tensor = torch.full((num_items,), user,
                            dtype=torch.long, device=device)
        item_tensor = torch.arange(num_items, device=device)

        predictions = model(user_tensor, item_tensor).squeeze()
```

```
        # Get top-k recommendations
        _, top_k_indices = torch.topk(predictions,
                                      k=min(k, len(predictions)))
        recommended_items = set(top_k_indices.cpu().numpy())

        # Calculate recall@k for this user
        relevant_recommended = len(relevant_items.intersection(recommended_
items))
        recall = relevant_recommended / len(relevant_items)
        user_recalls.append(recall)

    # Return average recall@k
    return np.mean(user_recalls) if user_recalls else 0.0
```

Listing 5.16 Recommendation Systems: Recall@k Function

Before we use this function, let's look at the Precision@k metric and how it works. Imagine you have a recommendation system that suggests ten movies (k = 10). Your goal is to assess how many of these are really good suggestions, so Precision@k calculates the proportion of relevant movies in the top-*k* recommendation list generated by the recommendation system. The formula is as follows:

$$Precision@k = \frac{Count\ of\ relevant\ items\ in\ Top-k\ recommendations}{Count\ of\ Top-k\ recommendations}$$

Let's look at an example. The system suggests movies A, B, F, G, H, I, J, K, L, and M. You like movies A, B, C, D, and E, so the relevant movies in the recommendations are movies A and B, which means there are two relevant movies. The total number of recommendations is ten, so the Precision@k is 2/10, or 20%. This means that 20% of the top ten recommendations are relevant to you. A higher Precision@k value is always more advantageous, as it indicates that the system's suggestions are of high quality and contain a smaller number of irrelevant items.

Unlike Recall@k, which aims to find as many relevant items as possible, Precision@k focuses on relevance within the limited recommendation list.

Listing 5.17 shows the implementation of the function for determining Precision@k. Its structure is similar to the Recall@k function in that the same parameters are passed. The function calculates the average Precision@k metric by evaluating the accuracy of the top-*k* recommendations for each user.

As with the recall function, we first identify the relevant items for each user by using a threshold value. The model then generates the top-*k* recommendations, and the function calculates the proportion of relevant items in this recommended list to finally return the average value across all users.

```
def calculate_precision_at_k(model, val_dataloader, k=10, threshold=4.0, de-
vice='cpu'):
    """
    Calculate Precision@k for the validation set.

    Args:
        model: Trained recommender system model
        val_dataloader: Validation data loader
        k: Number of top recommendations to consider
        threshold: Rating threshold to consider an item as relevant
        device: Device to run computations on

    Returns:
        float: Average Precision@k across all users
    """
    model.eval()
    user_precisions = []

    # Group validation data by user
    user_items = {}
    user_ratings = {}

    with torch.no_grad():
        for user_id, item_id, rating in val_dataloader:
            for i in range(len(user_id)):
                u = user_id[i].item()
                item = item_id[i].item()
                r = rating[i].item()

                if u not in user_items:
                    user_items[u] = []
                    user_ratings[u] = []

                user_items[u].append(item)
                user_ratings[u].append(r)

    # Calculate precision@k for each user
    for user in user_items.keys():
        if len(user_items[user]) < 2:  # Skip users with too few interactions
            continue

        # Get relevant items for this user (items with rating >= threshold)
        relevant_items = set([item for item, rating in zip(user_items[user],
        user_ratings[user])
```

5

```
                              if rating >= threshold])

        if len(relevant_items) == 0:
            continue

        # Get predictions for all items for this user
        user_tensor = torch.full((num_items,), user, dtype=torch.long,
        device=device)
        item_tensor = torch.arange(num_items, device=device)

        predictions = model(user_tensor, item_tensor).squeeze()

        # Get top-k recommendations
        _, top_k_indices = torch.topk(predictions, k=min(k, len(predictions)))
        recommended_items = set(top_k_indices.cpu().numpy())

        # Calculate precision@k for this user
        relevant_recommended = len(relevant_items.intersection(recommended_items))
        precision = relevant_recommended / len(recommended_items)
    if len(recommended_items) > 0 else 0.0
        user_precisions.append(precision)

    # Return average precision@k
    return np.mean(user_precisions) if user_precisions else 0.0

#%% Calculate and display recall@k and precision@k metrics
print("\n" + "="*50)
print("RECOMMENDATION SYSTEM EVALUATION")
print("="*50)

# Calculate metrics for different k values
k_values = [5, 10, 20, 50]
threshold = 4.0  # Consider items with rating >= 4 as relevant

for k in k_values:
    recall_k = calculate_recall_at_k(model, val_dataloader,
    k=k, threshold=threshold, device=DEVICE)
    precision_k = calculate_precision_at_k(model, val_dataloader, k=k,
    threshold=threshold, device=DEVICE)

    print(f"Recall@{k}: {recall_k:.4f}")
    print(f"Precision@{k}: {precision_k:.4f}")
    print("-" * 30)
```

Listing 5.17 Recommendation Systems: Precision@k Function

Now, we apply these functions to our model and our data in Listing 5.18. We use a threshold value of 4 and vary the number *k*. The corresponding Recall@k and Precision@k values are displayed on the console.

```
#%% Visualize recall@k and precision@k for different k values
threshold = 4.0
k_values = [5, 10, 20, 50, 100]
recalls = []
precisions = []

for k in k_values:
    recall_k = calculate_recall_at_k(model, val_dataloader, k=k, threshold=
    threshold, device=DEVICE)
    precision_k = calculate_precision_at_k(model, val_dataloader, k=k,
    threshold=threshold, device=DEVICE)
    recalls.append(recall_k)
    precisions.append(precision_k)


======================================================
RECOMMENDATION SYSTEM EVALUATION
======================================================
Recall@5: 0.0224
Precision@5: 0.0058
------------------------------
Recall@10: 0.1100
Precision@10: 0.0149
------------------------------
Recall@20: 0.1794
Precision@20: 0.0130
------------------------------
Recall@50: 0.4019
Precision@50: 0.0114
------------------------------
```

Listing 5.18 Recommendation Systems: Evaluation Results

Graphical representation is generally quicker and easier to understand, so we visualize the results with the code from Listing 5.19.

```
# Create visualization
fig, (ax1, ax2) = plt.subplots(1, 2, figsize=(12, 5))

# Recall@k plot
ax1.plot(k_values, recalls, 'bo-', linewidth=2, markersize=8)
ax1.set_xlabel('k (Number of Recommendations)')
```

```
ax1.set_ylabel('Recall@k')
ax1.set_title('Recall@k vs k')
ax1.grid(True, alpha=0.3)
ax1.set_xscale('log')

# Precision@k plot
ax2.plot(k_values, precisions, 'ro-', linewidth=2, markersize=8)
ax2.set_xlabel('k (Number of Recommendations)')
ax2.set_ylabel('Precision@k')
ax2.set_title('Precision@k vs k')
ax2.grid(True, alpha=0.3)
ax2.set_xscale('log')

plt.tight_layout()
plt.show()
```

Listing 5.19 Recommendation Systems: Visualization of Evaluation

Figure 5.6 shows the progression of `Recall@k` and `Precision@k` for different k values.

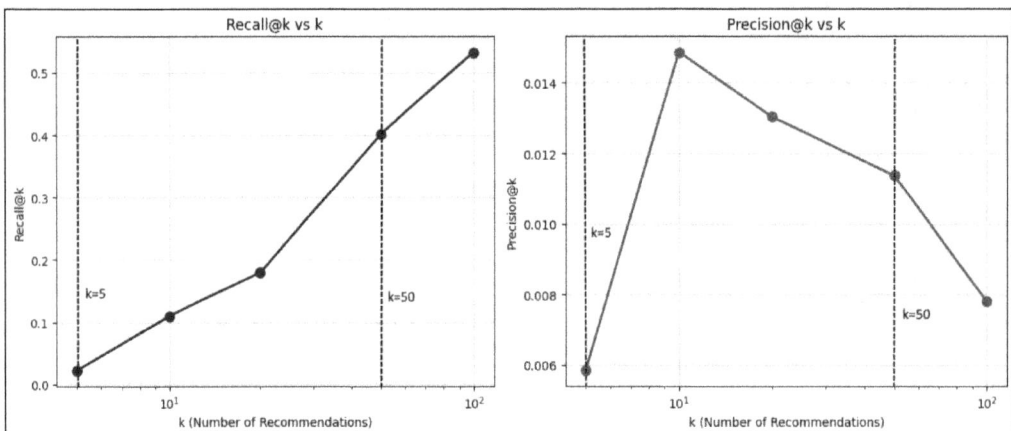

Figure 5.6 Recommendation Systems: Recall@k and Precision@k

The recall value increases significantly with increasing k. At Recall@50, it reaches 0.4, which means that the system finds an average of 40% of the items relevant to a user in the top fifty recommendations. This is a positive development, as the system identifies a greater amount of relevant content when it has a larger "list" to suggest.

The precision values are generally very low and decrease with increasing k (after a certain point). For Precision@5, the value is only about 0.58%, which means that only a very small proportion of the top five recommendations were relevant. For Precision@50, the value is even lower (0.0114), as the larger list of recommendations contains proportionally more irrelevant items.

There is a clear trade-off between recall and precision. The system is good at "remembering" relevant items (i.e., it has high recall), but it's less good at placing these items at the top of the list (i.e., it has low precision). The results indicate that the model can identify many relevant items in the entire dataset, but the relevance of the top recommendations can still be improved.

To make the system more useful for the end user, we should focus on increasing precision (i.e., the ability to place the best suggestions at the top of the recommendation list). We could do this by fine-tuning the model or including additional features (e.g., content features of items). However, this is outside the scope of this book, and we only offer it to you here as a suggestion for further elaboration.

Finally, let's look at the model's specific recommendations for a given player. Ultimately, it's all about bringing the model results into the real world.

Listing 5.20 shows the function for determining the top recommendations. The function receives the trained model (model), the idea of the user under consideration (user_id), the number of recommendations (*k*) and the optional parameter device as parameters. The function returns a list of the top recommendations.

```
def get_top_k_recommendations(model, user_id, k=10, device='cpu'):
    model.eval()
    with torch.no_grad():
        user_tensor = torch.full((num_items,), user_id,
 dtype=torch.long, device=device)
        item_tensor = torch.arange(num_items, device=device)
        predictions = model(user_tensor, item_tensor).squeeze()

        # Get top-k recommendations
        _, top_k_indices = torch.topk(predictions, k=min(k,
 len(predictions)))
        return top_k_indices.cpu().numpy().tolist()
```

Listing 5.20 Recommendation Systems: Function for Top-k Recommendations for Given User

In Listing 5.21, we apply the function to the user with ID 2 as an example and obtain the top ten recommendations for this player.

```
example_user_id = 2
top_10_recommendations = get_top_k_recommendations(model,
    example_user_id, k=10, device=DEVICE)

# Get the games this user has rated highly (rating >= 4)
user_liked_items = df[
    (df['user_id'] == example_user_id) &
    (df['rating'] >= 4)
]['item_title'].tolist()
```

```
print(f"\nUser {example_user_id}'s highly rated games:")
for item_title in user_liked_items:
    game_name = df[df['item_title'] == item_title]['item_title'].iloc[0]
    print(f"- Game ID: {game_name}")

print(f"\nTop 10 recommended games for user {example_user_id}:")
for item_id in top_10_recommendations:
    game_name = df[df['item_id'] == item_id]['item_title'].iloc[0]
    print(f"- Game ID: {game_name}")
```

```
User 2's highly rated games:
- Game ID: Ultra Street Fighter IV
- Game ID: FINAL FANTASY XIII
- Game ID: The Elder Scrolls V Skyrim

Top 10 recommended games for user 2:
- Game ID: Football Manager 2012
- Game ID: Football Manager 2013
- Game ID: Football Manager 2015
- Game ID: Football Manager 2014
- Game ID: Breezeblox
- Game ID: Movie Studio 13 Platinum - Steam Powered
- Game ID: DARK SOULS II
- Game ID: Football Manager 2016
- Game ID: Counter-Strike Global Offensive
- Game ID: Fallout 4
```

Listing 5.21 Recommendation Systems: Top-k Recommendations for Given User

5.3   Summary

In this chapter, we've dealt with the topic of recommendation systems. In Section 5.1, we learned about the basic concepts of recommendation systems, discussed content-based filtering and collaborative filtering, and became familiar with matrix factorization as a special form of collaborative filtering. In Section 5.2, we then applied this theoretical knowledge in practice by using a model for recommending computer games.

Chapter 6

Autoencoders

"Among competing hypotheses, the one with the fewest assumptions should be selected."
—*Wilhelm von Ockham*

In the field of deep learning, we often must deal with an overwhelming amount of data. This sheer complexity often seems insurmountable—and that's when a very old philosophical principle called *Occam's razor* comes to the rescue. It states that the simplest explanation is the best. Applied to our field, this means that often, we can present the true essence of a dataset in a much simpler, condensed form without losing essential elements.

This is precisely the central task of autoencoders. An *autoencoder* is a special form of neural network, and it consists of an encoder and a decoder. Its aim is to extract the most important features of the data and compress them into a minimal, "shaved" form (with the encoder). Then, at a later point in time, a user can restore the original data as accurately as possible from this compressed form (with the decoder). This network learns to remove the noise and redundancies from the data.

In this chapter, we'll look at how an autoencoder performs this task and how we can use this powerful tool to reduce dimensionality, detect anomalies, and generate new data. Section 6.1 explains the basic structure of an autoencoder, which consists of an encoder and a decoder network. It shows how these two parts work together to compress and restore data. Section 6.2 provides a practical guide to implementing an autoencoder. It covers the entire process from preparing the data to validating the trained model.

Section 6.3 introduces the variational autoencoder (VAE) as a further development of the classical autoencoder. It explains the fundamental difference between the two, which lies in the probabilistic nature of the latent space. Finally, in Section 6.4, we implement a VAE network and emphasize the difference between it and a classical autoencoder.

6.1   Architecture

An autoencoder generally consists of three main components:

- An *encoder*
- A *bottleneck*, also called a *latent space*
- A *decoder*

Figure 6.1 illustrates this general structure. The input data should be represented by less data from the autoencoder, so that the network provides reconstructed input data that can be represented by significantly fewer parameters.

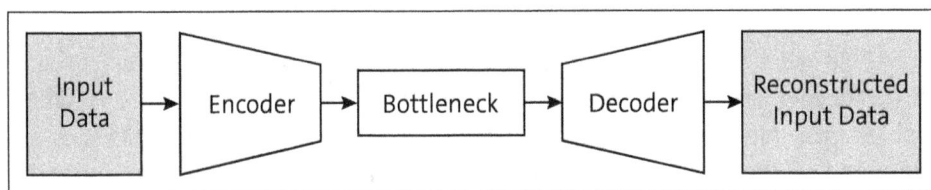

Figure 6.1 Autoencoder: General Structure

The input data can be tabular data with many columns or, for example, images. The encoder is responsible for compressing the input data in a latent space (i.e., representing it with fewer parameters). Compared to the input data, this latent space is a low-dimensional space that reflects the essential characteristics of the data. Its dimensions are described by the number of neurons in this layer, so if the latent space has 128 neurons, it's referred to as a 128-dimensional latent space. The decoder reconstructs the input data from the compressed display.

At the beginning of the network training, the model result is randomly initialized, so it corresponds to an image full of noise. As time progresses, the decoder becomes better at reconstructing the data so that the model result corresponds more and more closely to the original input data.

During training, the loss function must measure the deviation of the reconstructed image from the original image. A simple approach here would be to compare the pixel values in each case and square the deviations. We already learned about this mean squared error (MSE) in earlier chapters.

6.2   Coding: Autoencoder

In this section, we'll train an autoencoder model that can represent handwritten digits with a small number of latent dimensions. The dataset we'll use is called MNIST, and it contains a large number of handwritten digits. This classic dataset consists of 60,000 images, and each image is a grayscale image with 28 × 28 pixels describing the digits between 0 and 9. Figure 6.2 shows the original figures and the reconstructions created from the model.

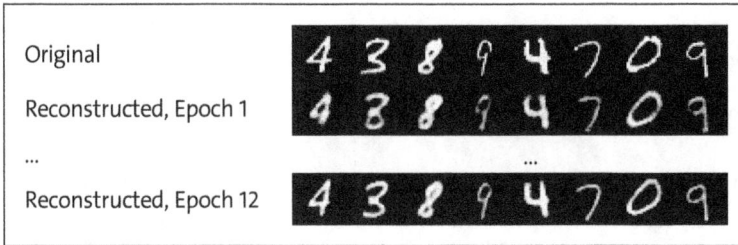

Figure 6.2 Original and Reconstructed Digits Based on Autoencoder

We'll develop an autoencoder model in this section, beginning with the preparation of the data in Section 6.2.1. Then, we'll develop our own model in Section 6.2.2 and train it based on the data in Section 6.2.3. Finally we'll check the model performance in Section 6.2.4.

6.2.1 Data Preparation

The complete script for this example can be found at *070_Autoencodermnist_autoencoder.py*. We'll start with the code in Listing 6.1, which imports the required packages. We'll provide the training data via torchvision and the MNIST class. Other packages (torch and all loaded submodules) are required for modeling, and still others (matplotlib and seaborn) are required for visualization.

```
import os
import torch
import torch.nn as nn
import torchvision.transforms as transforms
import torchvision.utils as vutils
from torchvision.datasets import MNIST
from torch.utils.data import DataLoader, random_split
import matplotlib.pyplot as plt
import seaborn as sns
```

Listing 6.1 Autoencoder: Loading of Required Packages

We control the model training via various hyperparameters, which we define as in Listing 6.2. In addition to the already known BATCH_SIZE, EPOCHS, and LR parameters, we define the LATENT_DIMS parameter here to determine how many latent dimensions the data will describe.

```
BATCH_SIZE = 32
EPOCHS = 12
```

```
LR = 0.001
LATENT_DIMS = 16
VAL_SPLIT = 0.1
OUT_DIR = "runs"
DEVICE = torch.device("cuda" if torch.cuda.is_available() else "cpu")
```

Listing 6.2 Autoencoder: Hyperparameter

Now, let's move on to loading the data in Listing 6.3. Before we load the data, we must transform it. We create these transformations with transforms.Compose, and we pass the preprocessing steps to this function in a list. At the beginning, we adjust the size of the images. They should all already have 28 × 28 pixels, but we're better off ensuring that they do with Resize. We then convert the images into tensors with ToTensor and normalize them with Normalize.

```
# %% transformations
my_transforms = transforms.Compose([
    transforms.Resize((28,28)),
    transforms.ToTensor(),
    # Normalize to [-1, 1] so a Tanh decoder output matches input scale (1
channel)
    transforms.Normalize(mean=[0.5], std=[0.5])
])

dataset = MNIST(root=os.path.join("data", "mnist"), train=True, download=True,
transform=my_transforms)
```

Listing 6.3 Autoencoder: Dataset Preparation

Next, we check the value range of the data and see that it's actually between –1.0 and +1.0, as follows:

```
x, _ = dataset[0]
print(x.min().item(), x.max().item())
```

-1.0 1.0

Now, let's now continue with the splitting of the data into training data (train_dataset) and validation data (val_dataset). The splitting takes place according to a proportion that we define in VAL_SPLIT. We can also use the random_split function for the split by giving it the dataset and the number of datasets via the lengths parameter, as follows:

```
val_size = int(len(dataset) * VAL_SPLIT)
train_size = len(dataset) - val_size
train_dataset, val_dataset = random_split(dataset=dataset, lengths=[train_size,
val_size])
```

At this point, we've created the `train_dataset` and `val_dataset` datasets, so we use them to can create the `DataLoader`s in Listing 6.4.

```
train_loader = DataLoader(dataset=train_dataset, batch_size=BATCH_SIZE, shuf-
fle=True)
val_loader = DataLoader(dataset=val_dataset, batch_size=BATCH_SIZE, shuffle=
False)
```

Listing 6.4 Autoencoder: Creating Training DataLoader and Validation DataLoader

You should check the assumptions from time to time. The initial dataset has 60,000 elements, and with a 10% share of the validation data, there should be 54,000 training data and 6,000 validation data—which is true in this case. In this example, we use this two-part split, but you're free to create a three-way split of training, validation, and test data.

```
print(f"Training data count: {len(train_dataset)}")
print(f"Validation data count: {len(val_dataset)}")
```

```
Training data count: 54000
Validation data count: 6000
```

We've now prepared the data enough to turn our attention to modeling.

6.2.2 Modeling

The autoencoder consists of an encoder and a decoder. For more complex models, having a separate class for the latent space could make sense—but in this simple case, we don't need one and can consider the output of the encoder as a latent space.

It also makes sense for us to define the encoder and decoder class separately so that we can use them individually later. Figure 6.3 shows the concrete structure of our network, which consists of an encoder and a decoder.

Listing 6.5 shows the structure of the encoder class. As we are working with images, it makes sense to use a `Conv2d` layer. This has an input channel `in_channels = 1`, as the image is in grayscale. For RGB images, there would be three channels.

We generate 6 output channels and use the standard kernel size of 3. We activate the convolutional layer with `ReLU` and "smooth" the results with `Flatten`. The dimensions have been reduced from 28 pixels to 26, due to the CNN layer, and the linear layer thus has 6 × 26 × 26 neurons. This last step ensures that the 4D tensor (`BATCH_SIZE`, feature channels, horizontal pixels, and vertical pixels) becomes a 2D tensor (`BATCH_SIZE` and `LATENT_DIMS`).

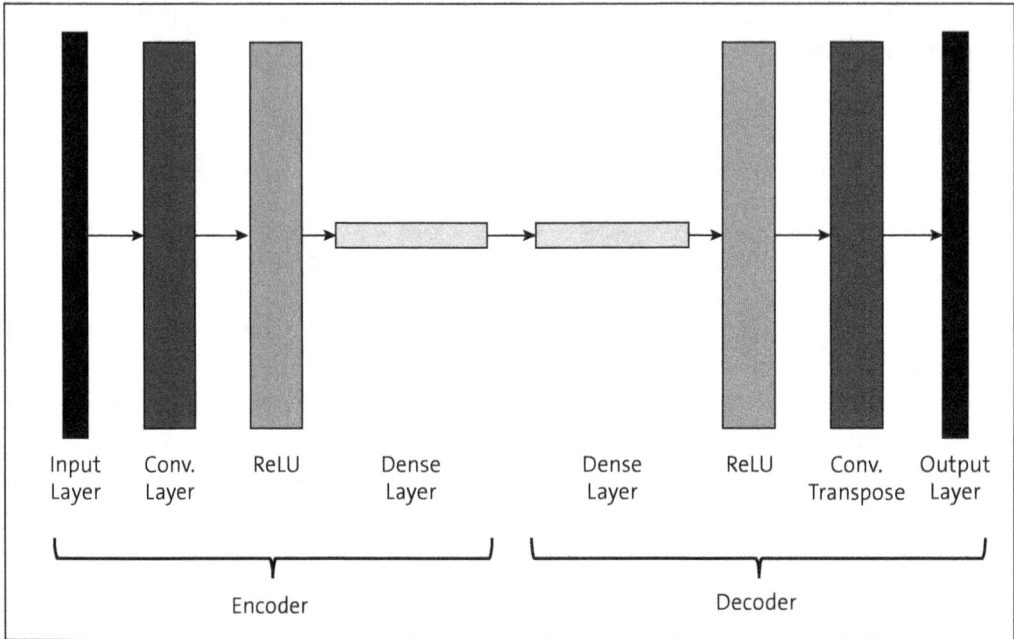

Figure 6.3 Autoencoder Network Setup

```
class Encoder(nn.Module):
    def __init__(self):
        super().__init__()
        self.conv1 = nn.Conv2d(in_channels=1,
                                out_channels=6, kernel_size=3)
        self.relu = nn.ReLU()
        self.flatten = nn.Flatten()
        self.fully_connected = nn.Linear(6*26*26, LATENT_DIMS)

    def forward(self, x):
        x = self.conv1(x)  # in: (BS, 1, 28, 28), out: (BS, 6, 26, 26)
        x = self.relu(x)  #
        x = self.flatten(x)  # out: (BS, 6*26*26)
        x = self.fully_connected(x)  # out: (BS, LATENT_DIMS)
        return x
```

Listing 6.5 Autoencoder: Encoder Class

The decoder class shown in Listing 6.6 reverses the process. It usually corresponds to the mirroring of the encoder class, and we use the following mirror functions for this: ConvTranspose2d (for reversing Conv2d) and unflatten (for reversing flatten).

We activate the output layer by using the hyperbolic tangent tanh. We do that because it normalizes the output to –1 to +1, which corresponds exactly with the value range of the input data. And so, we've managed to use this special activation function at least at this one point.

```
class Decoder(nn.Module):
    def __init__(self):
        super().__init__()
        self.fully_connected = nn.Linear(LATENT_DIMS, 6*26*26)
        self.unflatten = nn.Unflatten(dim=1, unflattened_size=(6, 26, 26))
        self.conv1 = nn.ConvTranspose2d(6, 1, 3)
        self.relu = nn.ReLU()
        self.tanh = nn.Tanh()

    def forward(self, x):
        x = self.fully_connected(x)
        x = self.relu(x)
        x = self.unflatten(x)  # out: (BS, 6, 26, 26)
        x = self.conv1(x)  # out: (BS, 1, 28, 28)
        x = self.tanh(x)  # match input scale [-1, 1]
        return x
```

Listing 6.6 Autoencoder: Decoder Class

The actual autoencoder class in Listing 6.7 consists of a connection between the encoder class and the decoder class.

```
class Autoencoder(nn.Module):
    def __init__(self):
        super().__init__()
        self.encoder = Encoder()
        self.decoder = Decoder()

    def forward(self, x):
        x = self.encoder(x)
        x = self.decoder(x)
        return x
```

Listing 6.7 Autoencoder: Autoencoder Class

To perform the postprocessing of the results, we need some auxiliary functions, which we'll define here. We want to visualize the reconstructed images at the end, and for this purpose, we'll create the denormalize and save_reconstruction_grid functions to display the original and reconstruction images.

Listing 6.8 first describes how we create the OUT_DIR output folder (if it doesn't already exist). We then create the denormalize function, obtain the tensors with the value range from −1 to +1, and scale back the value range to 0 to 255 so that we can display the tensors as images. We use the save_reconstruction_grid function in such a way that it is called in each epoch and returns the original representation and the corresponding reconstruction for ten sample digits.

```python
if not os.path.exists(OUT_DIR):
    os.makedirs(OUT_DIR, exist_ok=True)

def denormalize(img_tensor: torch.Tensor) -> torch.Tensor:
    # Input in [-1, 1] -> [0, 1]
    return (img_tensor + 1.0) / 2.0

def save_reconstruction_grid(model: nn.Module, images: torch.Tensor,
        epoch: int, out_dir: str) -> None:
    model.eval()
    with torch.no_grad():
        images = images.to(DEVICE)
        reconstructed = model(images)
        # Prepare a grid with originals (top) and reconstructions (bottom)
        original = denormalize(images).cpu()
        reconstructed_image = denormalize(reconstructed).cpu()
        grid = vutils.make_grid(torch.cat([original, reconstructed_image], dim=
0),
            nrow=images.size(0))
        plt.figure(figsize=(12, 6))
        plt.axis('off')
        plt.imshow(grid.permute(1, 2, 0).squeeze())
        plt.tight_layout()
        plt.savefig(os.path.join(out_dir, f"epoch_{epoch:03d}.png"))
        plt.close()
```

Listing 6.8 Autoencoder: Helper Functions

We can now carry out the training with the prepared data and the helper functions.

6.2.3 Model Training

Listing 6.9 describes how to instantiate the model instance model, the optimizer, and the loss function (loss_fun).

```
#%% Model, optimizer, loss
model = Autoencoder().to(DEVICE)
optimizer = torch.optim.Adam(model.parameters(), lr=LR)
loss_fun = nn.MSELoss()
```

Listing 6.9 Autoencoder: Model Instance, Optimizer, and Loss Function

We should always use the same images for validation so that we an see the progress of the training over several epochs. To do this, we use the next and iter functions in Listing 6.10 to extract images from the DataLoader objects, which we then use for validation.

```
if len(val_loader) > 0:
    fixed_images, fixed_labels = next(iter(val_loader))
else:
    fixed_images, fixed_labels = next(iter(train_loader))
# Use a small grid (up to 8 images)
fixed_images = fixed_images[:8]
```

Listing 6.10 Autoencoder: Ensure Consistent Validation Images

We use the training loop shown in Listing 6.11 to perform the model training. We train the model as usual, based on the training data, and then validate it with the validation data. To continuously monitor the progress of the training, we save the losses and create an image using the save_reconstruction_grid function to compare the original and reconstruction images. We then save these in the folder we define via the OUT_DIR variable.

```
#%% Training loop
loss_train, loss_val = [], []
for epoch in range(1, EPOCHS + 1):
    model.train()
    running_loss = 0.0
    for i, (X_batch, _) in enumerate(train_loader):
        X_batch = X_batch.to(DEVICE)
        y_pred = model(X_batch)
        loss = loss_fun(y_pred, X_batch)
        loss.backward()
        optimizer.step()
        optimizer.zero_grad()
        running_loss += loss.item() * X_batch.size(0)
    train_epoch_loss = running_loss / (len(train_loader.dataset)
    loss_train.append(train_epoch_loss)
```

```
# Validation loss
model.eval()
val_running_loss = 0.0
with torch.no_grad():
    for X_val, _ in val_loader:
        X_val = X_val.to(DEVICE)
        y_val = model(X_val)
        v_loss = loss_fun(y_val, X_val)
        val_running_loss += v_loss.item() * X_val.size(0)
val_epoch_loss = val_running_loss / (len(val_loader.dataset)
loss_val.append(val_epoch_loss)
print(f"Epoch {epoch:02d} | train_loss={train_epoch_loss:.4f} |
val_loss={val_epoch_loss:.4f}")

# Save reconstructions and latent space every epoch
save_reconstruction_grid(model, fixed_images, epoch, OUT_DIR)
```

Epoch 01 | train_loss=0.1127 | val_loss=0.0692
Epoch 02 | train_loss=0.0614 | val_loss=0.0562
...
Epoch 12 | train_loss=0.0384 | val_loss=0.0395

Listing 6.11 Autoencoder: Training Loop

After we finish training the model, we'll want to check how well we've trained it, and we'll cover that in the following section.

6.2.4 Model Validation

First, let's look at the training and validation losses over the training progress by using the code from Listing 6.12.

```
sns.lineplot(x=range(1, EPOCHS + 1), y=loss_train, label="train")
sns.lineplot(x=range(1, EPOCHS + 1), y=loss_val, label="val")
plt.title("Training- and Validation Loss")
plt.xlabel("Epoch [-]")
plt.ylabel("Loss [-]")
plt.show()
```

Listing 6.12 Autoencoder: Training Losses and Validation Losses

The losses are shown in Figure 6.4 over the epochs, for both the training data and the validation data. The loss looks very good—both the training and validation losses initially decrease sharply and then asymptotically approach a loss value.

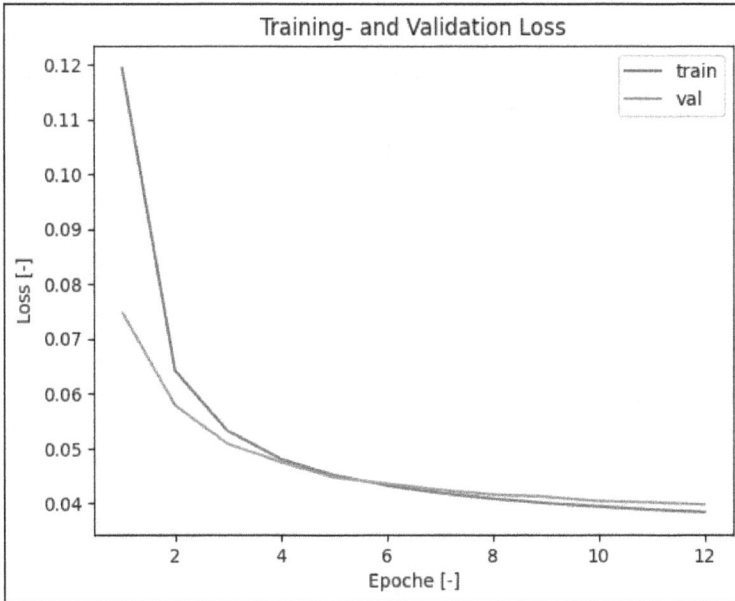

Figure 6.4 MNIST: Training Losses and Validation Losses

Finally, we want to check how well the model can reconstruct the original images at different points in time, as in Figure 6.5.

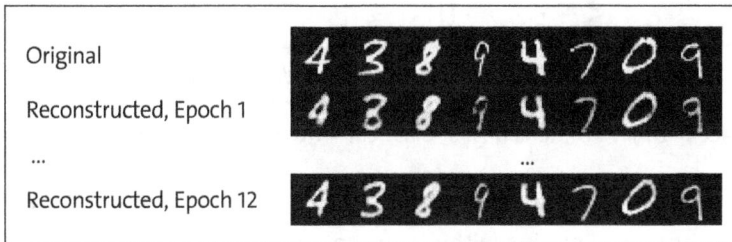

Figure 6.5 MNIST: Original and Reconstructed Images

After one epoch, the result already looks quite good but is still somewhat unclear. But after twelve epochs, we can clearly assign the reconstructed images to the numbers.

We must also consider that the original images were described by 28 × 28 (i.e.. 784) pixels. The reconstructed images were restored from the compressed latent space, which in our example has only 16 dimensions. We were therefore able to compress the data to a very high degree, namely by 2% of the original data. Data compression is also one of the main purposes of autoencoders, and we were able to demonstrate this functionality impressively.

VAEs, to which we'll turn in the following section, fulfill a completely different task. We can use them to generate new, similar data that differs from the training data.

6.3   Variational Autoencoders

Variational autoencoders (VAEs) are generative models because we can use them create new data. While a conventional autoencoder reduces the dimensionality of the data, VAEs go a decisive step further: we can use them to detect anomalies (i.e., deviations from "normal" behavior).

Whereas with classic autoencoders, we map the data to fixed points in latent space, with VAEs, we map the data points to a probability distribution in latent space. Figure 6.6 shows the latent space with the following two dimensions:

- **p(x)**
  This is the distribution of observed data.

- **p(z)**
  This is the distribution of latent data.

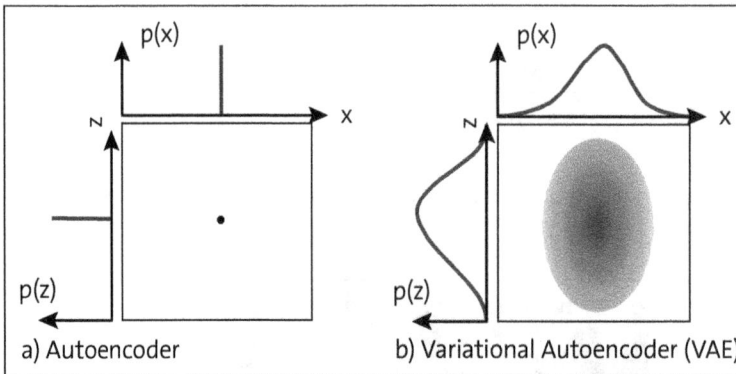

Figure 6.6 Latent Space with Two Dimensions: a) Classical Autoencoder; b) VAE

A classical autoencoder learns a compressed representation of data, which is the so-called latent space. In the representation in this latent space (here, with the dimensions X and Z), each observed data point p(x) is mapped to a single point.

There is no distribution of probabilities because the autoencoder is deterministic. It maps the input to a fixed, predicted point in latent space. The distribution of the latent data p(z) therefore consists of individual, discrete points, which are often represented as a delta function (a vertical line at a certain point that is infinitely high and whose integral is 1).

In contrast, the VAE doesn't model a single point for each observed data point p(x); instead, it models a probability distribution in latent space. This distribution is typically assumed to be a Gaussian curve (a normal distribution). This means that the VAE estimates a mean value (μ) and a standard deviation (σ) for each dimension of the latent variable z, resulting in a range of possible values instead of a single point.

The distribution of the latent data p(z) in the shared space therefore looks like a hilltop that slopes in all directions. The density is highest in the center and decreases as one moves toward the outside, which means that the VAE "knows" which points in the latent space are most likely to represent the original data. This property allows the VAE to generate new, plausible data points by sampling from this distribution.

6.4   Coding: Variational Autoencoder

Now, we'll implement a VAE in practice. The complete script can be found at *070_Auto-encoder\mnist_vae.py*. The data preparation is analogous to the previous script, so we can go straight into the modeling of the network in Section 6.4.1. In this case, the loss function requires special attention, and we'll cover it in Section 6.4.2.

6.4.1   Network Architecture

We'll develop a network consisting of an encoder and a decoder. Figure 6.7 shows the actual implementation.

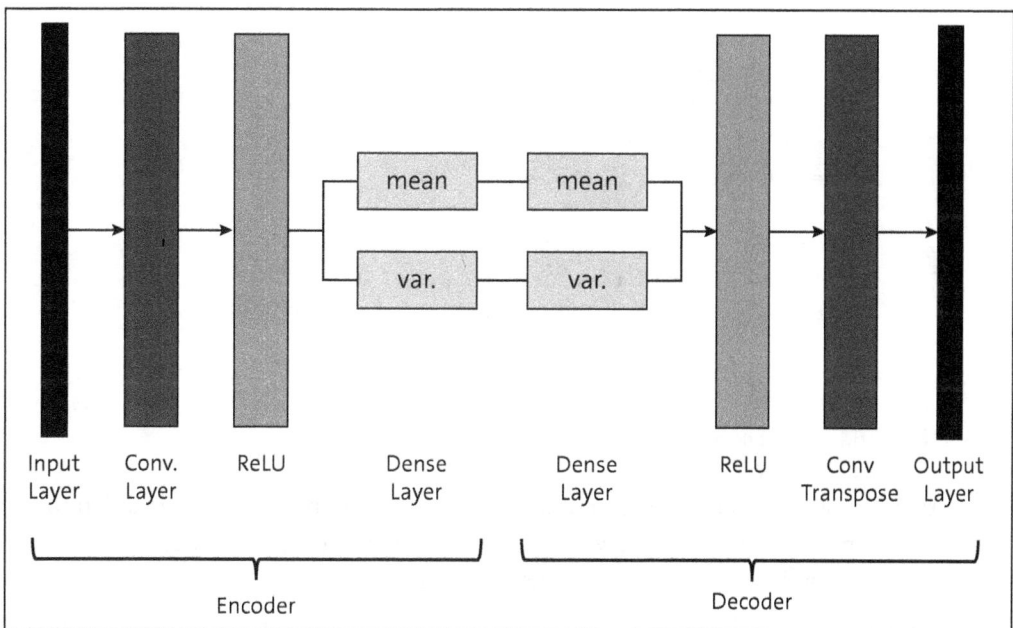

Figure 6.7 VAE: Network Structure

The decoder corresponds to the mirror image of the encoder. In contrast to the network of the classical autoencoder, there are two vectors for the dense layers: one for the mean values and one for the variances.

We start by implementing the encoder in Listing 6.13. This encoder doesn't differ from the encoder of the classical autoencoder, for the most part. We only find differences at the end of the model, as the mean values (fc_mean) and the variances (fc_logvar) are now predicted.

```python
class VAEEncoder(nn.Module):
    def __init__(self):
        super().__init__()
        self.conv1 = nn.Conv2d(in_channels=1,
                               out_channels=6, kernel_size=3)
        self.relu = nn.ReLU()
        self.flatten = nn.Flatten()
        self.fc = nn.Linear(6*26*26, 128)
        self.relu2 = nn.ReLU()

        # Separate layers for mean and log variance
        self.fc_mean = nn.Linear(128, LATENT_DIMS)
        self.fc_logvar = nn.Linear(128, LATENT_DIMS)

    def forward(self, x):
        x = self.conv1(x)  # in: (BS, 1, 28, 28), out: (BS, 6, 26, 26)
        x = self.relu(x)
        x = self.flatten(x)  # out: (BS, 6*26*26)
        x = self.fc(x)  # out: (BS, 128)
        x = self.relu2(x)

        # Get mean and log variance
        mu = self.fc_mean(x)  # out: (BS, LATENT_DIMS)
        logvar = self.fc_logvar(x)  # out: (BS, LATENT_DIMS)

        return mu, logvar
```

Listing 6.13 VAE: Encoder Class

What's new here is how the mean values (mu) and the variances (logvar) are determined. The output from the flatten layer is now fed into two parallel branches, one for mu and the other for the layer with the variances (logvar).

The decoder class shown in Listing 6.14 mirrors the encoder. There are differences in the inversion functions (unflatten and conv1 [which uses ConvTranspose2d]), as well as in the activation of the output with tanh.

```python
class VAEDecoder(nn.Module):
    def __init__(self):
        super().__init__()
        self.fc = nn.Linear(LATENT_DIMS, 128)
```

```
    self.relu = nn.ReLU()
    self.fc2 = nn.Linear(128, 6*26*26)
    self.relu2 = nn.ReLU()
    self.unflatten = nn.Unflatten(dim=1, unflattened_size=(6, 26, 26))
    self.conv1 = nn.ConvTranspose2d(6, 1, 3)
    self.tanh = nn.Tanh()

def forward(self, x):
    x = self.fc(x)   # in: (BS, LATENT_DIMS), out: (BS, 128)
    x = self.relu(x)
    x = self.fc2(x)   # out: (BS, 6*26*26)
    x = self.relu2(x)
    x = self.unflatten(x)   # out: (BS, 6, 26, 26)
    x = self.conv1(x)   # out: (BS, 1, 28, 28)
    x = self.tanh(x)   # match input scale [-1, 1]
    return x
```

Listing 6.14 VAE: Decoder Class

The VAE class in Listing 6.15 now brings together both elements: the encoder and the decoder. We need to mention a few special features here. There is a `reparameterize` method that uses the reparameterization trick, which makes the training of a VAE possible in the first place.

Reparameterization Trick in VAE

We have the following problem with VAEs. A VAE must draw random samples from a learned probability distribution in the latent space, but to train a neural network, we need backpropagation, which, as we know, calculates the gradients to then adjust the weights. The problem here, however, is that the process of random sampling is not differentiable. The solution is to separate randomness from the learnable parameters of the network, and we do this as follows:

1. The encoder of the VAE continues to learn and output the mean and standard deviation of a distribution.

2. Instead of directly drawing a random sample from this distribution, we take a simple random number from a fixed standard normal distribution (mean 0; standard deviation 1).

3. We convert this simple random number (ϵ) into a new variable (z) by using the formula $z = \mu + \sigma \times \epsilon$.

This transformation is deterministic and differentiable. Backpropagation can now flow through this formula without any problems and adjust the parameters for the mean and standard deviation of the encoder so that the VAE is trained effectively.

Another method was written to draw random numbers from the number range 0 to 1 in a normally distributed manner and pass this number through the decoder.

```python
class VAE(nn.Module):
    def __init__(self):
        super().__init__()
        self.encoder = VAEEncoder()
        self.decoder = VAEDecoder()

    def reparameterize(self, mu, logvar):
        std = torch.exp(0.5 * logvar)
        eps = torch.randn_like(std)
        return mu + eps * std

    def forward(self, x):
        mu, logvar = self.encoder(x)
        z = self.reparameterize(mu, logvar)
        return self.decoder(z), mu, logvar

    def generate(self, num_samples=1):
        z = torch.randn(num_samples, LATENT_DIMS).to(DEVICE)
        return self.decoder(z)
```

Listing 6.15 VAE: VAE Class

The code in Listing 6.16 starts by ensuring that the output folder for storing the images is available. The denormalize function allows us to return the tensor to a form that we can then display as an image.

```python
if not os.path.exists(OUT_DIR):
    os.makedirs(OUT_DIR, exist_ok=True)

def denormalize(img_tensor: torch.Tensor) -> torch.Tensor:
    # Input in [-1, 1] -> [0, 1]
    return (img_tensor + 1.0) / 2.0
```

Listing 6.16 VAE: Denormalization Function

Listing 6.17 shows two further functions. The save_reconstruction_grid function allows us to visualize the capabilities of the model reconstructed data, while the save_generated_samples function lets us generate digits with the help of the trained VAE.

```python
def save_reconstruction_grid(model: nn.Module,
    images: torch.Tensor, epoch: int, out_dir: str) -> None:
    model.eval()
```

```
with torch.no_grad():
    images = images.to(DEVICE)
    reconstructed, _, _ = model(images)
    # Prepare a grid with originals (top) and reconstructions (bottom)
    original = denormalize(images).cpu()
    reconstructed_image = denormalize(reconstructed).cpu()
    grid = vutils.make_grid(torch.cat([original, reconstructed_image], dim=
0),
    nrow=images.size(0))
    plt.figure(figsize=(12, 6))
    plt.axis('off')
    plt.imshow(grid.permute(1, 2, 0).squeeze())
    plt.tight_layout()
    plt.savefig(os.path.join(out_dir, f"reconstruction_epoch_{ep-
och:03d}.png"))
    plt.close()

def save_generated_samples(model: nn.Module, epoch: int,
    out_dir: str, num_samples=16) -> None:
    """Generate and save new samples from the VAE"""
    model.eval()
    with torch.no_grad():
        generated = model.generate(num_samples)
        generated_images = denormalize(generated).cpu()
        grid = vutils.make_grid(generated_images, nrow=4)
        plt.figure(figsize=(8, 8))
        plt.axis('off')
        plt.imshow(grid.permute(1, 2, 0).squeeze())
        plt.tight_layout()
        plt.savefig(os.path.join(out_dir, f"generated_epoch_{epoch:03d}.png"))
        plt.close()
```

Listing 6.17 VAE: Function for Digit Creation

6.4.2   Loss Function

Listing 6.18 shows how to calculate the loss function for the VAE. We calculate the losses from the following two components:

- **Reconstruction loss (recon_loss)**
  This measures how well the VAE reconstructs the original input data. Our goal is for the reconstructed output (recon_x) to be as similar as possible to the original input (x), and we achieve this by using the MSE, which calculates the average of the squared differences between the two tensors.

235

- **KL divergence loss** (kl_loss)

  This measures the difference between the probability distribution learned by the encoder (in the mu and logvar variables) and a standard normal distribution. The VAE penalizes this if the learned distributions are too far removed from a simple, well-organized, normal distribution. This restriction is necessary to ensure that the latent space is continuous and orderly, and it enables the VAE to generate new and realistic data points later.

```
def vae_loss_function(recon_x, x, mu, logvar):
    # Reconstruction loss (MSE)
    recon_loss = nn.MSELoss(reduction='sum')(recon_x, x)

    # KL divergence loss
    kl_loss = -0.5 * torch.sum(1 + logvar - mu.pow(2) - logvar.exp())

    return recon_loss + kl_loss
```

Listing 6.18 VAE: Loss Function

Listing 6.19 takes us back to familiar territory. We create a model instance (model) and an instance of the optimizer.

```
#%% Model, optimizer
model = VAE().to(DEVICE)
optimizer = torch.optim.Adam(model.parameters(), lr=LR)
```

Listing 6.19 VAE: Model Instance and Optimizer

To validate the reconstruction, we use a fixed selection of images and create the corresponding fixed_images object in Listing 6.20.

```
# Fixed validation batch for consistent visualization
if len(val_loader) > 0:
    fixed_images, fixed_labels = next(iter(val_loader))
else:
    fixed_images, fixed_labels = next(iter(train_loader))
# Use a small grid (up to 8 images)
fixed_images = fixed_images[:8]
```

Listing 6.20 Fixed Selection of Images for Validation

Now, we've come to the actual training loop. Listing 6.21 shows the various steps that we're already familiar with, and the only unusual aspect here is the loss function (vae_loss_function). Since we're training several tensors (recon_batch, mu, and logvar), calling the loss function is also somewhat more complex. In addition to these tensors, we must pass the real values (X_batch).

At the end of each epoch, we generate digits by using the save_generated_samples function and save the reconstructions as an image by using save_reconstruction_grid.

```
#%% Training loop
loss_train, loss_val = [], []
for epoch in range(1, EPOCHS + 1):
    model.train()
    running_loss = 0.0
    for i, (X_batch, _) in enumerate(train_loader):
        X_batch = X_batch.to(DEVICE)

        # Forward pass
        recon_batch, mu, logvar = model(X_batch)

        # Calculate loss
        loss = vae_loss_function(recon_batch, X_batch, mu, logvar)

        # Backward pass
        optimizer.zero_grad()
        loss.backward()
        optimizer.step()

        running_loss += loss.item()

    train_epoch_loss = running_loss / len(train_loader)
    loss_train.append(train_epoch_loss)

    # Validation loss
    model.eval()
    val_running_loss = 0.0
    with torch.no_grad():
        for X_val, _ in val_loader:
            X_val = X_val.to(DEVICE)
            recon_val, mu_val, logvar_val = model(X_val)
            v_loss = vae_loss_function(recon_val, X_val, mu_val, logvar_val)
            val_running_loss += v_loss.item()

    val_epoch_loss = val_running_loss / len(val_loader)
    loss_val.append(val_epoch_loss)

    print(f"Epoch {epoch:02d} | train_loss={train_epoch_loss:.4f} |
        val_loss={val_epoch_loss:.4f}")
```

```
# Save reconstructions and generated samples every epoch
save_reconstruction_grid(model, fixed_images, epoch, OUT_DIR)
save_generated_samples(model, epoch, OUT_DIR)
```

Listing 6.21 VAE: Training Loop

Using the code shown in Listing 6.22, we plot the loss for the training and validation data.

```
#%% Visualize train, val loss
plt.figure(figsize=(10, 6))
sns.lineplot(x=range(1, EPOCHS + 1), y=loss_train, label="train")
sns.lineplot(x=range(1, EPOCHS + 1), y=loss_val, label="val")
plt.title("VAE Training- and Validation Loss")
plt.xlabel("Epoch [-]")
plt.ylabel("Loss [-]")
plt.show()
```

Listing 6.22 VAE: Training Losses and Validation Losses

Figure 6.8 shows the loss curve, which indicates successful model training.

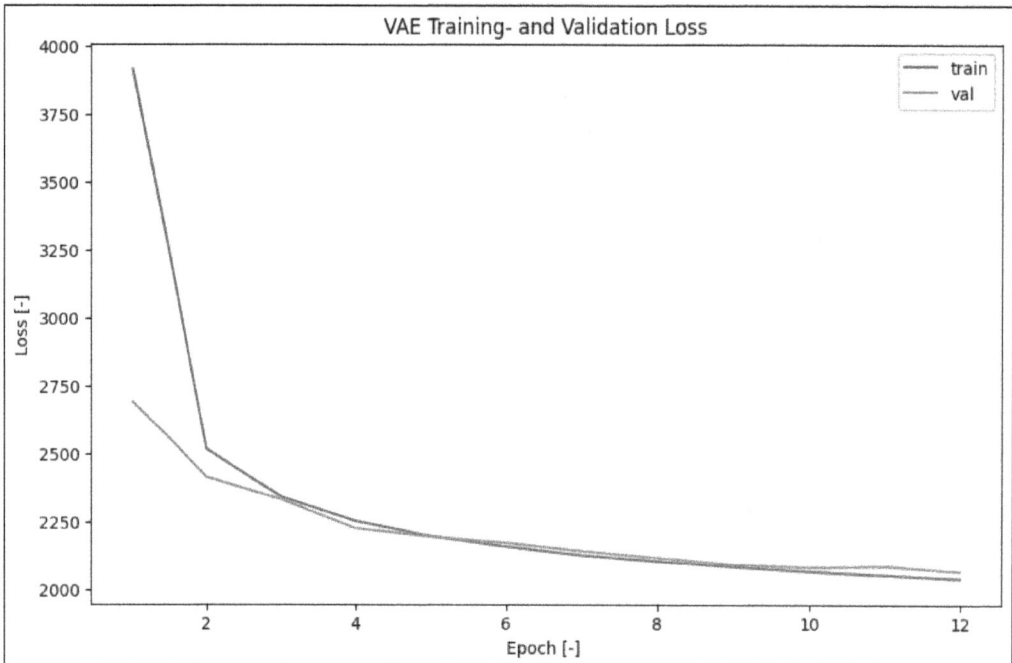

Figure 6.8 VAE: Training and Validation Losses

Finally, we want to test the capabilities of the model by generating sixteen digits. Listing 6.23 shows the corresponding code. We use the model's generate method, which randomly selects sixteen points from the latent space and generates digits on this basis that were not part of the dataset.

```
#%% Generate new digits
model.eval()
with torch.no_grad():
    # Generate 16 new samples
    new_digits = model.generate(16)
    new_digits_denorm = denormalize(new_digits).cpu()

    # Create a grid and display
    grid = vutils.make_grid(new_digits_denorm, nrow=4)
    plt.figure(figsize=(10, 10))
    plt.axis('off')
    plt.imshow(grid.permute(1, 2, 0).squeeze(), cmap='gray')
    plt.title("Generated digits ")
    plt.tight_layout()
    plt.show()
```

Listing 6.23 VAE: Generation of Digits

Figure 6.9 shows the digits generated by the VAE. Most of them are easily recognizable as familiar digits, but some digits show that longer training may be necessary.

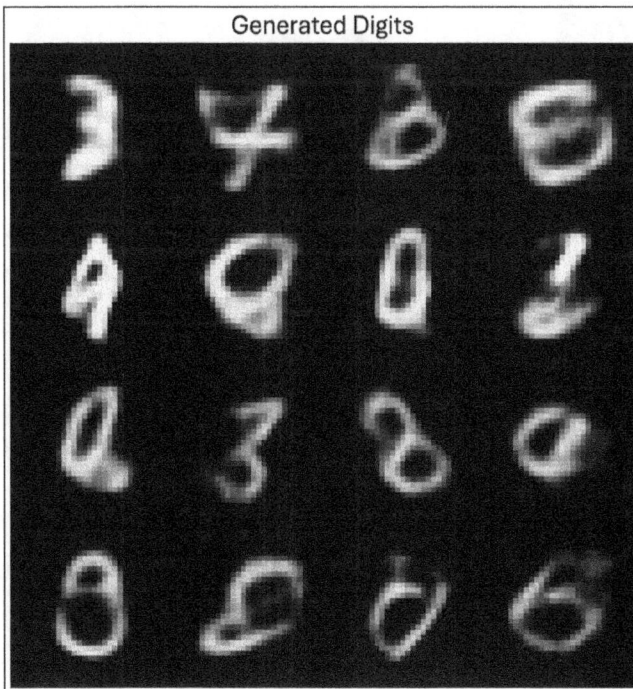

Figure 6.9 VAE: Generated Digits

6.5 Summary

In this chapter, we've learned about autoencoders and their two main representatives: classical autoencoders and variational autoencoders. The main fields of application of autoencoders and VAEs include dimensionality reduction, anomaly detection, and data cleansing, with VAEs also being used extensively in the field of generative modeling.

A classical autoencoder is a neural network that is trained to reconstruct its input. It consists of an encoder, which converts the data into a compressed representation (the so-called latent space) and a decoder, which restores the data from this representation.

The special feature of the classical autoencoder is that it learns a deterministic mapping for each data point. In latent space, a data point becomes a fixed point, and that makes it very suitable for tasks such as compression or noise suppression. VAE goes one step further. It's not a classic reconstruction algorithm but a generative model. Instead of mapping each data point to a single point in latent space, the encoder learns to represent it as a probability distribution (typically, a Gaussian distribution). We examined both representatives in theory and then implemented them in practice.

Now that you know how autoencoders learn efficient representations for flat data such as images, in the next chapter, you'll learn about graph neural networks that specialize in learning representations for complex structured data.

Chapter 7

Graph Neural Networks

"The whole is greater than the sum of its parts."
—Aristotle

Conventional models generally view data points as isolated units, and this approach reaches its limits as soon as the data points are no longer independent of each other but are connected to each other. *Graph neural networks* (GNNs) recognize that the full information content is only hidden when the connections among the data points are considered. This makes GNNs suitable, for example, for analyzing social networks, forecasting traffic, and detecting fraudulent patterns in the financial sector. A GNN can learn the structure of the whole by looking not only at the properties of the individual parts but also at the way they interact with each other.

Section 7.1 provides an introduction to graph theory, in which we learn about basic concepts such as graphs, adjacency matrices, and their characteristics. We pay special attention to message passing within graphs, which is a central concept for GNNs, and then, we highlight various use cases. Section 7.2 presents the practical part, the coding, in which we deal with the structure of a graph in PyTorch Geometric—a Python library for deep learning on graphs that's based on PyTorch. Section 7.3 looks at the training of a GNN.

7.1 Introduction to Graph Theory

A *graph* is a mathematical structure that represents a set of objects and the relationships among those objects. A graph always consists of two basic components:

- **Nodes**
  These are the objects in the graph. For example, the nodes in a social network can be people.

- **Edges**
  These represent the connections between or among nodes. In a social network, an edge would represent a friendly relationship between two people.

There are two types of edges: directional and nondirectional. Directional edges have a direction, like a one-way street. In the example of the social network, a directional edge can reflect the interest of person A in person B. If person A likes person B and person B does not like person A, then that's an example of a directional edge. If the two people like each other, then that's an example of a nondirectional or bidirectional edge.

Graphs are used to model complex systems in many areas. In addition to social networks, they can model biological networks, transportation routes, and citations between scientific papers.

We'll gain an understanding of graphs and adjacency matrices in Section 7.1.1. We can equip graphs and their nodes with features, and we'll find out how in Section 7.1.2. Model training of a graph neural network differs from the model training of the networks we've learned so far in the way information is processed via message passing, and we'll learn about that in Section 7.1.3. Finally, graph neural networks can be useful for different applications, and we'll learn about that in Section 7.1.4.

7.1.1   Graphs and the Adjacency Matrix

Figure 7.1 shows a simple network structure with five nodes (A to E), which are connected to one another.

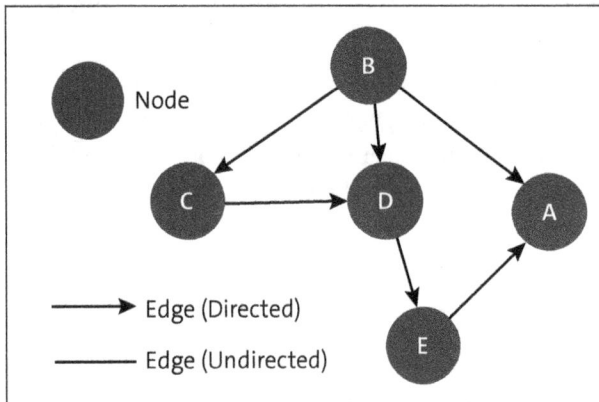

Figure 7.1 Example Structure of GNN

How can we model the connections between the nodes? This is quite simple with the help of an *adjacency matrix*, which is a square matrix that represents the connections in a graph. The matrix is structured in such a way that both the rows and the columns represent the nodes. The rows can represent the source node, and the columns can represent the target node. The values in the matrix indicate whether there is a connection between two nodes.

Figure 7.2 represents the adjacency matrix for the graph shown above. For example, a connection points from node B to node C, and we can read this connection in the matrix by searching in the second row (corresponding to the source node B) for the column with the target node (the third column for node C). At the intersection of both, there is the value 1, which stands for the directed connection. The following list shows how to create the entries in the matrix in Figure 7.2:

- An entry of $X_{ij} = 1$ means that there is an edge between node i and node j.
- If the edge is undirected, $X_{ji} = 1$ automatically applies as well.
- An entry of $X_{ij} = 0$ means that there is no edge between i and j.

	Target				
	A	B	C	D	E
A	0	0	0	0	0
B	1	0	1	1	0
C	0	0	0	1	0
D	0	0	1	0	1
E	1	0	0	1	0

Figure 7.2 Adjacency Matrix

For an undirected graph, the adjacency matrix is always symmetrical since the connection between A and B is the same as the connection between B and A. But for directed graphs, the matrix is asymmetrical, as shown in the example.

7.1.2 Features

The nodes can also contain certain characteristics. If the nodes represent people, then the characteristics can represent specific properties of those people. Figure 7.3 shows an example of a feature matrix for the five nodes.

		Age	Place
	A	25	Hamburg
	B	33	Munich
Node	C	47	Berlin
	D	19	Stuttgart
	E	82	Cologne

Figure 7.3 Characteristics Matrix

So how can we train a graph neural network? To do it, we need to understand the concept of passing messages.

7.1.3   Passing Messages

The starting point, as shown in Figure 7.4, is a graph that consists of five nodes. A GNN processes the information from the graph to generate a new representation for the target node (B).

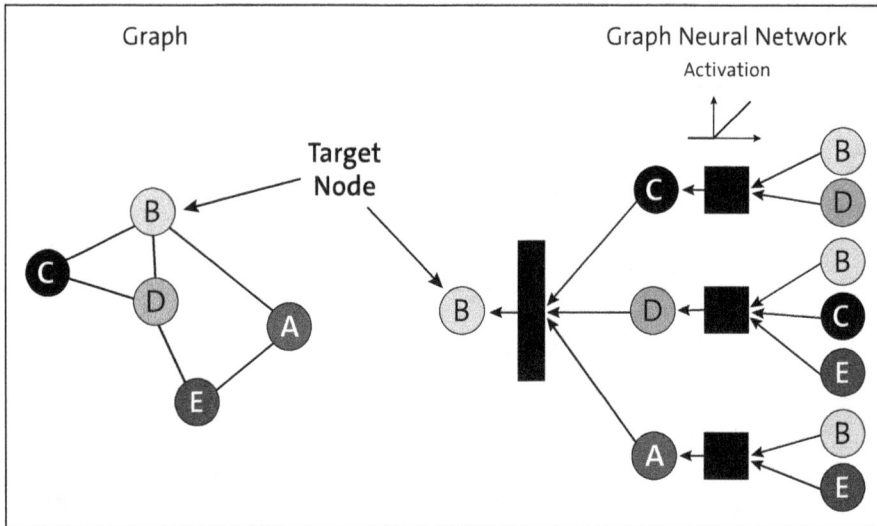

Figure 7.4 Passing Messages in GNNs

This process takes place in the following steps:

1. The direct neighbors of B are identified. In this case, these are nodes C, D, and A.

2. The *GNN layer* collects the properties (known as *messages*) of these neighboring nodes, which are the feature vectors. Each of the brown boxes in the figure represents an aggregation function that processes the features of the neighbors, and this function can be an average, a sum, or a more complex operation.

3. The aggregated information from the neighbors is combined with the original features of the target node (B). This combination takes place in the dark blue box, and the result is a new, updated feature representation for node B, which now contains knowledge about its neighbors.

This process is repeated for each node in the graph. The information flows through several layers of a GNN over longer distances, so that a node incorporates the features of its "neighbors of the neighbors" and so on. This creates a representation (known as an *embedding*) that reflects not only the local structure but also the more global structures of the graph.

With each GNN layer, the embedding of a node becomes richer as it picks up information from increasingly distant neighbors. At the end of the process, the final embeddings have the entire graph structure encoded in them, which makes them very useful for downstream tasks such as classification.

7.1.4   Use Cases

GNNs are very special neural networks that are ideal for special applications.

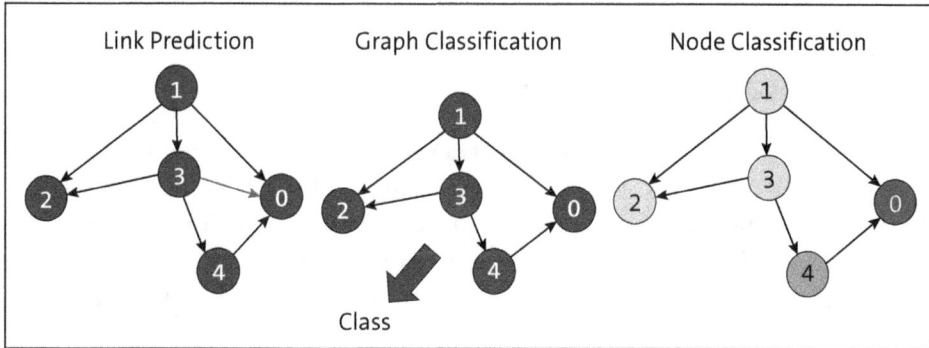

Figure 7.5 GNN Use Cases

Figure 7.5 shows some of their most common use cases, as follows:

- **Node classification**
  The goal of node classification is to predict a category or characteristic for a single node in a graph. The GNN uses not only the characteristics of the respective node but also the information on its direct neighbors to obtain a more accurate classification. We can use node classification to detect fraudsters or spam bots in social networks because the GNN analyzes their connection patterns to other suspicious accounts. Similarly, we can use node classification to detect conspicuous or fraudulent accounts in a network of money flows.

- **Link prediction**
  Link prediction aims to predict the probability of a connection between two nodes that doesn't exist in the graph. The model learns the patterns that lead to existing connections and then applies this knowledge to new pairs. One such use case is in recommendation systems. A GNN can predict whether a user might buy a product by analyzing connections among similar users and their purchasing decisions. We can use link prediction to generate friendship and partner suggestions on social networks or dating platforms, and link prediction is also used extensively in knowledge graphs to complete missing facts (e.g., to predict the country of origin of a particular company), thereby enriching and validating large databases. It's also indispensable in biological networks, for example, for predicting new protein-protein interactions in genome research.

- **Graph classification**
  Graph classification involves assigning a category to the entire graph, not just to individual nodes. The GNN aggregates the information from all nodes and their connections to create a global representation that characterizes the entire graph. Graph classification is very commonly used in chemistry to represent molecules as graphs,

with atoms as nodes and chemical bonds as connections between atoms. A GNN can classify the entire molecular graph to then predict, for example, whether the molecule is toxic or has a certain property.

7.2   Coding: Developing a Graph

In this section, you'll learn how to create a graph from scratch. The complete script can be found at *080_GraphNN\create_graph.py*. Listing 7.1 shows the required packages. We're now working with a new package: PyTorch Geometric (`torch_geometric`), a specialized package that builds on PyTorch to simplify the writing and training of graph neural networks. It provides efficient data structures for graphs, which we'll use in this script.

We'll use the well-known matplotlib package for visualization, and we'll also use `networkx`, which is suitable for displaying graphs.

```
# data preparation
from torch_geometric.data import Data
# modeling
import torch
from torch_geometric.utils.convert import to_networkx
# visulalization
import networkx as nx
import matplotlib.pyplot as plt
```

Listing 7.1 Graph Creation: Required Packages

Now, we need a function to create our network, so we'll build a social network consisting of a few people with certain characteristics. Our goal in this script is not to train a model but instead to gain an understanding of how to work with graphs. The `create_matchmaking_dataset` function, which we create in Listing 7.2, receives as parameters a dictionary consisting of people and their interests.

With the `name_to_idx` object, we'll create a dictionary from a name and index. We'll form the undirected edges as (i, j) pairs, and we'll also form all edges of the clique consisting of Bert, Lea, Elisa, and Kiki. There will also be a special treatment for the connection between Bert and Steve.

```
def create_matchmaking_dataset(person_interests: dict) -> Data:
    person_names = list(person_interests.keys())
    x = torch.tensor(list(person_interests.values()), dtype=torch.float)

    name_to_idx = {name: idx for idx, name in enumerate(person_names)}
```

```
    # Create all edges
    edges = []

    # Clique among Bert, Lea, Elisa, and Kiki (if present in the dict)
    clique = [n for n in ["Bert", "Lea", "Elisa", "Kiki"]
                 if n in name_to_idx]
    for ai in range(len(clique)):
        for bi in range(ai + 1, len(clique)):
            i = name_to_idx[clique[ai]]
            j = name_to_idx[clique[bi]]
            edges.append([i, j])

    # Only Bert and Steve know each other
    if "Bert" in name_to_idx and "Steve" in name_to_idx:
        i = name_to_idx["Bert"]
        j = name_to_idx["Steve"]
        if i < j:
            edges.append([i, j])
        elif i > j:
            edges.append([j, i])

    # Convert to edge_index (2, E). Empty if there are no edges.
    edge_index = torch.tensor(edges, dtype=torch.long).T if edges else
torch.empty((2, 0), dtype=torch.long)

    # Return graph
    data = Data(x=x, edge_index=edge_index)
    return data
```

Listing 7.2 Graph Creation: Function for Graph Creation

First, we create the dictionary with the persons and interests in Listing 7.3, and then, we create the graph using our previously created function.

```
person_interests = {
    "Bert": [0.8, 0.2, 0.2],
    "Lea": [0.1, 0.9, 0.8],
    "Elisa": [0.0, 0.5, 0.8],
    "Kiki": [0.0, 1.0, 0.0],
    "Steve": [0.2, 0.3, 0.5],
}
matchmaking_data = create_matchmaking_dataset(person_interests=person_
interests)
```

Listing 7.3 Graph Creation: Creation of Our First Graph

At this point, we look at the graph shown in Figure 7.6.

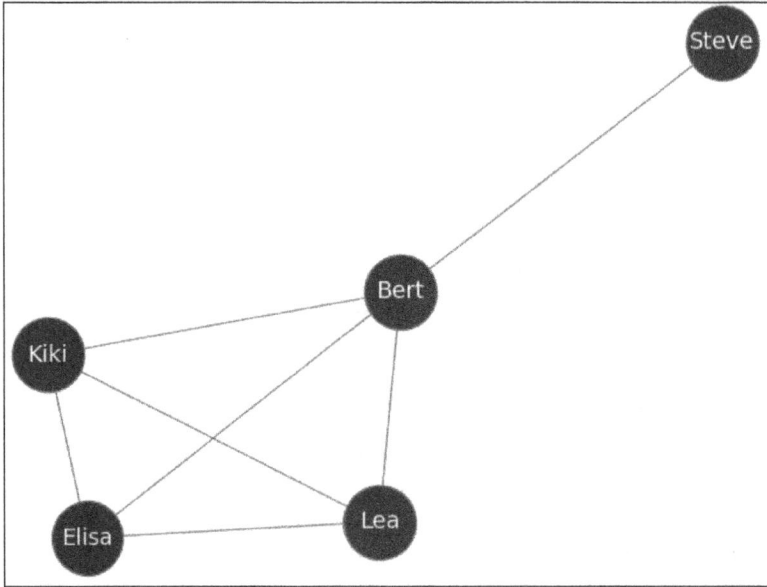

Figure 7.6 Graph of Persons

The graph shows which persons are directly connected (known) to each other. Using the code from Listing 7.4, we now check the constructed graph. We can determine the number of nodes by using the num_nodes property, and we can obtain the number of node features with node_features. The total number of connections is given by the num_edges property.

The feature matrix for each person was stored in matchmaking_data.x, and finally, edge_index contains the connections that the GNN will use for learning.

```
print(f"Number of nodes (people): {matchmaking_data.num_nodes}")
print(f"Number of node features (interests): {matchmaking_data.num_node_fea-
tures}")
print(f"Number of existing friendships (edges): {matchmaking_data.num_edges}")

print("\n--- Main components of the data object ---")
print(f"data.x: shape {matchmaking_data.x.shape} - The feature matrix for each
person.")
print(f"data.edge_index: shape {matchmaking_data.edge_index.shape}")
person_names = list(person_interests.keys())
```

```
Number of nodes (people): 5
Number of node features (interests): 3
Number of existing friendships (edges): 7
```

```
--- Main components of the data object ---
data.x: shape torch.Size([5, 3]) - The feature matrix for each person.
data.edge_index: shape torch.Size([2, 7])
```

Listing 7.4 Graph Creation: Graph Review

Graphs constitute a versatile and extremely helpful data type because we can visualize and thus easily check them for correctness. For this purpose, in Listing 7.5, we create a visualize_graph function that maps the graph and the features. We use the matplotlib and networkx packages for this, and we pass the graph to the function as the parameter data and the node names (node_names).

In the first step, we plot the nodes with nx.draw_networkx_nodes and then plot the connections with nx.draw_networkx_edges. The graph consists of two subplots, with the graph on the left and the characteristics as a bar chart on the right.

```python
def visualize_graph(data: Data, node_names=None):
    G = to_networkx(data, to_undirected=True)

    # Create subplots - one for graph, one for interests
    fig, (ax1, ax2) = plt.subplots(1, 2, figsize=(15, 6))

    # Graph visualization
    pos = nx.spring_layout(G, seed=42)
    nx.draw_networkx_nodes(G, pos, node_color="#2C2C54",
                           node_size=2000, edgecolors="#C05C37", ax=ax1)
    nx.draw_networkx_edges(G, pos, alpha=0.5, ax=ax1)

    # Labels
    if node_names is not None and len(node_names) == data.num_nodes:
        labels = {i: node_names[i] for i in range(data.num_nodes)}
    else:
        labels = {i: str(i) for i in range(data.num_nodes)}
    nx.draw_networkx_labels(G, pos, labels=labels, font_size=14,
                            font_color="#FFFFFF", ax=ax1)

    ax1.set_title("Freundschaftsgraph")
    ax1.axis("off")

    # Interests visualization
    interests = ["Lesen", "Laufen", "Kochen"]
    x = range(len(node_names))
    width = 0.25

    for i, interest in enumerate(interests):
```

```
        interests_values = data.x[:, i].numpy()
        ax2.bar([xi + i*width for xi in x], interests_values, width,
                label=interest, alpha=0.7)

    ax2.set_ylabel("Merkmalswert [-]")
    ax2.set_title("Merkmale der Personen")
    ax2.set_xticks([xi + width for xi in x])
    ax2.set_xticklabels(node_names, rotation=45)
    ax2.legend()
    ax2.set_ylim(0, 1.2)

    plt.tight_layout()
    plt.show()

visualize_graph(matchmaking_data, person_names)
```

Listing 7.5 Graph Creation: Visualization of Graph and Its Features

Figure 7.7 shows the result of our efforts, with the graph on the left and the characteristics as a bar chart on the right.

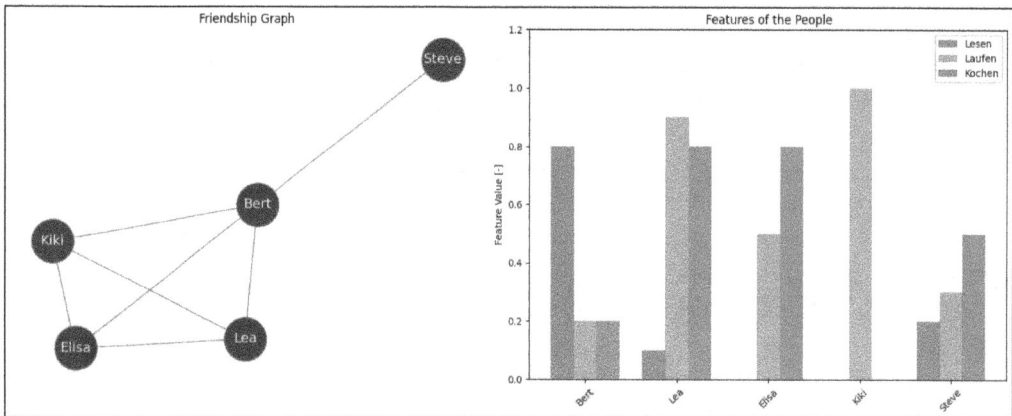

Figure 7.7 Graph and Features

This gives you an impression of how you can create a graph. In the following section, we'll use a much larger dataset to train a PyTorch model.

7.3   Coding: Training a Graph Neural Network

We'll work with the Cora dataset, which is a widely used dataset for the development of GNNs.

Cora Dataset

This dataset is a citation network of scientific publications in which each publication is assigned to one of seven research categories, such as "Neural Networks" or "Probabilistic Methods."

We'll train a node classification model, and our goal will be to predict the category of a given paper, considering its own characteristics and its connections in the network. The characteristics describe a *bag-of-words* approach, which means that the characteristics of the node (e.g., the text of a document or profile description) are converted into a vector that simply counts how often certain words occur, ignoring the order of the words. Each dimension of the feature vector corresponds to a specific, unique word from a given dictionary. If the value is 1, the word occurs; if it's 0, it doesn't.

7

The complete script can be found at *080_GraphNN\node_classification.py*. We load the required packages via the code in Listing 7.6. One new item at this point is our use of t-distributed stochastic neighbor embedding (t-SNE), which is a dimension reduction technique that transforms high-dimensional data into a low-dimensional representation (usually 2D or 3D) to visualize similarities and clusters in the dataset.

We load the dataset directly via the torch_geometric package and its Planetoid class. Planteoid refers to a collection of three commonly used academic graph datasets (Cora, CiteSeer, and PubMed) that are used for benchmarking GNNs and typically focus on node classification of scientific documents.

The model will use the GATConv network layer, which is a GNN layer from the PyTorch Geometric package that uses self-aware weights to dynamically and adaptively determine the importance of each neighbor node's information during message passing, rather than assigning equal importance to all neighbors.

```
import os
import torch
import torch.nn.functional as F
from sklearn.manifold import TSNE
from torch_geometric.datasets import Planetoid
from torch_geometric.transforms import NormalizeFeatures
from torch_geometric.nn import GATConv
import matplotlib.pyplot as plt
import seaborn as sns
```

Listing 7.6 GNN Training: Required Packages

If possible, we want to carry out the training on the GPU and determine the device as follows:

```
device = torch.device('cuda' if torch.cuda.is_available() else 'cpu')
print(f"Device: {device}")
```

Now, let's move on to importing the data in Listing 7.7. We download the data to the dataset_path folder via the Planetoid class and load it onto the device by using data.to(). We then display some properties of the dataset.

```
dataset_path = 'data/Planetoid'
if not os.path.exists(dataset_path):
    dataset = Planetoid(root=dataset_path, name='Cora', transform=NormalizeFea-
tures())
else:
    dataset = Planetoid(root=dataset_path, name='Cora', transform=NormalizeFea-
tures(), force_reload=False)
data = dataset[0]
data = data.to(device)

print(f'Number of graphs: {len(dataset)}')
print(f'Number of features: {dataset.num_features}')
print(f'Number of classes: {dataset.num_classes}')
print(f'Graph object: {data}')
print(f'Is the graph directed: {data.is_directed()}')
print(f'Number of nodes: {data.num_nodes}')
print(f'Number of edges: {data.num_edges}')
print(f'Number of training nodes: {data.train_mask.sum()}')
print(f'Number of validation nodes: {data.val_mask.sum()}')
print(f'Number of test nodes: {data.test_mask.sum()}')
print(f"Number of nodes in the graph: {data.num_nodes}")
```

```
Number of graphs: 1
Number of features: 1433
Number of classes: 7
Graph object: Data(x=[2708, 1433], edge_index=[2, 10556], y=[2708], train_mask=[
2708], val_mask=[2708], test_mask=[2708])
Is the graph directed: False
Number of nodes: 2708
Number of edges: 10556
Number of training nodes: 140
Number of validation nodes: 500
Number of test nodes: 1000
Number of nodes in the graph: 2708
```

Listing 7.7 GNN Training: Data Import

The seven categories to which the articles can belong are as follows:

- Neural networks
- Probabilistic methods
- Genetic algorithms
- Theory
- Case based
- Reinforcement learning
- Rule learning

Before we get to modeling, let's gain an understanding of how many elements are contained in each of the seven categories and the accuracy of a classifier that always predicts the most frequent category. Listing 7.8 provides an analysis of class counts and how to derive the accuracy of the dummy classifier.

```
class_counts = torch.bincount(data.y)
for class_idx, count in enumerate(class_counts):
    print(f'Class {class_idx}: {count.item()} nodes')

# Calculate dummy classifier accuracy (majority class)
majority_class = torch.argmax(class_counts)
dummy_predictions = torch.full_like(data.y, majority_class)
dummy_acc = (dummy_predictions[data.test_mask] == data.y[data.test_
mask]).float().mean()
print(f'\nAccuracy of the Dummy Classifier: {dummy_acc:.4f}')
```

```
Class 0: 351 nodes
Class 1: 217 nodes
Class 2: 418 nodes
Class 3: 818 nodes
Class 4: 426 nodes
Class 5: 298 nodes
Class 6: 180 nodes
Accuracy of the Dummy Classifier: 0.3190
```

Listing 7.8 GNN Training: Class Elements and Dummy Classifier

The dummy classifier suggests an accuracy of 31.9%.

The data is already well prepared by the package, so we don't have to go through any further preparation steps, and we can describe the model class as a graph attention network directly with the code in Listing 7.9. We're building a *graph attention transformer* (GAT) network, so we've named the model class GAT. It's a neural network that's characterized using attention mechanisms.

We also use GATConv layers, which are the core elements of the graph attention network and implement the message passing mechanism with a transformer-based approach. A GATConv layer learns an individual weighting for each of its neighboring nodes for each node, and it assesses which neighbors are most important for updating the features of a node. It gives a node that's more relevant for the classification of the target node a higher weight so that the more relevant node contributes more to the new node representation.

In addition to these core elements, we use dropout layers to avoid overfitting and use an *exponential linear unit* (ELU) for activation. Activation with elu is a variation of ReLU activation, and for positive values, it corresponds to the input value and is identical to the normal ReLU function. However, for negative values, the elu function doesn't output a zero but instead outputs an exponentially decreasing curve. This avoids the problem of dying neurons.

The model activates the last layer with log_softmax because it's numerically more stable than a combination of softmax and log.

```
class GAT(torch.nn.Module):
    def __init__(self, num_features, num_hidden, num_classes, heads=8):
        super().__init__()
        self.conv1 = GATConv(num_features, num_hidden, heads=heads,
                             dropout=0.6)
        self.conv2 = GATConv(heads * num_hidden, num_classes, heads=1,
                             concat=False, dropout=0.6)

    def forward(self, x, edge_index):
        x = F.dropout(x, p=0.6, training=self.training)
        x = self.conv1(x, edge_index)
        x = F.elu(x)
        x = F.dropout(x, p=0.6, training=self.training)
        x = self.conv2(x, edge_index)
        x = F.log_softmax(x, dim=1)
        return x
```

Listing 7.9 GNN Training: Model Class

We can now create an instance of the model, the optimizer, and the loss function as shown in Listing 7.10.

```
model = GAT(num_features=dataset.num_features, num_hidden=8, num_classes=data-
set.num_classes, heads=8).to(device)
optimizer = torch.optim.Adam(model.parameters(), lr=0.01, weight_decay=5e-4)
criterion = torch.nn.NLLLoss()
```

Listing 7.10 GNN Training: Model Instance, Optimizer, and Loss Function

We carry out the individual training step by using the train function from Listing 7.11, which returns the current loss value. We also define the test function used for the validation and test data that returns the accuracy.

```python
def train(model, data, optimizer, criterion):
    """Performs a single training step."""
    model.train()
    optimizer.zero_grad()
    X_train = data.x
    y_true = data.y
    y_pred = model(X_train, data.edge_index)
    train_mask = data.train_mask
    loss = criterion(y_pred[train_mask], y_true[train_mask])
    loss.backward()
    optimizer.step()
    return loss.item()


def test(model, data, mask):
    """Evaluates the model on a specific mask (e.g., test or validation)."""
    model.eval()
    with torch.no_grad():
        X_test = data.x
        y_true = data.y
        logits = model(X_test, data.edge_index)
        y_pred = logits[mask].argmax(dim=1)
        correct = (y_pred == y_true[mask]).sum().item()
        total = mask.sum().item()
        accuracy = correct / total
    return accuracy
```

Listing 7.11 GNN Training: Training Function and Test Function

We carry out the training in Listing 7.12. The loop is quite short because a large part of the code was previously outsourced to the test and train function.

```python
num_epochs = 200
loss_list = []
train_acc_list = []
val_acc_list = []


for epoch in range(num_epochs):
    epoch_loss = train(model, data, optimizer, criterion)
    train_acc = test(model, data, data.train_mask)
    val_acc = test(model, data, data.val_mask)
```

```
    loss_list.append(epoch_loss)
    train_acc_list.append(train_acc)
    val_acc_list.append(val_acc)

    if (epoch + 1) % 20 == 0:
        print(f'Epoch: {epoch + 1:03d} | Verlust: {epoch_loss:.4f} | '
              f'Training Acc: {train_acc:.4f} |
                Validierungs-Accuracy: {val_acc:.4f}')
```

```
--- Starting Model Training ---
Epoch: 020 | Loss: 1.5871 | Train Acc: 0.9429 | Val Acc: 0.7540
...
Epoch: 200 | Loss: 0.6110 | Train Acc: 1.0000 | Val Acc: 0.7960
--- Training Complete ---
```

Listing 7.12 GNN Training: Training Loop

We can visualize the training progress with the code from Listing 7.13.

```
sns.set_style("whitegrid")
fig, (ax1, ax2) = plt.subplots(2, 1, figsize=(10, 8), sharex=True)

# Plot loss
sns.lineplot(data=loss_list, ax=ax1, color='blue', label='Training Loss')
ax1.set_ylabel('Verlust [-]')
ax1.set_title('Modell-Training-Performance')

# Plot accuracies
sns.lineplot(data=train_acc_list, ax=ax2, color='orange', linestyle='--', label=
'Train Accuracy')
sns.lineplot(data=val_acc_list, ax=ax2, color='green', linestyle='--', label=
'Validation Accuracy')
ax2.set_xlabel('Epoch [-]')
ax2.set_ylabel('Accuracy [-]')
plt.tight_layout()
plt.show()
```

Listing 7.13 GNN Training: Visualization of Losses and Accuracy

Figure 7.8 shows the model training performance with two subplots. The top half of the figure shows the loss change over epochs, and the bottom half shows the training and validation accuracy.

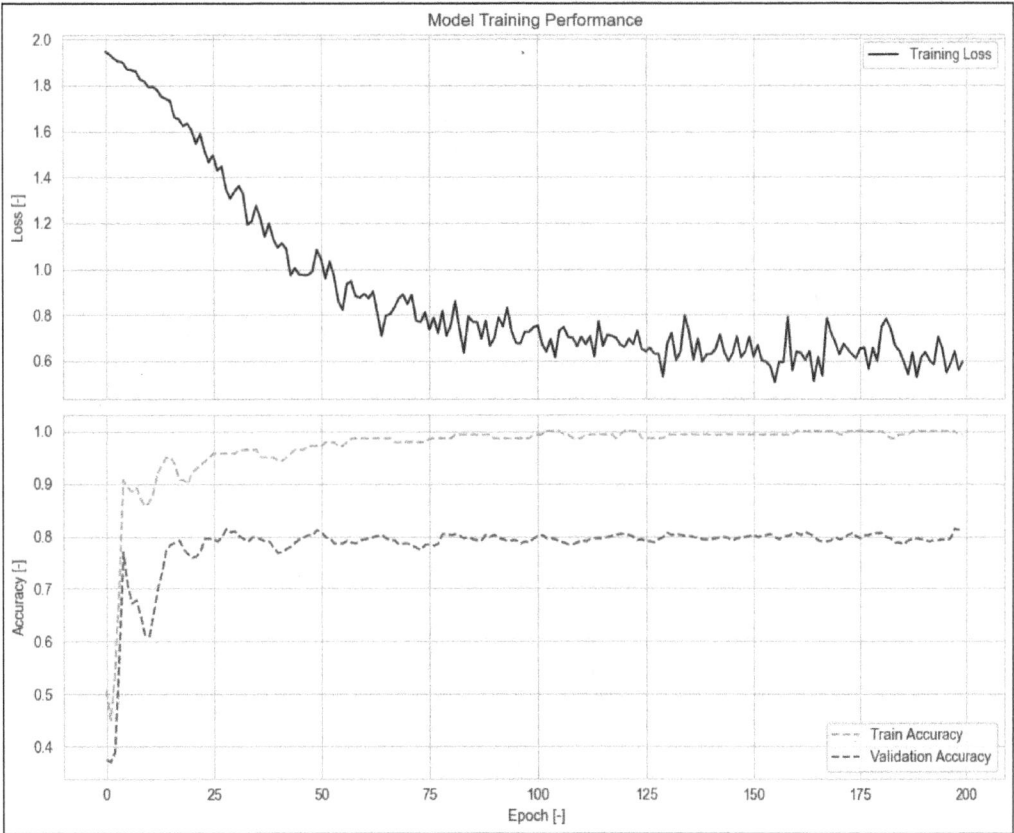

Figure 7.8 GNN Training: Visualization of Losses and Accuracy

The training losses decrease very well and approach an asymptotic value, and at the same time, the training and validation accuracy level off at a high value.

Now, let's look at the final accuracy based on the test data (see Listing 7.14). This is over 82% and therefore far higher than the dummy classifier. Our model therefore delivers very good results.

```
final_test_acc = test(model, data, data.test_mask)
print(f'\nFinale Test-Accuracy: {final_test_acc:.4f}')
```

Final Test Accuracy: 0.8210

Listing 7.14 GNN Training: Test Function

Finally, we want to visualize the embeddings of the data using the code from Listing 7.15. The t-SNE algorithm is suitable for this because it lets us reduce higher dimensions to two dimensions. We predict the test data (z) and then reduce it to a 2D object (z_tsne).

```
model.eval()
with torch.no_grad():
    # Get the final node embeddings (the output of the model)
    z = model(data.x, data.edge_index).detach().cpu().numpy()
    z_tsne = TSNE(n_components=2, perplexity=30, random_state=42).fit_trans-
form(z)

# Visualize the embeddings using Seaborn
sns.set_style("white")
plt.figure(figsize=(10, 8))

# Color the nodes by their ground-truth class
sns.scatterplot(
    x=z_tsne[:, 0], y=z_tsne[:, 1],
    hue=data.y.cpu().numpy(),
    palette=sns.color_palette("hsv", n_colors=dataset.num_classes),
    s=50,
    alpha=0.8
)
plt.title('t-SNE Visualization of the Node Embeddings')
plt.xlabel('t-SNE Component 1')
plt.ylabel('t-SNE Component 2')
plt.legend(title='Class')
plt.show()
```

Listing 7.15 GNN-Training: Visualization of Embeddings

Figure 7.9 shows the visualization of the node embeddings. It's very clear that the different classes could be clearly separated from each other in large parts, and that's a sign that the model clusters data of the same class and that different clusters are further separated from each other.

We've shown with the model that the thematic category (one of the seven classes) is possible for each scientific publication (each node) in the Cora citation network. We can also predict the correct thematic category of a publication by using the characteristics and linkage of the publication.

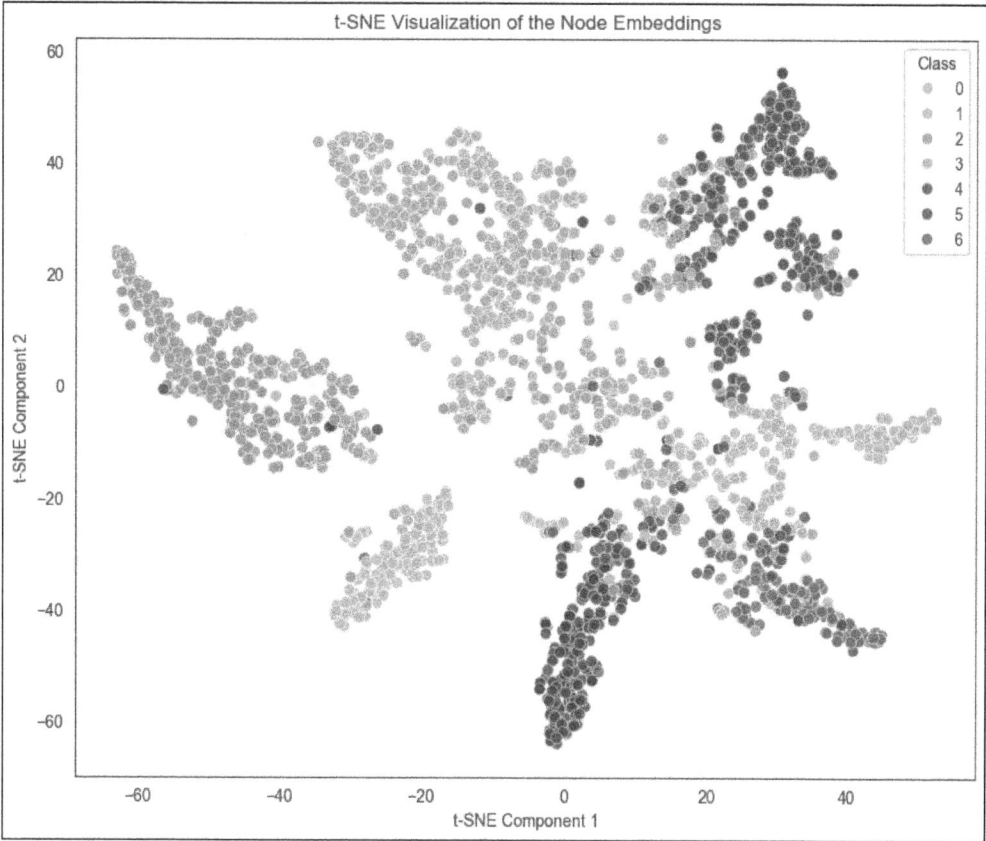

Figure 7.9 GNN Training: Visualization of Embeddings

7.4   Summary

In this chapter, we've learned about the basic principles of GNNs, familiarized ourselves with the theoretical foundations of graph theory, and gained an understanding of how message transmission works on graphs.

In the practical part, we started by creating a simple graph with five nodes and their characteristics. Then, we applied what we had learned by training GNN to solve the complex task of node classification on the real Cora dataset. This training demonstrated the ability of GNNs to extract valuable information from the structure and attributes of graphs.

In this chapter, we've assumed that the data structure of the graph is static. However, in many real-life application scenarios, the data is not only spatial but also evolves over time. This is where time series analysis comes in, and we'll look at it in the next chapter.

Chapter 8
Time Series Forecasting

"Predictions are difficult, especially when they concern the future."
—Mark Twain, Winston Churchill, Niels Bohr, and Karl Valentin

This famous quote, whose origin is not entirely clear, sums up a central challenge: the uncertainty of the future. This problem is omnipresent in the field of time series analysis. The data—which could be stock prices, weather data, or sales figures for your company—is available in chronological order, and the aim of time series analysis is to recognize patterns in the past to make predictions about the future.

In Section 8.1, you'll start by learning about the special features of modeling time series data. Then, you'll learn about various methods that are particularly suitable for modeling time series with correlated data points. In Section 8.2, you'll move on to implementing these features and methods in practical code, and we'll illustrate the theoretical approaches we discussed in Section 8.1 by using a small example dataset. In Section 8.3, we'll take a close look at PyTorch Forecasting, which is the kind of large framework that you'll often use (instead of proprietary network models) when modeling time series.

8.1 Modeling Approaches

Before we start coding, we need to gain a better understanding of time series data and why models for this kind of task need specific treatment. To do that, we'll learn about some special features of time series models in Section 8.1.1. We'll need to model the data in specific ways that we'll learn about in Section 8.1.2. Then, we'll cover three different modeling approaches: long short-term memory (LSTM) in Section 8.1.3, one-dimensional CNNs in Section 8.1.4, and transformers in Section 8.1.5.

8.1.1 Special Features of Time Series Models

There's a crucial difference between time series models and other models we've already seen: the order of the data points. While the sequence doesn't play a role in the tabular data—the observations are statistically independent of each other—this is different in a time series model. The data points of a time series are sequential and interdependent because the value of an observation depends on the values of the previous observations. This *autocorrelation* (also known as *serial correlation*) is the fundamental characteristic of time series data.

This means that when modeling, not only is the value itself important, but so is the time at which it was observed. This aspect requires special modeling approaches that take this dependency into account, and this has serious consequences for our data preparation and modeling, as follows:

- **No random sampling**
  We must use the data in its natural, chronological order. Random division into training and test data is strongly discouraged, as that would destroy the temporal structure.

- **Trends and seasonality**
  We can think of time series as a composition of underlying *trends*, superimposed *seasonality*, and randomness.

8.1.2   Data Modeling

Figure 8.1 illustrates how we model the data for time series analysis. The diagrams on the left ❶ show multivariate time series because in many cases, there will be not just one time series but several correlated time series. Let's take an example from my professional experience: I worked with colleagues who modeled the stock market energy price. Due to the complex energy mix involved in intra-European electricity trading and many renewable energies, this modeling quickly becomes complex and results in a model with dozens of features. For example, a summer heat wave will have a major impact on the feed-in of photovoltaics (which will be high), the feed-in of nuclear energy (which will be low due to shutdowns), electricity demand (which will be high due to air conditioning), etc.

The example shown in Figure 8.1 is simplified. Only two features (temperature and wind speed) are plotted over time.

To maintain temporal consistency, we divide the time series into sequences of equal length. For example, we record the first ten measurement points in the first row of the table on the right ❷. Then, we shift the entire window under consideration by one time step and enter it into the second row, and we generate a corresponding table for each individual feature.

The resulting tensor is thus 3D and contains the following dimensions:

- The dimension for the samples or batches
- The dimension for the contiguous sequences
- The dimension for the features

It's crucial to bring the data into such a form, but which models are suitable for the time series? I will introduce three common models: LSTM, CNNs, and transformers.

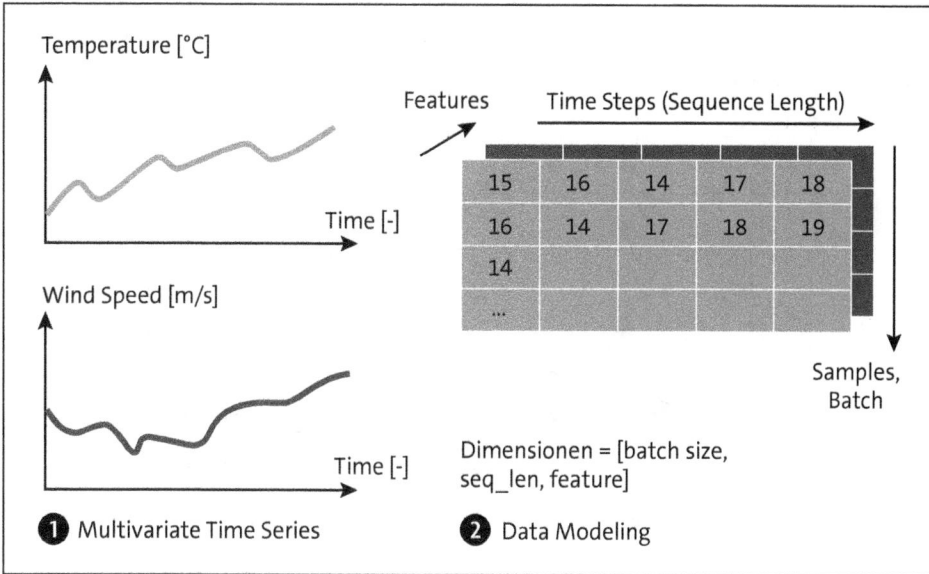

Figure 8.1 Data Modeling

8.1.3   Long Short-Term Memory

A *long short-term memory* (LSTM) is a type of neural network designed to remember sequential information over long periods of time. Think of it as a network with an intelligent memory that selectively decides what to keep and what to forget. Conventional networks have difficulty remembering information that lies far in the past, and that's because the information is gradually lost when the sequence is processed. This is known as the *vanishing gradient problem*. LSTMs were developed to solve this problem and retain important patterns that are relevant for prediction, even if they appeared in the time series a long time ago.

Figure 8.2 shows the structure of an LSTM cell. The core of an LSTM is the *cell state line*, which runs through the time series like a continuous memory chain. At each step of the sequence, meaning each data point, there are three gates that control the flow of information in the following ways:

- **Forget gate**
  This gate decides which old information should be removed from the cell state. It considers the current data point and the previous output, and it decides which information is no longer relevant.

- **Input gate**
  This gate decides which new information should be added to the cell state. It evaluates the current data point and updates the cell state with the relevant new information.

■ **Output gate**

This gate decides which information from the current cell state should be used for the LSTM's output. It selects a filtered version of the cell state and passes it on as the final output.

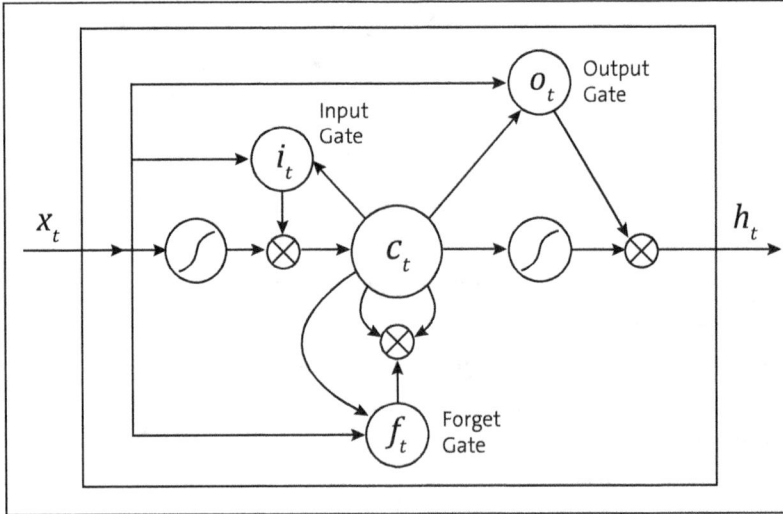

Figure 8.2 Structure of LSTM Module[1]

These three gates allow the LSTM to learn selectively and remember the most important information. This enables the LSTM to recognize complex and longer-term dependencies in the time series.

8.1.4   One-Dimensional Convolutional Neural Networks

In Chapter 4, we looked at 2D CNNs in detail, where the filter slides across two spatial dimensions (width and height), typically for image analysis. In contrast, we use 1D CNNs for time series prediction. The "1D" dimension in this context represents time.

The convolutional filter (or kernel) is a one-dimensional array that operates like a sliding window over the entire time series. At each position, a mathematical operation—the convolution—is performed. This involves element-by-element multiplication between the filter's weights and the corresponding segment of the time series, followed by summation to produce a single value in the resulting feature map. With that process, local patterns and temporal features are extracted from the sequence.

Each number in this feature map represents the extent to which the pattern searched by the filter was present in the original data at various points in time. We can also use several filters simultaneously to detect different patterns, such as sudden rises or falls.

---

1   BiObserver, "Long short-term memory units," Wikimedia, *https://upload.wikimedia.org/wikipedia/commons/d/d5/Long_Short_Term_Memory.png.*

Figure 8.3 shows an example of this process using a time series that's connected to a 1D convolutional layer. Each element of the feature map is only connected to a part of the time series, and just as we know it from CNN in the field of image recognition, convolutional layers can also be connected in series. This is shown in Figure 8.3, where the first 1D convolutional layer is followed by another.

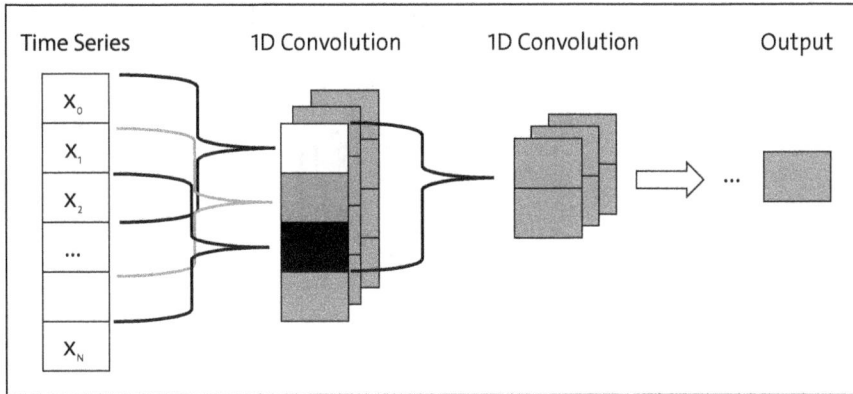

Figure 8.3 1D Convolution Process

We can also use a pooling layer that compresses the data. This makes the sequence shorter, but it also retains the most important pattern information. This makes the model more efficient and robust.

8.1.5 Transformer

Transformer models form the basis of today's language models, so we'll discuss transformers in detail in Chapter 9 and only provide a brief outline of them here. The central element of a transformer is the *self-attention mechanism*. Imagine that the transformer reads the entire time series all at once and then decides for each data point how important all other data points are for its meaning. This mechanism calculates a weighting for each data point in the time series that reflects how relevant each other data point in the time series is for that specific point in time. The model therefore assigns a score to each data point that allows the model to recognize, for example, that a strong increase in sales in December is closely linked to the sales figures in November and October while another increase in April is of little significance.

In contrast to LSTMs, which build up a kind of memory of the past, the transformer establishes the relationship between distant points in time directly and in parallel. This allows it to better capture long-term dependencies, and the calculation can take place in parallel.

Now that we've given you this brief theoretical introduction, we'll help you train your own model in the next section and implement the three model approaches we've learned about.

8.2   Coding: Custom Model

We'll use a simple time series representing the number of passengers on flights in the US from 1949 to 1960. Figure 8.4 graphically displays the evolution of passenger numbers over time, and it indicates an upward trend and strong seasonality.

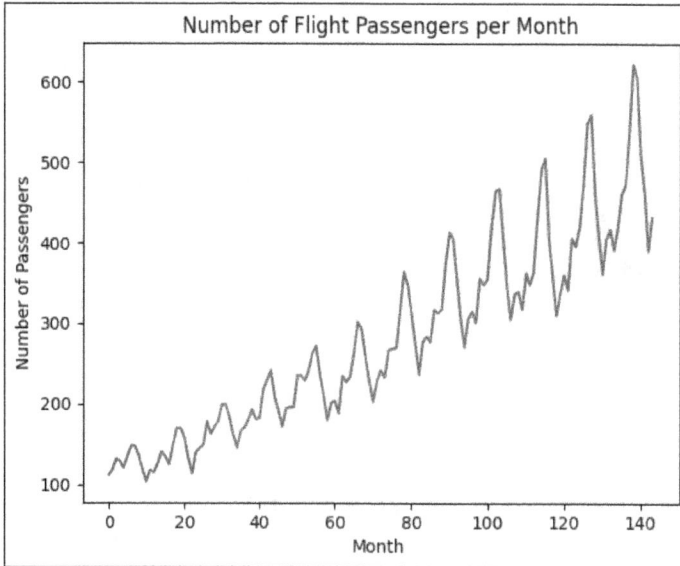

Figure 8.4 Passenger Numbers in Thousands in USA from 1949 to 1960

We'll train various models, but first, we must prepare the data accordingly. We'll start with the preparation of the data in Section 8.2.1 and implement the LSTM model in Section 8.2.2. We'll implement the competing CNN model in Section 8.2.3, and we'll cover the transformers model in Section 8.2.4.

8.2.1   Data Preparation

The complete script for data preparation can be found at *090_TimeSeries\data_prep.py*. Listing 8.1 shows the packages we use, which we divide into data preparation, modeling, and visualization for better understanding.

```
# data preparation
import numpy as np
import pandas as pd
from sklearn.preprocessing import MinMaxScaler

# modeling
import torch
from torch.utils.data import Dataset, DataLoader
```

```
# visualization
import matplotlib.pyplot as plt
import seaborn as sns
```

Listing 8.1 Data Preparation: Package Import

We only need to define the following two hyperparameters in Listing 8.2 in data preparation:

- **BATCH_SIZE**
  We need to use this when creating the DataLoader instances.

- **seq_len**
  We need to use the sequence length to describe how long the contiguous temporal sequence should be. Here, we've selected ten months.

```
BATCH_SIZE = 4
seq_len = 10
```

Listing 8.2 Data Preparation: Hyperparameter Definition

Now, let's move on to the actual data import in Listing 8.3. This is a classic dataset that's provided via the seaborn package. The flights dataframe has the year, month, and passengers features.

```
flights = sns.load_dataset("flights")
print(f'Number of Entries: {len(flights)}')
flights.head(2)
```

	year	month	passengers
0	1949	Jan	112
1	1949	Feb	118

Listing 8.3 Data Preparation: Data Import

Listing 8.4 shows the code for visualizing the time series, which was already shown in Figure 8.4.

```
sns.lineplot(x=range(len(flights)), y='passengers', data=flights)
plt.title('Number of Flight Passengers per Month')
plt.xlabel('Month')
plt.ylabel('Number of Passengers')
plt.show()
```

Listing 8.4 Data Preparation: Data Visualization

So far, the years and months have been displayed in separate columns, but we need to apply a timestamp that's coded as datetime. To do this, we create a new year_month column that receives this data type, as follows:

```
flights["year_month"] = pd.to_datetime(flights["year"].astype(str) + "-" +
flights["month"].astype(str))
```

It makes sense to scale the data as shown in Listing 8.5, and to do that, we use the Min-MaxScaler and first create the scaler instance, which then adjusts the data with fit_transform. We then obtain the scaled data (Xy_scaled), which has been transformed from a row vector into a column vector with the reshape(-1, 1) function.

```
scaler = MinMaxScaler()
Xy = flights.passengers.values.astype(np.float32)
Xy_scaled = scaler.fit_transform(Xy.reshape(-1, 1))
```

Listing 8.5 Data Preparation: Data Scaling

In Section 8.1.1, you learned how to reshape the data so that you can process it in the context of time series models. In Listing 8.6, we restructure the data by placing a sliding window over Xy_scaled. For the newly structured independent data (X_restruct), we create an input sequence of successive time steps, starting from each start index (i).

The target variable (y_restruct) behaves in a similar way. For this, we select the immediately following element as the target value, and the resulting objects have a structure of [N - seq_len, seq_len, number of features] for X_restruct and [N - seq_len, number of features] for y_restruct.

```
X_restruct = np.array([Xy_scaled[i:i+seq_len] for i in range(len(Xy_scaled) -
seq_len)])
y_restruct = np.array([Xy_scaled[i+seq_len] for i in range(len(Xy_scaled) -
seq_len)])
print(f'X_restruct shape: {X_restruct.shape}')
print(f'y_restruct shape: {y_restruct.shape}')
```

X_restruct shape: (134, 10, 1)
y_restruct shape: (134, 1)

Listing 8.6 Data Preparation: Data Restructuring

This transformation step from a 1D vector to a 3D object is the most difficult part of the data preparation. Listing 8.7 shows the next step—splitting the data into training and test data— which is simpler. In it we specify that the last twelve months (last_n_months) are reserved for the test. In the clip_point variable, we record the point in time in the time series, and we allocate all periods before that to training and all periods after that to testing.

```
last_n_months = 12
clip_point = len(X_restruct) - last_n_months
X_train = X_restruct[:clip_point]
X_test = X_restruct[clip_point:]
y_train = y_restruct[:clip_point]
y_test = y_restruct[clip_point:]

print(f"X_train shape: {X_train.shape}")
print(f"y_train shape: {y_train.shape}")
print(f"X_test shape: {X_test.shape}")
print(f"y_test shape: {y_test.shape}")

X_train shape: (122, 10, 1)
y_train shape: (122, 1)
X_test shape: (12, 10, 1)
y_test shape: (12, 1)
```

Listing 8.7 Data Preparation: Train/Test Split Creation

Listing 8.8 shows the code for creating the figure that displays the time series and the time of the split.

```
plt.figure(figsize=(10, 6))
sns.lineplot(data=flights, x='year_month', y='passengers', label='All data)
# The indices need to be adapted to time split_date = flights['year_
month'].iloc[clip_point + seq_len]
plt.axvline(x=split_date, color='red', linestyle='--', label='Train/Test
Split')
plt.title('Number of Flight Passengers per Month')
plt.legend()
plt.show()
```

Listing 8.8 Data Preparation: Train/Test Split Visualization

Figure 8.5 shows the resulting mapping of the time series with the cutoff date, which reflects the time when the training and test data were separated.

We create the FlightDataset class with the code in Listing 8.9. The squeeze(-1) method in __getitem__ is worth mentioning here because it reduces the dimension of the tensor, which is necessary since we're only dealing with univariate data here. This means that the input sequence with the [seq_len, 1] dimensions becomes [seq_len] instead.

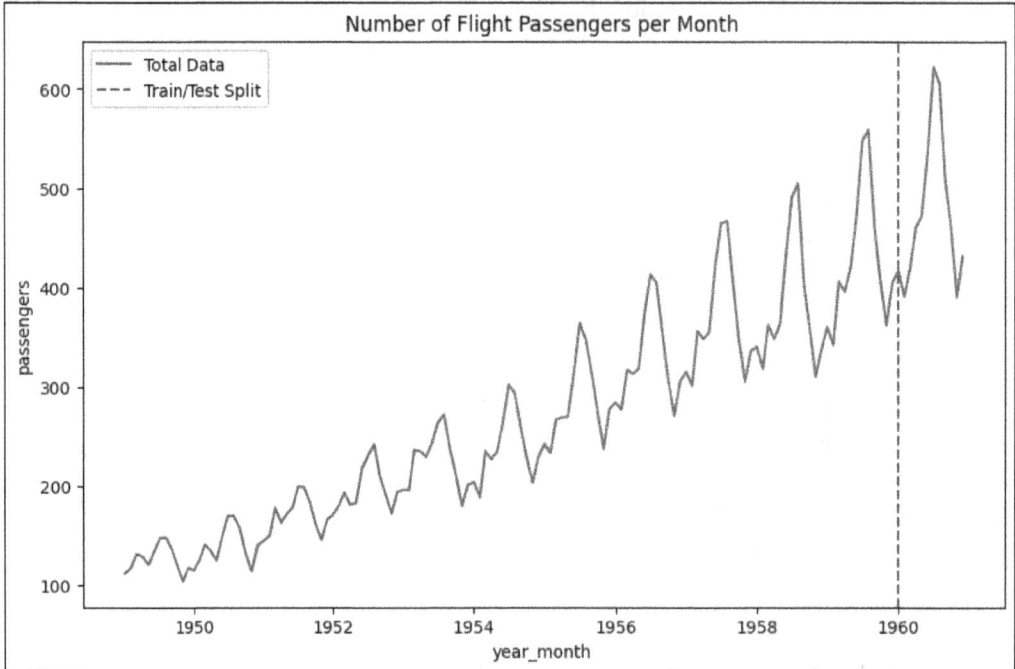

Figure 8.5 Time Series with Cutoff Date for Separation of Training and Test Data

```
class FlightDataset(Dataset):
    def __init__(self, X, y):
        self.X = torch.tensor(X, dtype=torch.float32)
        self.y = torch.tensor(y, dtype=torch.float32)

    def __len__(self):
        return len(self.X)

    def __getitem__(self, idx):
        return self.X[idx].squeeze(-1), self.y[idx]
#%% Dataloader
train_loader = DataLoader(FlightDataset(X_train, y_train),
                    batch_size=BATCH_SIZE)
test_loader = DataLoader(FlightDataset(X_test, y_test),
                    batch_size=len(y_test))
```

Listing 8.9 Data Preparation: Dataset and DataLoader Classes

The data is available in the form of train_loader and test_loader. Now, we are going to train the model.

8.2.2 Long Short-Term Memory

We start with the training of the LSTM model, and we can find the script at *090_TimeSeries\flight_lstm.py*. In Listing 8.10, we load the packages for modeling (torch), the packages for visualizing the results (seaborn and matplotlib), and the DataLoader train_loader and test_loader from our data preparation script.

```
import torch
from torch import nn
# visualization
import matplotlib.pyplot as plt
import seaborn as sns
from data_prep import train_loader, test_loader
```

Listing 8.10 LSTM: Package Import

We also need the number of training EPOCHS as a hyperparameter:

```
EPOCHS = 100
```

In Listing 8.11, we create the FlightModel model class. The class should be generic and work with other data, so the parameters are input_size, hidden_size and output_size. The structure of the model is simple, as the complicated part (the LSTM module) is provided directly by PyTorch.

The sequence is processed by self.lstm, which returns the output (x) with these dimensions: [batch, seq_len, hidden_size]. The object (x[:, -1, :]) takes the hidden state of the last timestep, and the subsequent ReLU activation performs nonlinearity and then passes the data on to the output layer with self.fc1.

```
class FlightModel(nn.Module):
    def __init__(self, input_size, hidden_size=50, output_size=1):
        super(FlightModel, self).__init__()
        self.lstm = nn.LSTM(input_size, hidden_size, num_layers=1,
                        batch_first=True)
        self.fc1 = nn.Linear(hidden_size, output_size)
        self.relu = nn.ReLU()

    def forward(self, x):
        x = x.unsqueeze(-1)
        x, _ = self.lstm(x)
        x = self.relu(x[:, -1, :])
        x = self.fc1(x)

        return x
```

Listing 8.11 LSTM: Model Class

In Listing 8.12, we create the model instance, the loss function, and the optimizer.

```
model = FlightModel(input_size=1)
loss_fun = nn.MSELoss()
optimizer = torch.optim.Adam(model.parameters())
```

Listing 8.12 LSTM: Model Instance, Loss Function, and Optimizer

Listing 8.13 shows the model training, which follows the same pattern as before.

```
loss_train = []
for epoch in range(EPOCHS):
    loss_epoch = 0
    for j, train_batch in enumerate(train_loader):
        X_batch, y_batch = train_batch

        optimizer.zero_grad()
        y_pred = model(X_batch)
        loss = loss_fun(y_pred, y_batch)
        loss.backward()
        optimizer.step()
        loss_epoch += loss.item()
    loss_train.append(loss_epoch/len(train_loader))

    print(f"Epoch: {epoch}, Loss: {loss.data}")
```

Listing 8.13 LSTM: Model Training

Now let's take a look at the training losses using the code from Listing 8.14.

```
sns.lineplot(x=range(EPOCHS), y=loss_train)
plt.xlabel('Epoch [-]')
plt.ylabel('Loss [-]')
plt.title('Loss')
```

Listing 8.14 LSTM: Loss Visualization

Figure 8.6 shows the training losses over the training epoch. The loss decreases sharply and reaches a convergence after 100 epochs.

Now, we want to find out how well the predictions match the actual passenger numbers in 1960. To do this, we create the predictions in Listing 8.15.

```
X_test_torch, y_test_torch = next(iter(test_loader))
with torch.no_grad():
    y_pred = model(X_test_torch)
```

```
y_act = y_test_torch.numpy().squeeze()
x_act = range(y_act.shape[0])
```

Listing 8.15 LSTM: Creating Predictions for Test Data

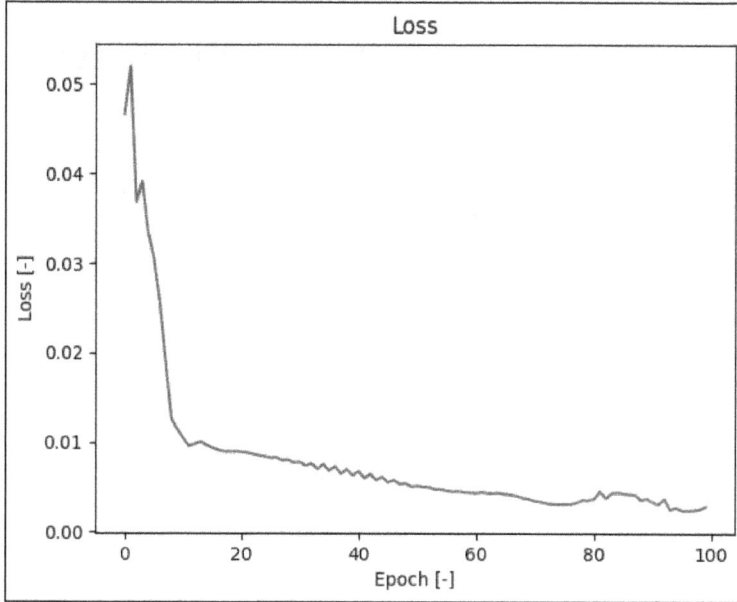

Figure 8.6 LSTM: Training Losses

With the code from Listing 8.16, we can visualize the result and have a direct comparison between actual and predicted passenger numbers.

```
sns.lineplot(x=x_act, y=y_act, label = 'tatsächlich',color='black')
sns.lineplot(x=x_act, y=y_pred.squeeze(), label = 'vorhergesagt',color='red')
plt.ylabel('Normierte Passagierzahlen [-]')
plt.xlabel('Monat [-]')
plt.title('Vorhersage vs. tatsächlicher Wert')
```

Listing 8.16 LSTM: Visualization of Actual and Predicted Values

The two time-series are shown in Figure 8.7, and they show relatively good agreement between the two curves.

We can be satisfied with the result that we see in the following:

```
rmse = np.sqrt(np.mean((y_act - y_pred.squeeze().numpy())**2))
print(f"RMSE: {rmse:.2f}")
```

RMSE: 0.08

```
#%% calculate mape error
mape = np.mean(np.abs((y_act - y_pred.squeeze().numpy()) / y_act)) * 100
print(f"MAPE: {mape:.2f}%")
```

MAPE: 8.79%

We'll now turn our attention to the next type of model: convolutional neural networks.

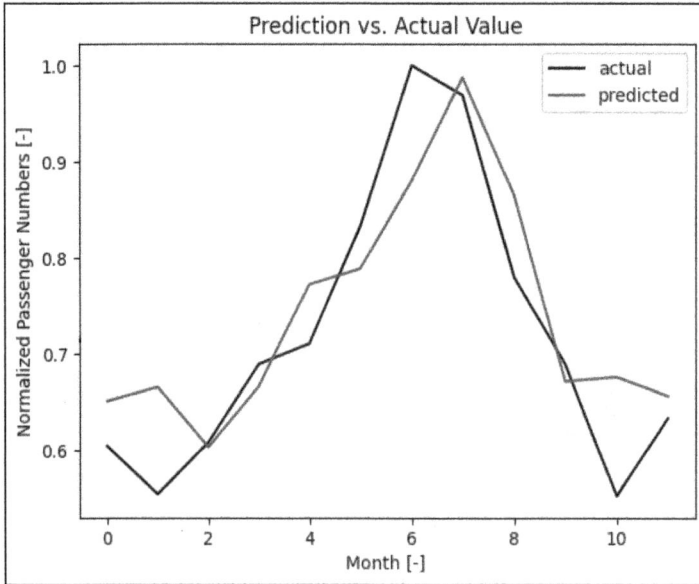

Figure 8.7 LSTM: Actual and Predicted Passenger Numbers

8.2.3 Convolutional Neural Network

The script can be found at *090_TimeSeries\flight_cnn.py*. First, we need the packages shown in Listing 8.17.

```
# modeling
import torch
from torch import nn
# visualization
import matplotlib.pyplot as plt
import seaborn as sns
from data_prep import train_loader, test_loader
```

Listing 8.17 1D CNN: Loading Packages

Listing 8.18 shows the hyperparameters we use. The model uses a convolutional layer internally, and it's based on the OUT_CHANNELS and KERNEL_SIZE parameters. The hidden layer receives the number of HIDDEN_SIZE parameters.

```
EPOCHS = 100
BATCH_SIZE = 4
SEQ_LEN = 10
OUT_CHANNELS = 16
KERNEL_SIZE = 3
HIDDEN_SIZE = 50
```

Listing 8.18 1D CNN: Hyperparameters

Listing 8.19 shows how the model class is structured. PyTorch expects the data in the form of [batch_size, channels, seq_len] for Conv1d. Our input has the form of [batch_size, seq_len], and we have to transform it.

We process the data in the conv1 layer and activate it with relu. At the transition from the convolutional layer to the linear layer, we must flatten the tensor by using torch.flatten to have the correct dimension.

We then process the data further in the fc1 and fc2 linear layers.

```
class FlightModel(nn.Module):
    def __init__(self, in_channels, out_channels, kernel_size, seq_len,
                 hidden_size, output_size):
        super(FlightModel, self).__init__()
        self.conv1 = nn.Conv1d(in_channels, out_channels, kernel_size)
        conv_output_size = out_channels * (seq_len - kernel_size + 1)

        self.fc1 = nn.Linear(conv_output_size, hidden_size)
        self.fc2 = nn.Linear(hidden_size, output_size)
        self.relu = nn.ReLU()

    def forward(self, x):
        x = x.unsqueeze(1)
        x = self.conv1(x)
        x = self.relu(x)
        x = torch.flatten(x, start_dim=1)
        x = self.fc1(x)
        x = self.relu(x)
        x = self.fc2(x)
        x = self.relu(x)
        return x
```

Listing 8.19 1D CNN: Model Class

Listing 8.20 shows the creation of the model instance, the loss function, and the optimizer. We transfer the previously defined hyperparameters to the model instance, we use MSE as the loss function, and we choose Adam as the optimizer.

```
model = FlightModel(
    in_channels=1,
    out_channels=OUT_CHANNELS,
    kernel_size=KERNEL_SIZE,
    seq_len=SEQ_LEN,
    hidden_size=HIDDEN_SIZE,
    output_size=1
)
loss_fun = nn.MSELoss()
optimizer = torch.optim.Adam(model.parameters())
```

Listing 8.20 1D CNN: Model Instance, Loss Function, and Optimizer

Now, let's move on to the model training in Listing 8.21.

```
loss_train = []
for epoch in range(EPOCHS):
    loss_epoch = 0
    for j, train_batch in enumerate(train_loader):
        X_batch, y_batch = train_batch

        optimizer.zero_grad()
        y_pred = model(X_batch)
        loss = loss_fun(y_pred, y_batch)
        loss.backward()
        optimizer.step()
        loss_epoch += loss.item()
    loss_train.append(loss_epoch/len(train_loader))

    print(f"Epoch: {epoch}, Loss: {loss.data}")
```

Listing 8.21 1D CNN: Model Training

The training losses of our CNN model are shown in Figure 8.8.

We can generate the model predictions for the test data by using the following code:

```
X_test_torch, y_test_torch = next(iter(test_loader))
with torch.no_grad():
    y_pred = model(X_test_torch)
```

Now, we can use the predictions and actual values to determine metrics such as root mean square error (RMSE) and mean average percentage error (MAPE):

```
rmse = np.sqrt(np.mean((y_act - y_pred.squeeze().numpy())**2))
print(f"RMSE: {rmse:.2f}")
```

RMSE: 0.09

```
mape = np.mean(np.abs((y_act - y_pred.squeeze().numpy()) / y_act)) * 100
print(f"MAPE: {mape:.2f}%")
```

MAPE: 11.49%

The results are similar to those we obtained from the previously trained LSTM model.

And that brings us to the third and final model class that I would like to introduce to you in the context of time series modeling: transformers.

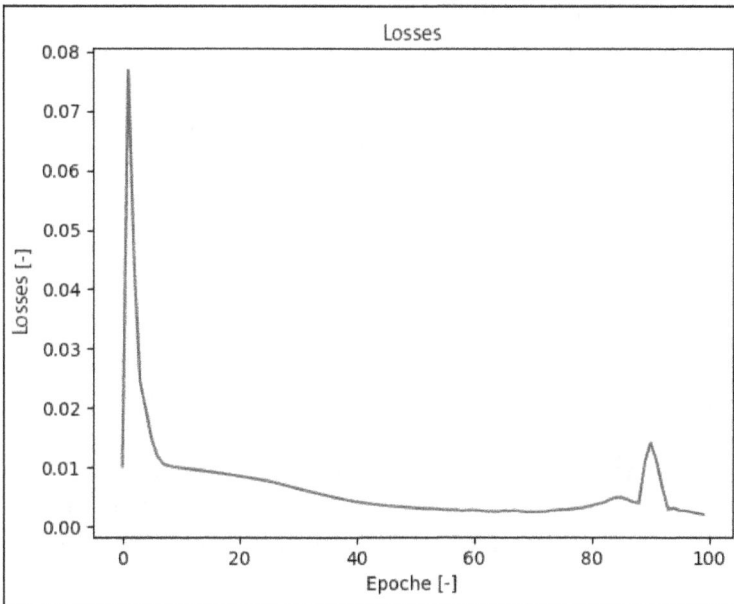

Figure 8.8 CNN: Training Losses

8.2.4   Transformers

The script can be found at *090_TimeSeries\flight_transformer.py*. Listing 8.22 shows which packages are required to train this model.

```
# modeling
import torch
from torch import nn
# visualization
import matplotlib.pyplot as plt
import seaborn as sns
# own scripts
from data_prep import train_loader, test_loader
```

Listing 8.22 Transformer: Import Packages

The model requires the hyperparameters listed in Listing 8.23. We use several attention heads (NHEAD) here. The more heads the model has, the better it can recognize different patterns (short versus long term, trend versus seasonality, etc.). Each head can learn its own view on the same inputs.

We control the dimensionality of the time series embedding and the output of the encoder layers via the D_MODEL parameter. We must also ensure that the D_MODEL number is an integer divided by NHEAD. The NUM_LAYERS parameter defines the number of stacked transformer encoder layers, and a higher number increases the representational power of the model but entails the risk of overfitting. The HIDDEN_SIZE parameter describes the size of the hidden layer in the feed-forward layers.

```
EPOCHS = 100
BATCH_SIZE = 4
SEQ_LEN = 10
NHEAD = 4
D_MODEL = 48
NUM_LAYERS = 2
HIDDEN_SIZE = 50
```

Listing 8.23 Transformer: Hyperparameter

Now, we can create the FlightModel model class in Listing 8.24. We must extend the data by one dimension by using unsqueeze(-1) so that the transformer can process it. We then embed the data, with the embedding layer converting each value of the sequence into a vector of size d_model. The transformer_encoder layer with self-attention processes the sequence and learns the relationships between the points in time. Then, we come to the prediction. We calculate the target value from the last timestep of the encoder output via a linear layer (fc1).

```
class FlightModel(nn.Module):
    def __init__(self, input_size, d_model, nhead, num_layers, output_size):
        super(FlightModel, self).__init__()
        self.embedding = nn.Linear(input_size, d_model)
        self.encoder_layer = nn.TransformerEncoderLayer(d_model=d_model, nhead=
nhead, batch_first=True)
        self.transformer_encoder = nn.TransformerEncoder(self.encoder_layer,
num_layers=num_layers)
        self.fc1 = nn.Linear(d_model, output_size)

    def forward(self, x):
        x = x.unsqueeze(-1)
        x = self.embedding(x)
        x = self.transformer_encoder(x)
```

```
    x = self.fc1(x[:, -1, :])
    return x
```

Listing 8.24 Transformer: Model Class

Now, we can create the model instance, the loss function, and the optimizer, as shown in Listing 8.25. When instantiating the model, we set the input_size and output_size parameters to 1 since we're dealing with a univariate time series. We define the other parameters via the corresponding variables.

```
model = FlightModel(
    input_size=1,
    d_model=D_MODEL,
    nhead=NHEAD,
    num_layers=NUM_LAYERS,
    output_size=1
)
loss_fun = nn.MSELoss()
optimizer = torch.optim.Adam(model.parameters())
```

Listing 8.25 Transformer: Model Instance, Loss Function, and Optimizer

The model training is based on the code in Listing 8.26, which includes no surprises.

```
loss_train = []
for epoch in range(EPOCHS):
    loss_epoch = 0
    for j, train_batch in enumerate(train_loader):
        X_batch, y_batch = train_batch

        optimizer.zero_grad()
        y_pred = model(X_batch)
        loss = loss_fun(y_pred, y_batch)
        loss.backward()
        optimizer.step()
        loss_epoch += loss.item()
    loss_train.append(loss_epoch/len(train_loader))

    print(f"Epoch: {epoch}, Loss: {loss.data}")
```

Listing 8.26 Transformer: Model Training

The model will be trained after 100 epochs.

Now, let's examine the RMSE and MAPE metrics to check the model quality and compare the different models with each other, as follows:

```
rmse = np.sqrt(np.mean((y_act - y_pred.squeeze().numpy())**2))
print(f"RMSE: {rmse:.2f}")
```

RMSE: 0.11
```
mape = np.mean(np.abs((y_act - y_pred.squeeze().numpy()) / y_act)) * 100
print(f"MAPE: {mape:.2f}%")
```

MAPE: 12.12%

We've thus become familiar with the third model class for solving time series problems. Modeling time series can quickly become complex, and for that reason, various frameworks have been created to make your life easier. One very popular such framework is PyTorch Forecasting, which we'll get to know better in the next section.

8.3   Coding: Using PyTorch Forecasting

The *PyTorch Forecasting* package simplifies the prediction of time series and is suitable for both data scientists and researchers. For a long time, classic methods such as the autoregressive integrated moving average (ARIMA) were commonly used in time series forecasting. In contrast to computer vision and natural language processing, deep learning has only recently gained a foothold in this field, and it can now also deliver results in this area.

We've already seen that preparing the dataset can be a headache, and for that reason, the developers of this package have designed the data interface to be as simple as possible. We can simply pass a DataFrame to the framework, and we'll be ready to go.

The framework makes several models available, such as the Temporal Fusion Transformer and NBeats.

The implementation follows the typical structure of preparing the data in Section 8.3.1, performing the model training in Section 8.3.2, and evaluating the results in Section 8.3.3.

8.3.1   Data Preparation

We load the required packages in Listing 8.27. The pytorch_forecasting package provides its own handling of the dataset via TimeSeriesDataSet, and it also allows us to access powerful time series models. In this example, we use TemporalFusionTransformer.

The *Temporal Fusion Transformer* (TFT) is a deep learning model that was developed specifically for multivariate time series prediction. It stands out because it can process static, known-dynamic, and unknown-dynamic inputs simultaneously. It's also a *transformer model*, which means it's based on the attention mechanism. A key feature is the *gated residual network* (GRN) block, which enables selective processing of input data to filter out noise and irrelevant information. In addition, it has a variable selection layer that determines which variables are most important for the prediction. This also makes the model interpretable—a property that's crucial in many use cases. The complete script can be found under *090_TimeSeries\flight_py_forecasting.py*.

```
import numpy as np
import lightning.pytorch as pl
from pytorch_forecasting import TimeSeriesDataSet, TemporalFusionTransformer
import seaborn as sns
import matplotlib.pyplot as plt
```

Listing 8.27 PyTorch Forecasting: Package Import

We import the dataset via seaborn in Listing 8.28, and then, we check the existing data and display some information on the console to check what form the dataset has or what time period it covers.

```
flights = sns.load_dataset("flights")

print(f"Form of the original dataset: {flights.shape}")
print(f"Date range: {flights['year'].min()} - {flights['year'].max()}")
print(f"Number of months: {len(flights)}")
```

```
Form of the original dataset: (144, 3)
Date Range: 1949 - 1960
Number of Months: 144
```

Listing 8.28 PyTorch Forecasting: Data Import

Next, we must prepare the features in Listing 8.29. Here, we create a zero-based time index (time_idx), and we implement the month and year as categorical features. We should question these features in case of doubt, as there are several ways to model the data at this point. It's ensured that the target variable passengers are available as a float.

```
# This is a crucial step for pytoch-forecasting
flights["time_idx"] = range(len(flights)).astype(int)
flights["month"] = flights["month"].astype("category")
flights["year"] = flights["year"].astype(str).astype("category")
flights["series"] = "flights"
flights["passengers"] = flights["passengers"].astype(float)
```

Listing 8.29 PyTorch Forecasting: Features Preparation

The code in Listing 8.30 prepares data for a time series prediction by setting the maximum length of the encoder and the prediction to twelve time units (i.e., one year). It divides the flights dataset into training and validation data, with the training area extending up to TRAINING_CUTOFF and the rest being used for validation.

```
MAX_ENCODER_LENGTH = 12
MAX_PREDICTION_LENGTH = 12
TRAINING_CUTOFF = len(flights) - 12

print(f"Training split point: {TRAINING_CUTOFF}")
print(f"Training data points: {TRAINING_CUTOFF}")
print(f"Validation data points: {len(flights) - TRAINING_CUTOFF}")
train_data = flights.iloc[:TRAINING_CUTOFF].copy()
val_data = flights.iloc[TRAINING_CUTOFF:].copy()
```

Training Split Point: 132
Training data points: 132
Validation data points: 12

Listing 8.30 PyTorch Forecasting: Separation of Training and Validation Data

Before proceeding further, we check the dimensions of the datasets in Listing 8.31. Then, we perform the training based on 11 years (132 time units) to predict the last year of the dataset.

```
print(f"Form of the training data: {train_data.shape}")
print(f"Form of the validation data: {val_data.shape}")
print(f"Time index range of the training data: {train_data['time_idx'].min()} -
{train_data['time_idx'].max()}")
print(f"Time index range of the validation data: {val_data['time_idx'].min()} -
{val_data['time_idx'].max()}")
```

Form of training data: (132, 5)
Form of validation data: (12, 5)
Time Index Range of training data: 0 - 131
Time Index Range of Validation Data: 132 - 143

Listing 8.31 PyTorch Forecasting: Review of Training Data and Validation Data

Then, we can visualize the two time series, training and validation, with the code in Listing 8.32.

```
plt.figure(figsize=(10, 6))
sns.lineplot(x="time_idx", y="passengers", data=train_data, label='train')
sns.lineplot(x="time_idx", y="passengers", data=val_data, label='val')
plt.title('Training- and Validation data)
plt.xlabel(Time index [-]')
plt.ylabel('Passengers [-]')
plt.show()
```

Listing 8.32 PyTorch Forecasting: Visualization of Training Data and Validation Data

Figure 8.9 clearly shows the connected training time series and the subsequent validation time series.

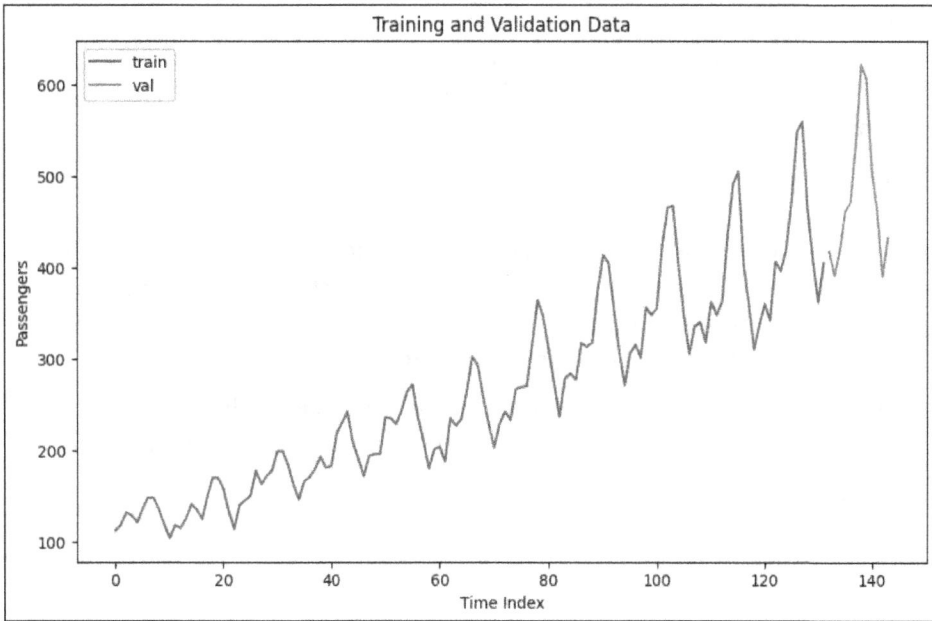

Figure 8.9 PyTorch Forecasting: Visualization of Training Data and Validation Data

In Listing 8.33, we create a `TimeSeriesDataSet` object that serves as the central data container for training the time series model. The following parameters require explanation:

- `max_encoder_length`
  This parameter determines how many historical data points the model should use for prediction.

- `max_prediction_length`
  This parameter specifies how many future time steps the model should predict.

- `group_ids`
  You can use this parameter to identify individual time series in the dataset.

- `time_varying_unknown_reals`
  You can use this parameter to specify variables that change over time and whose future values are not known in advance.

- `time_varying_known_categoricals`
  This parameter works like `time_varying_unknown_reals`, but for categorical data.

```
full_dataset = TimeSeriesDataSet(
    flights,
    time_idx="time_idx",
```

```
    target="passengers",
    group_ids=["series"],
    max_encoder_length=MAX_ENCODER_LENGTH,
    max_prediction_length=MAX_PREDICTION_LENGTH,
    time_varying_unknown_reals=["passengers"],
    time_varying_known_categoricals=["month"],
    static_categoricals=["year"])
```

Listing 8.33 PyTorch Forecasting: Creating Complete Dataset

In Listing 8.34, we create the training data and validation data from the complete dataset. We use an already configured TimeSeriesDataset object (full_dataset) as a template to create a new dataset, and we use the from_dataset method to ensure that settings such as the target size or group_ids are applied directly.

We use min_prediction_length to guarantee that the model predicts at least one step into the future, and we use min_encoder_length to determines the minimum length of historical data that the model requires for a prediction.

```
training = TimeSeriesDataSet.from_dataset(
    dataset=full_dataset,
    data=train_data,
    min_prediction_length=1,
    min_encoder_length=MAX_ENCODER_LENGTH//2
)
validation = TimeSeriesDataSet.from_dataset(
    dataset=full_dataset,
    data=val_data,
    min_prediction_length=1,
    min_encoder_length=MAX_ENCODER_LENGTH//2
)
```

Listing 8.34 PyTorch Forecasting: Training Dataset and Validation Dataset

The datasets are now prepared, and we can continue with the training of the model.

8.3.2 Model Training

Listing 8.35 shows the hyperparameters for the training. We use the transformer-specific ATTENTION_HEAD_SIZE parameter, which defines the number of attention heads. The higher this number is, the more possibilities the model has to capture different aspects of the input data simultaneously. We also use DROPOUT, a regularization technique to avoid overfitting. Specifically, the value of 0.1 means that 10% of the neurons are randomly switched off during training.

```
BATCH_SIZE = 4
LEARNING_RATE = 0.01
HIDDEN_SIZE = 4
ATTENTION_HEAD_SIZE = 4
DROPOUT = 0.1
HIDDEN_CONTINUOUS_SIZE = 4
```

Listing 8.35 PyTorch Forecasting: Hyperparameter

We create the `train_dataloader` and `val_dataloader` data loader instances by using the `.to_dataloader` method, as follows:

```
train_dataloader = training.to_dataloader(train=True, batch_size=BATCH_SIZE)
val_dataloader = validation.to_dataloader(train=False, batch_size=BATCH_SIZE)
```

Listing 8.36 describes the creation of the model instance. Since PyTorch Forecasting is based on the PyTorch Lightning framework, the actual model training and the creation of the model instance here look different from what we know from the previous chapters.

At this point, we'll take the modified structure for granted. Chapter 11 will be dedicated to PyTorch Lightning.

We call the `TemporalFusionTransformer` class with its `from_dataset` method, and we transfer the hyperparameters described earlier.

```
model = TemporalFusionTransformer.from_dataset(
    training,
    hidden_size=HIDDEN_SIZE,
    attention_head_size=ATTENTION_HEAD_SIZE,
    dropout=DROPOUT,
    hidden_continuous_size=HIDDEN_CONTINUOUS_SIZE,
    optimizer="adam",
    learning_rate=LEARNING_RATE,
)
```

Listing 8.36 PyTorch Forecasting: Model Class

We create a trainer instance in Lightning and transfer the model class to it, and Listing 8.37 describes exactly how we do this. The advantage is that the trainer can separate the model training from the CPU or GPU during the actual implementation. We use a GPU via the `accelerator="auto"` parameter if available; otherwise, we use the CPU. We deactivate the automatic creation of checkpoints with `enable_checkpointing` as well as logging with `logger=False`.

We also use the `fit` method to adapt the model to the data. The API is therefore strongly oriented toward scikit-learn.

```
trainer = pl.Trainer(
    max_epochs=10,
    accelerator="auto",
    enable_checkpointing=False,
    logger=False,
)
trainer.fit(model, train_dataloaders=train_dataloader, val_dataloaders=val_
dataloader)
```

Listing 8.37 PyTorch Forecasting: Trainer Instance and Training Kickoff

After a short time, the model training will be complete and we'll be able to turn to evaluation.

8.3.3   Evaluation

We create the model predictions on the validation with the code in Listing 8.38. We do this by using model.predict, and we also extract the actual values (actual_values) from the val_dataloader. We then transfer the data so that it's available as a one-dimensional array.

```
predictions = model.predict(val_dataloader, return_x=False).cpu().numpy()

actual_values = []
for batch in val_dataloader:
    batch_actual = batch[1][0].numpy().flatten()
    # Filter out padding values (usually 0 or NaN)
    batch_actual = batch_actual[batch_actual != 0]
    if len(batch_actual) > 0:
        actual_values.extend(batch_actual)

actual_values = np.array(actual_values)

actual_flat = actual_values.flatten()
predicted_flat = predictions[:, 0]

min_length = min(len(actual_flat), len(predicted_flat))
actual_flat = actual_flat[:min_length]
predicted_flat = predicted_flat[:min_length]

print(f"Shape of actual data: {actual_flat.shape}")
print(f"Shape of predicted data: {predicted_flat.shape}")
```

Shape of actual data: (11)
Shape of predicted data: (11)

Listing 8.38 PyTorch Forecasting: Creating Predictions

Finally, we want to make a comparison between the actual and predicted passenger numbers. We use the code from Listing 8.39 for this purpose.

```
plt.figure(figsize=(10, 6))
x_act = range(len(actual_flat))
y_act = actual_flat
y_pred = predicted_flat
sns.lineplot(x=x_act, y=y_act, label='actual', color='black')
sns.lineplot(x=x_act, y=y_pred.squeeze(), label='predicted', color='red')
plt.ylabel('Passenger Numbers [-]')
plt.xlabel('Month [-]')
plt.title(Prediction vs. Actual Value')
plt.grid(True, alpha=0.3)
plt.tight_layout()
plt.show()
```

Listing 8.39 PyTorch Forecasting: Comparison of Actual and Predicted Values

The results are shown in Figure 8.10. The model also predicts the basic trend, although it consistently underestimates the total volume. In general, the more complex the dataset, the more the model complexity should match the dataset's complexity. Models such as TFT are powerful for tiny univariate datasets like the one used here.

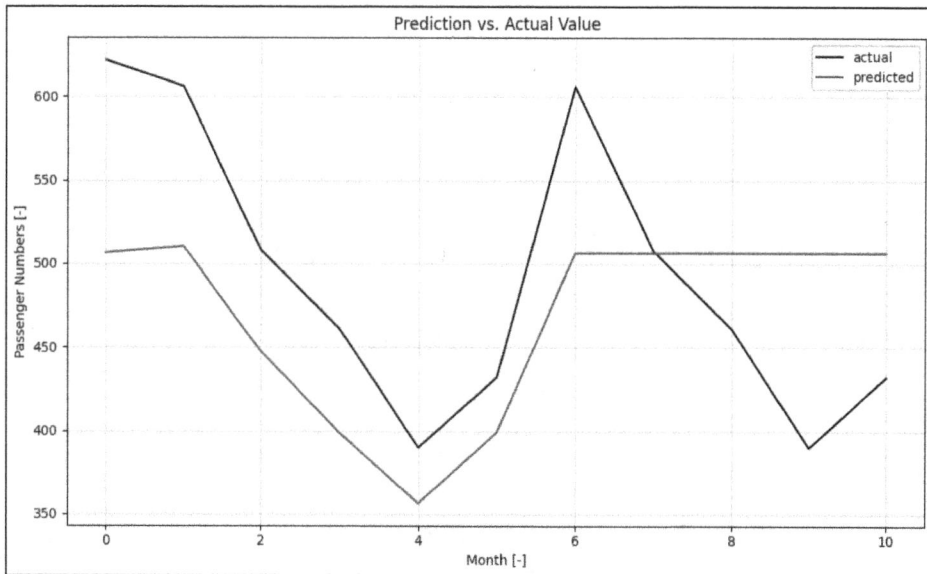

Figure 8.10 PyTorch Forecasting: Comparison of Actual and Predicted Values

I leave it to you to familiarize yourself with the many capabilities of PyTorch Forecasting. In this section, you have only taken the first steps into this vast area.

8.4   Summary

In Section 8.1 of this chapter on time series analysis with PyTorch, we gave you an introduction to the special features of time series data. Then, we presented various network architectures (such as 1D-CNN, LSTM, and transformers) that are particularly suitable for modeling time series.

We dedicated Section 8.2 to the practical implementation of these theoretical concepts. We translated the approaches presented in Section 8.1 into practical code and illustrated them by using a small sample dataset. In Section 8.3, we presented an alternative to this because it's common practice to use specialized frameworks instead of developing one's own models. This section helped you get to know the widely used PyTorch Forecasting framework.

Chapter 9
Language Models

"The limits of my language mean the limit of my world."
—Ludwig Wittgenstein

Our language is not only a tool for communication but also the medium through which we structure our thinking, perception, and understanding. If we lack the words to describe something, our perspective narrows—and when we expand languages, new ways of understanding the world automatically open up.

This is precisely where large language models come in. They are classified as *foundation models*, which are versatile basic models that have been trained on huge data sets. They serve as a foundation that can be adapted for a variety of different tasks.

Language models can do more than just generate texts. By making patterns in language visible, they expand our range of expressions—so it's important for us to understand not only how language models work but also how they change our world.

Large language models (LLMs) have established themselves as a transformative technology that fundamentally redefines machines' understanding and generation of texts. At their core, LLMs are complex neural networks that have been trained on unimaginably large amounts of text data. This enables the models to recognize complex linguistic patterns and correlations, which in turn gives them the ability to create coherent texts with appropriate content, answer questions, carry out translations, and produce creative content. The remarkable capabilities of LLMs are based on advances in model architecture, particularly the *transformer* architecture—a special type of network that makes it possible to efficiently process long-term dependencies in texts.

In Section 9.1, we jump right into the deep end and learn how to use language models directly with Python. In Section 9.2, we look at model parameters that help us influence the model response, and we'll also learn what parameters such as temperature, top-p, and top-k are all about. In Section 9.3, we look at the parameters that we can use to select the right model from the huge selection of language models that are available nowadays.

Section 9.4 covers message types, and Section 9.5 covers prompt templates, both of which form the basis of *chains*, which are the basic infrastructure we use to connect LLMs with other components such as data sources and tools. To map complex workflows, it makes sense to use chains, as they help to modularize the code and offer a great deal of flexibility. Section 9.6 covers chains.

In Section 9.7 we'll learn about a specific form of chain that returns structured outputs, which can be extremely helpful if we want to store the model result in a database or use it as input for a subsequent process. In Section 9.8, we conclude the chapter by looking at the underlying architecture of language models. Without the transformer architecture, today's language models would be inconceivable, so we take a technical deep dive in this section to help us understand this pioneering network type.

Now, let's get started and learn how we can use language models directly from Python.

9.1   Using Large Language Models with Python

Due to the complexity and technical requirements of training language models, we'll start at a higher level of abstraction. We'll use language models that have already been trained, and above all, we'll learn how to use them efficiently.

Figure 9.1 shows several popular language models and model families. The most powerful models include the flagship models from OpenAI (e.g., GPT-5), Google (Gemini 2.5 Pro), Anthropic (Claude Sonnet 4.5), and X-AI (Grok 4). Worth mentioning here is Mistral with LeChat as a European provider that also works in compliance with the General Data Protection Regulation (GDPR). Some model providers only grant access to their proprietary models via API, but there are also open weight or open-source models. More on this in Section 9.3.4.

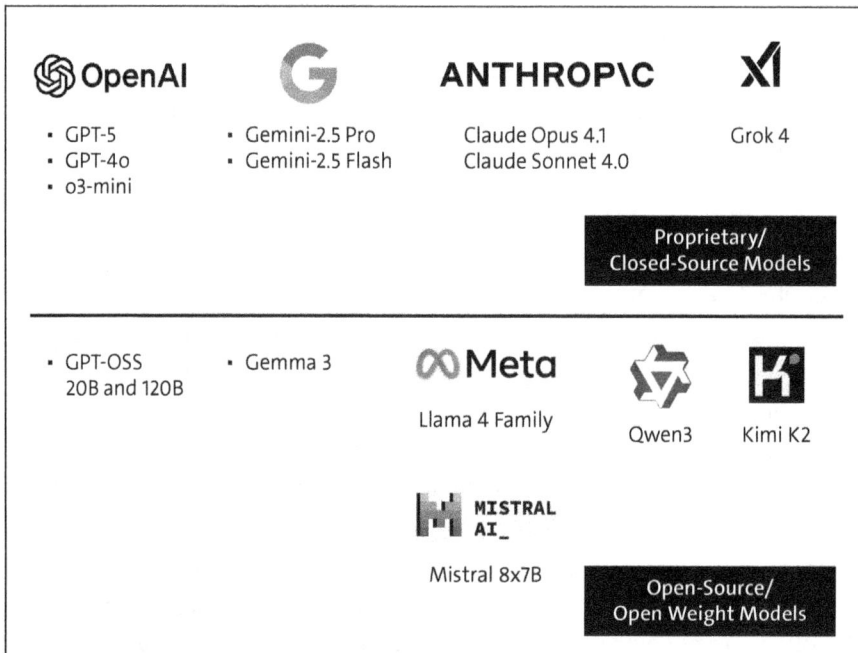

Figure 9.1 Popular Language Models and Model Families

For many people, the topic of language models is synonymous with ChatGPT. This is because OpenAI got the ball rolling with the release of ChatGPT at the end of 2022, and since then, the GPT model series has been the most famous one for language models. It's a very powerful model, but you must pay for it. However, since there are other extremely capable LLMs that are open source, I'll show you how you can use LLMs for free via Groq, and you'll see that interacting with the models is very easy thanks to the Python LangChain framework. After these two examples, you'll be able to connect to any other LLM provider.

In Section 9.1.1, we'll take our first steps by interacting with LLMs via Python script and work with OpenAI. In Section 9.1.2, we'll work with Groq. In Section 9.1.3 we learn how to use multimodal models to pass images alongside text to a model. In Section 9.1.4, we'll learn how to run LLMs locally, for privacy reasons.

9.1.1   Coding: Using OpenAI

To use OpenAI, you first need an *API key*. These models are not free, so you'll be asked to pay to use them. Prices are constantly decreasing over time, and you can check the current prices at *https://openai.com/api/pricing/*.

If you want to use a web service, you normally need a username and password to access it—but if you want to use a service programmatically, you need an API key. An API key is therefore like a combination of a username and password. How can you get an API key if you don't have one yet? Please follow these steps:

1. Navigate to *https://platform.openai.com/*.
2. Create an account.
3. Activate billing and add some money to your account.
4. Go to the API keys section and create a new API key. The name you specify in the web frontend is only relevant for later recognition. In practice, you only need the key.
5. Copy the key to the clipboard.
6. Paste it into a file named *.env*. It should look like this: `sk-proj....`

It's generally good practice to separate code from credentials, so you should save the API key in a separate file. A common approach is to save it in a file called *.env* and place it in the working folder. In this file, you save the API key and possibly many other keys if necessary. Listing 9.1 shows what an environment file should look like.

```
OPENAI_API_KEY = sk-proj...
```

Listing 9.1 Sample Content of .env File

The API keys are treated as *environment variables*, which are typically variables that are used by your operating system. Our variable is called `OPENAI_API_KEY`, and it has a value that must be defined on the right side of the equal sign. It's important to use the same key name in the code script.

If you're hesitant to enter your bank details online, skip this lesson and go to the next one to work directly with Groq, which offers free access to the models. Also, don't confuse Groq with Grok. *Groq* is an AI startup that focuses on developing chips for fast inference of LLMs, while *Grok* is an LLM that was launched as an initiative by Elon Musk.

Let's start with our coding script. You can find it at *100_LLM/10_model_chat_openai.py*. Here, it's good practice to place all the required packages and functions at the beginning of the file. Let's go through what we need here.

We need the os package to retrieve and load the environment variables. All major model providers offer packages for integration in LangChain, so here, we use langchain_openai. We need the dotenv package to work with the environment variable file, and its load_dotenv() function loads the content of the *.env* file and makes it available as environment variables, as follows:

```
#%% packages
import os
from langchain_openai import ChatOpenAI
from dotenv import load_dotenv
load_dotenv('.env')
```

You can check whether the API key is available by running os.getenv('OPENAI_API_KEY'), which should display the API key on the screen.

Now, we create an instance of the model that we'll use by using the ChatOpenAI class. For this, we need a model name, so we've chosen gpt-4o-mini here. Another important parameter is the temperature (more on model parameters in Section 9.2), which controls the creativity of the model. And you need to pass the API key to authenticate yourself and allow OpenAI to calculate your costs based on your usage, as follows:

```
MODEL_NAME = 'gpt-4o-mini'
model = ChatOpenAI(model_name=MODEL_NAME,
                   temperature=0.5,
                   api_key=os.getenv('OPENAI_API_KEY'))
```

The model object has a very important method: invoke(), which allows you to execute the model based on certain parameters. In our first example, we ask the model for information on "What is LangChain?" The result is stored in an object and is an object of type AIMessage. We can find out what information we got from the model call by looking at the output of its model_dump() method. The model execution and the generated result are shown in Listing 9.2.

```
res = model.invoke("What is a LangChain?")
res.model_dump()
```

```
{'content': 'LangChain is ...',
 'additional_kwargs': {'refusal': None},
 'response_metadata': {'token_usage': {'completion_tokens': 290,
   'prompt_tokens': 13,
   'total_tokens': 303,
   'completion_tokens_details': {'accepted_prediction_tokens': 0,
    'audio_tokens': 0,
    'reasoning_tokens': 0,
    'rejected_prediction_tokens': 0},
   'prompt_tokens_details': {'audio_tokens': 0, 'cached_tokens': 0}},
  'model_name': 'gpt-4o-mini-2024-07-18',
  'system_fingerprint': 'fp_560af6e559',
  'id': 'chatcmpl-C5oA3wS5xLKygiq66IWHdOWbFigRC',
  'service_tier': 'default',
  'finish_reason': 'stop',
  'logprobs': None},
 'type': 'ai',
 'name': None,
 'id': 'run--95a59d13-84a9-4d1b-838a-38fa7b8b0d92-0',
 'example': False,
 'tool_calls': [],
 'invalid_tool_calls': [],
 'usage_metadata': {'input_tokens': 13,
  'output_tokens': 290,
  'total_tokens': 303,
  'input_token_details': {'audio': 0, 'cache_read': 0},
  'output_token_details': {'audio': 0, 'reasoning': 0}}}}
```

Listing 9.2 OpenAI Model Invocation and Output

There's a lot of information that comes back from the model, so let's start with the most important information: the content. This property contains the actual model output prompt. Of the other properties, I only want to mention the response_metadata, which contains information about token usage. You're asked to pay for input tokens and output tokens, and in the response_metadata , you can see how many tokens were used in the request.

You can familiarize yourself with the different models in the OpenAI model family by studying the model overview at *https://platform.openai.com/docs/models/overview*. OpenAI has created a model family (*https://platform.openai.com/docs/models*) consisting of the following models that are suitable for different tasks:

- **Language models**
  These include the GPT family (e.g., GPT-5), which can process and generate text, and some of them can also generate images.

- **Text-to-image generation**
  These include GPT Image 1 and DALL-E 3, which are models that can generate and edit images.

- **Text-to-speech (TTS) models**
  These include several models (such as GPT-4o mini TTS) that can convert text into natural, spoken audio.

- **Real-time models**
  These allow you to create text, audio inputs, and audio outputs in real time.

- **Text embeddings**
  Embeddings are numerical representations of text that are also the foundation of natural language processing.

Of course, you're not limited to the OpenAI model family. You can work with many other LLMs. Now, we'll discover how to work with open-source LLMs that you can run for free via Groq.

9.1.2 Coding: Using Groq

Groq is a company that develops AI hardware that enables fast inference. For developers, the company offers access to LLMs, especially open-source LLMs. You can use the service for free, but you need to authenticate yourself with an API key. So, the first step is to go to *https://console.groq.com/*, set up an account, and create an API key that you can use in your code. Then, copy this API key and save it in a file called *.env* in the working folder. The content of the file should look like Listing 9.3.

```
GROQ_API_KEY = gsk_...
```

Listing 9.3 Sample Content of .env File

The script, which you can find at *100_LLM/20_model_chat_groq.py*, starts by loading the relevant packages. The main package is langchain_groq, which is the interface to use the models from the Groq model family. You use the os and dotenv packages to set up and retrieve environment variables that contain the Groq API key, as follows:

```
#%% packages
import os
from langchain_groq import ChatGroq
from dotenv import load_dotenv
load_dotenv('.env')
```

We need to select a model, and details on specific models can be found in the overview of Groq models at *https://console.groq.com/docs/models*. Here, we choose a model from the Llama family. It's an open-source model, or more precisely, an open-weight model. This means that the model is made available to the public for free use and the training process is publicly available. However, not all details of the datasets used are publicly available.

Once we've decided on a model, we can create an instance of the ChatGroq class. In the following instantiation, we pass the name of the model as a parameter, and we also need to pass the API key that we created earlier. These two parameters are mandatory. Among many other available parameters, we only set the temperature parameter, which controls the creativity of the model. We'll learn more about the model parameters in Section 9.2 so that we'll have everything ready to interact with the LLM.

```
MODEL_NAME = 'llama-3.3-70b-versatile'
model = ChatGroq(model_name=MODEL_NAME,
                 temperature=0.5,
                 api_key=os.getenv('GROQ_API_KEY'))
```

We ask the model, "What is Hugging Face?" via the invoke() method, as follows:

```
# %% Run the model
res = model.invoke("What is a Hugging Face?")
```

The model_dump() method shown in Listing 9.4 gives us an overview of the model output.

```
# %% find out what is in the result
res.model_dump()
```

```
{'content': 'Hugging Face is a popular open-source library and platform for nat-
ural language processing (NLP) and machine learning (ML) ...',
'additional_kwargs': {},
'response_metadata': {'token_usage': {'completion_tokens': 314,
'prompt_tokens': 42,
'total_tokens': 356,
'completion_time': 1.032163328,
'prompt_time': 0.010863083,
'queue_time': 0.085254578,
'total_time': 1.043026411},
'model_name': 'llama-3.3-70b-versatile',
'system_fingerprint': 'fp_2ddfbb0da0',
'service_tier': 'on_demand',
'finish_reason': 'stop',
logprobs': None},
'type': 'ai',
'name': None,
'id': 'run--982093cc-280a-49cc-9568-a9c9e6d37943-0',
```

```
'example': False,
'tool_calls': [],
'invalid_tool_calls': [],
'usage_metadata': {'input_tokens': 42,
'output_tokens': 314,
'total_tokens': 356}}
```

Listing 9.4 Groq Model Invocation and Output

The most important output here is again the content. Typically, we access it directly via its property, as shown in Listing 9.5.

```
print(res.content)
```

Hugging Face is a popular open-source library and platform for natural language processing (NLP) and machine learning (ML) tasks. It was founded in 2016 by Ju-lien Chaumond, Clement Delangue, and Thomas Wolf. Hugging Face is known for its Transformers library, which provides pre-trained models and a simple interface for using and fine-tuning transformer-based models for various NLP tasks, such as text classification, language translation, question answering, and more.
…

Listing 9.5 Groq Model Output: Only Content (Source: 03_LLMs/10_model_chat_groq.py)

We've successfully called up a model from Groq, and you can find all models provided by Groq at *https://console.groq.com/docs/models*. The model ID information is provided for each model, and that's the string you need to use in your script. The developer and a limiting factor are also displayed, and for LLMs, the latter is the context window. The maximum number of tokens is also displayed, and you can find more information on that in the next info box. If you want to dive deeper into the model, you can view the model card and be redirected to the model's developer page.

Special features include the whisper models, which provide a speech-to-text model. This means you can upload an MP3 file (up to a certain size) and receive the corresponding transcription. Most of the available models are LLMs that are all open-source or open-weight models. Prominent models here include members of the OpenAI OSS family, the Llama family (from Meta), Gemma models (from Google), DeepSeek models, Kimi (from Moonshot AI), and Qwen (from Alibaba).

In the model overview, you may have seen one of the most important parameters: the *context window*, which refers to the maximum number of input tokens that a model can process at once. This is an important aspect because the model's ability to generate relevant outputs depends on how much information it can retain and use in a single prompt. Each LLM has a fixed limit on how many tokens it can process simultaneously in its context window.

It's important to note that an LLM breaks down the input text into smaller units called *tokens*. A token can be a word, part of a word, or even a punctuation mark. For example, the sentence "Language models are very powerful" can be broken down into the following tokens: *Language, models, are, very,* and *powerful.* A large context window allows the model to process more information, which improves its ability to understand longer texts or "remember" long conversations. On the other hand, as the size of the context window increases, so do the computing resources required to process the text. In addition, the latency of the model increases—it takes longer for the model to deliver a response.

9.1.3   Coding: Multimodal Models

In this section, we'll mainly work with text input and output and use traditional LLMs. But the demand for models that can understand and interact with more complex and diverse forms of information has increased, and that's why large multimodal models (LMMs) have been developed. These models are trained to understand and generate multiple types or modalities of input and output formats. Modalities typically include text, images, audio, and video.

In contrast to traditional LLMs, which work in a single modality (i.e., text), LMMs can process information in different formats. To achieve this, they combine advances in NLP with innovations in computer vision and audio processing. This means that LMMs can do the following:

- Analyze and describe images in text form.
- Generate images based on text descriptions.
- Transcribe audio recordings.
- Interpret audio recordings and provide responses based on them.

Let's use a few multimodal models to learn how to work with them. We'll find out how we can use an image as input for a model, and then, we'll interact with the LMM to find out if it understands what it "sees" in the image.

Figure 9.2 depicts a flowchart describing the process of training a deep neural network, and we'll pass this image to the multimodal model.

You can find the code for this script at *100_LLM\30_multimodal.py*, and it's mainly based on the Groq documentation (*https://console.groq.com/docs/vision*). In Listing 9.6, we start with the import of the required packages.

```
#%% packages
from groq import Groq
from dotenv import load_dotenv, find_dotenv
load_dotenv(find_dotenv(usecwd=True))
import base64
```

Listing 9.6 Multimodal Model: Required Packages

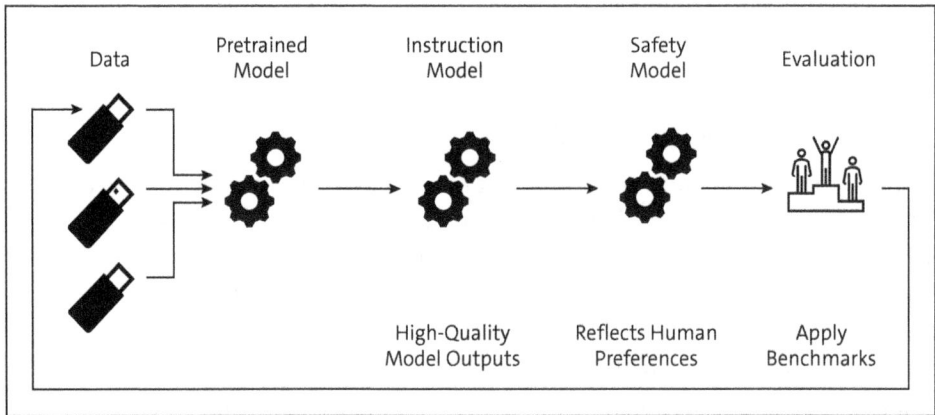

Figure 9.2 Training Process of LLM

It's good practice to define the constants at the beginning of the script, and here, we specify which model we choose, where the image is located, and what the user's input is, as follows:

```
MODEL = "meta-llama/llama-4-maverick-17b-128e-instruct"
IMAGE_PATH = "TrainingProcess.png"
USER_PROMPT = "What is shown in this image? Answer in a paragraph and in German."
```

Since we're working with a local image and we needs to send it to Groq's API, we need to load the image and convert it into a format so that we can send it as a text string. For this functionality, in Listing 9.7, we define a function called encode_image that loads the image and converts it directly into the base64 format. *Base64* is a binary-to-text coding method that converts binary data into an ASCII string format.

```
#%% Function to encode the image
def encode_image(image_path):
  with open(image_path, "rb") as image_file:
    return base64.b64encode(image_file.read()).decode('utf-8')

base64_image = encode_image(IMAGE_PATH)
```

Listing 9.7 Multimodal Model: Function for Image Encoding

Now, we can set up a `groq` instance in Listing 9.8. This is the native implementation of the `groq` package, so the chat request has a different format compared to the LangChain interaction with models. But we can recognize many elements, such as the messages. In the messages object, we define a user message. In it, we pass a dictionary with text content (the user prompt) and the image content (the image we want to interact with).

```python
#%% Getting the base64 string
client = Groq()

chat_completion = client.chat.completions.create(
    messages=[
        {
            "role": "user",
            "content": [
                {"type": "text", "text": USER_PROMPT},
                {
                    "type": "image_url",
                    "image_url": {
                        "url": f"data:image/jpeg;base64,{base64_image}",
                    },
                },
            ],
        }
    ],
    model=MODEL,
)
```

Listing 9.8 Multimodal Model: Model Instance Setup

We've received an answer in the `chat_completion` object, which we can display on the screen, as shown in Listing 9.9.

```python
#%% analyze the output
print(chat_completion.choices[0].message.content)

The image shows a process for developing a model based on artificial intelli-
gence. The process begins with data collection, which is shown in the form of
three USB sticks, and leads through various stages such as pre-trained model,
instruction model, and safety model to evaluation. Each step is symbolized by a
cogwheel graphic that represents the processing and refinement of the model.
```

Listing 9.9 Multimodal Model: Model Output

The model can provide valuable answers, and it understands what it sees. Try changing the user request to see if the model can answer more detailed questions about the image.

9.1.4   Coding: Using Large Language Models Locally

So far, we've used LLMs via API calls from software-as-a-service (SaaS) providers. Sometimes, you want to run a model locally because privacy is important and you want to avoid transmitting confidential information over the internet. In such cases, you can run a model on your local computer, and ideally, you'll have a powerful GPU that can run a decent-sized model. But a small model can also run on your CPU, and a powerful platform that makes this process very easy is *Ollama*, with which you can run an LLM on your laptop or desktop computer without the need for an internet connection. An alternative provider is LM Studio (*https://lmstudio.ai/*).

Local use of voice models provides privacy and full control by allowing you to interact with an LLM directly on your hardware. First, you must install the Ollama software locally on your computer by visiting *https://ollama.com/*, as shown in Figure 9.3.

Figure 9.3 Ollama Software Download

Here, you can download the software that matches your operating system. Ollama offers the software for macOS, Linux, and Windows. Once you've done this, you need to find out which model is suitable for your hardware and project requirements. You can find a list of available models at *https://ollama.com/library*. In this section, we'll work with gemma3. Figure 9.4 shows the gemma3 model class.

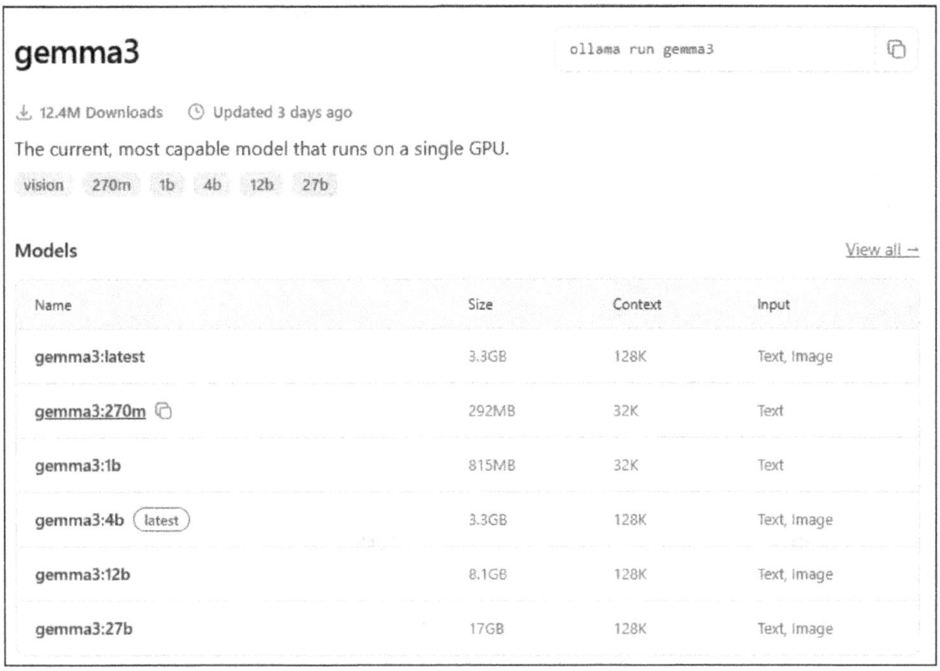

Name	Size	Context	Input
gemma3:latest	3.3GB	128K	Text, Image
gemma3:270m	292MB	32K	Text
gemma3:1b	815MB	32K	Text
gemma3:4b (latest)	3.3GB	128K	Text, Image
gemma3:12b	8.1GB	128K	Text, Image
gemma3:27b	17GB	128K	Text, Image

Figure 9.4 Ollama: Using gemma3 Model Class

This is a model with relatively few model parameters. It is provided by Google but is nevertheless very powerful. There are several different variants, from a tiny 270 million (270m) parameter model to a large 27 billion (27b) parameter model. If you click on the model version (which is actually a link), you'll find more information (e.g., the file sizes of the models, the actual name of the model).

A special feature of this model class is that it can also be multimodal. Look at the last column, where you'll see that models from 4b onward can process both text and images. We'll use the model with the "`gemma3:4b`" name, and Listing 9.10 shows how you can download a model by retrieving it via Ollama by executing the initial command.

```
ollama pull gemma3:4b

pulling manifest
pulling aeda25e63ebd: 100% ????????? 3.3 GB
pulling e0a42594d802: 100% ????????? 358 B
pulling dd084c7d92a3: 100% ????????? 8.4 KB
pulling 3116c5225075: 100% ????????? 77 B
pulling b6ae5839783f: 100% ????????? 489 B
verifying sha256 digest
writing manifest
success
```

Listing 9.10 Ollama: Model Download

At this point, the model has been downloaded to your hard disk and is available. You can check this via the following:

```
ollama list
```

```
NAME ID SIZE MODIFIED
gemma3:4b a2af6cc3eb7f 3.3 GB 2 minutes ago
```

The last step at the operating system level is to add the langchain-ollama Python package, which you can add via the following:

```
uv add langchain-ollama
```

You can also add it using pip via the following:

```
pip install langchain-ollama
```

This completes the preparations. Now, we can interact with the local model directly from a Python script.

In our Python script, which you can find at *100_LLMs\40_ollama.py*, the packages must first be imported. As in the previous section, we'll use the model multimodally and require the base64 package to encode the image. We access the model via ChatOllama, and we pass the user request as a HumanMessage (more on messages in Section 9.4), as follows:

```
import base64
from langchain_ollama import ChatOllama
from langchain_core.messages import HumanMessage
```

Now, we must define some variables that we'll access later. The model's name (MODEL_NAME) refers to gemma3, and the actual request consists of a question in USER_PROMPT as well as the image that we pass with its path IMAGE_PATH, as follows:

```
MODEL_NAME = "gemma3:4b"
USER_PROMPT = "What does this image show? Answer in a short paragraph!"
IMAGE_PATH = "TrainingProcess.png"
```

We can't pass the image as a path but must encode it with base64. For this, we use the encode_image auxiliary function, to which we pass the path to the image and from which we receive the encoded image back (see Listing 9.11).

```
def encode_image(image_path: str) -> str:
    with open(image_path, "rb") as image_file:
        return base64.b64encode(image_file.read()).decode("utf-8")

base64_image = encode_image(IMAGE_PATH)
```

Listing 9.11 Ollama: Image Encoding

Interaction with the model takes place via a model instance model, which we create with ChatOllama, as follows:

```
model = ChatOllama(model=MODEL_NAME, temperature=0.2)
```

We must transfer the request, which consists of a text and an image, via a message object. We create the object with HumanMessage, and it consists of the text and the image (see Listing 9.12).

```
message = HumanMessage(
    content=[
        {"type": "text", "text": USER_PROMPT},
        {"type": "image_url", "image_url": f"data:image/png;base64,{base64_im-
age}"},
    ]
)
```

Listing 9.12 Ollama: Message Object

Everything is prepared so that we can now transfer the data to the model, and to do this, we use invoke again, as follows:

```
res = model.invoke([message])
```

The nice thing about the model integrations via LangChain is that they all follow the same scheme. We can therefore access the model response via the following content property, as in the previous sections:

```
res.content
```

```
The image represents a process for evaluating language models. It starts with
data that is fed into a pre-trained model. This model then produces high-quali-
ty output that is aligned with human preferences. Finally, this output is evalu-
ated against benchmarks to assess the quality and appropriateness of the model.
It is an iterative process that aims to develop models that are not only techni-
cally capable but also aligned with human values and expectations.
```

Isn't that great? You can use an LLM, even with confidential information, without revealing any data over the internet.

Now, we come back to cover a little more theory on how to control the behavior of the model. In the following section, you'll get to know the most important model parameters.

9.2 Model Parameters

There are some very important parameters that you can adjust to control the generated model outputs. In Section 9.2.1, we'll cover the temperature parameter, and in Section 9.2.2 we'll cover the top-p and top-k parameters. All three of them play an important role, and with their help, you can control the creativity, randomness, and focus of the generated outputs. In Section 9.2.3, I provide some best practices for these parameters.

9.2.1 Model Temperature

You can use the model temperature to control the randomness of the results. Typical values are 0 (low temperature) and 1 or even higher (high temperature). Low temperatures keep the model very focused and provide more deterministic results, which means you get the same answer over and over again.

The model prefers extremely probable tokens, but high temperatures increase the randomness of token selection. They cause the model to select a wider distribution of tokens, allowing for more creative or unexpected outputs. Temperatures should not normally exceed 1 because that can lead to chaotic and incoherent outputs. Figure 9.5 shows the influence of temperature on the model result.

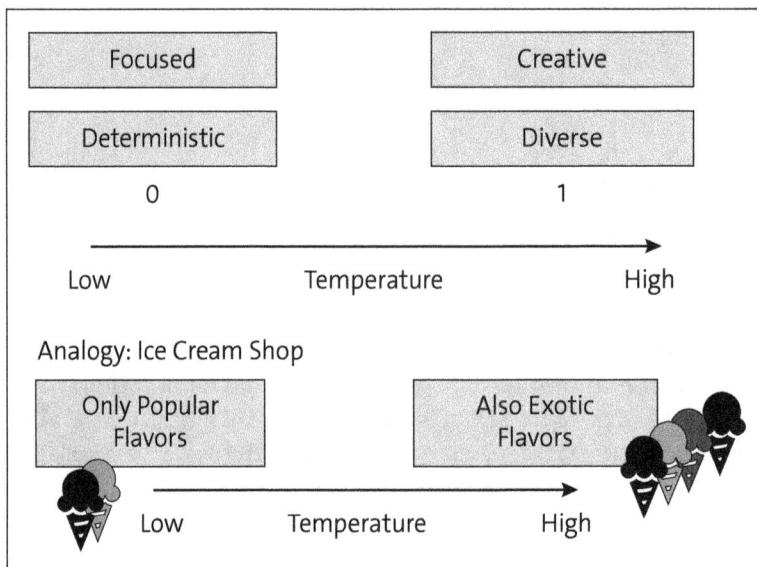

Figure 9.5 Model Parameter: Temperature Analogy

At low temperatures, you get very focused and deterministic answers, whereas at higher temperatures, the answers become more creative and varied. Imagine you own an ice cream parlor. When the (ambient) temperature is low, fewer customers come to your store and it might be a good business decision to offer only the most popular flavors. But when the temperature rises, demand increases and it's a good decision to offer more exotic flavors.

The temperature is directly linked to the probability distribution of the tokens. Let me explain this with an artificial example. You have the "Bert likes <MASK>." prompt, and the model has the task of filling in the missing word. There are a huge number of possible words. To simplify, we'll use and show only three words: read, run, and program.

The model has an underlying probability for these words based on its training data. At very low temperatures, the model amplifies the differences between the probabilities, and at very high temperatures, these differences disappear and all words have the same probability. Figure 9.6 shows the example probability distributions for a given user request and different temperature values.

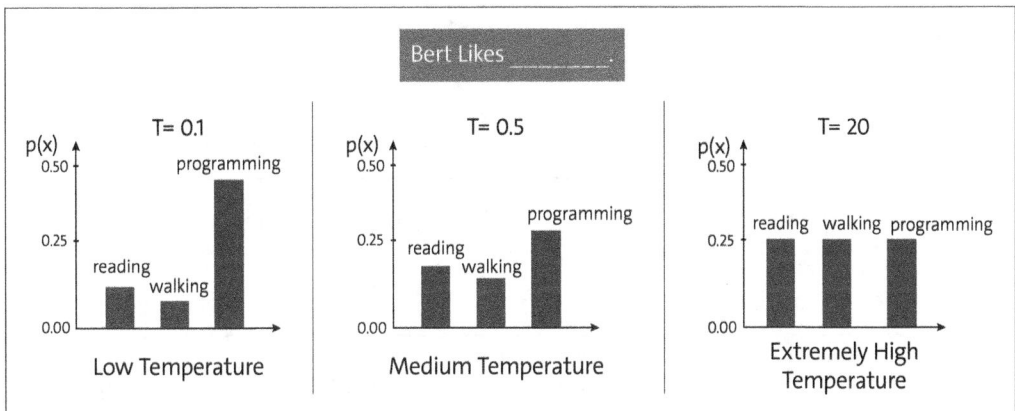

Figure 9.6 Model Parameter: Temperature and Probability Distribution

The diagram on the left shows a low temperature of 0.1, for example, the middle diagram shows the influence of a medium temperature on the probability distribution, and the diagram on the right shows an extremely high temperature.

Can you see how the differences become smaller as the temperature rises? This is exactly what this parameter does. Now, let's look at two other parameters that work together with the temperature: top-p and top-k.

9.2.2 Top-p and Top-k

Top-p sampling (also known as *nucleus sampling*) controls the probability of considering the next token. It does this by dynamically adjusting the number of possible tokens from which the model can choose. Let's take an example with top-p = 0.9. In this case, the model considers a cumulative probability of tokens that adds up to 90% and chooses the smallest set of tokens within these limits. This approach balances deterministic and creative outputs. If you set top-p = 1, then there's no filtering, and effectively, the model considers all possible tokens. If you define a small value such as top-p < 0.5, then the outputs of the model tend to be more focused and predictable because the model considers only the best tokens.

Top-k sampling controls how many of the most probable tokens the model considers when it generates the next word. If you select a value of top-k = 1, then the model will select only the most probable token. This gives you a completely deterministic result. On the other hand, if top-k = 50, then the model samples from the 50 most likely tokens for each step. This increases variety and allows for more creative and varied outputs. Top-k specifies a fixed number of tokens to choose from, regardless of the cumulative probability represented by the top-k tokens.

Figure 9.7 shows an example of top-p and top-k parameters.

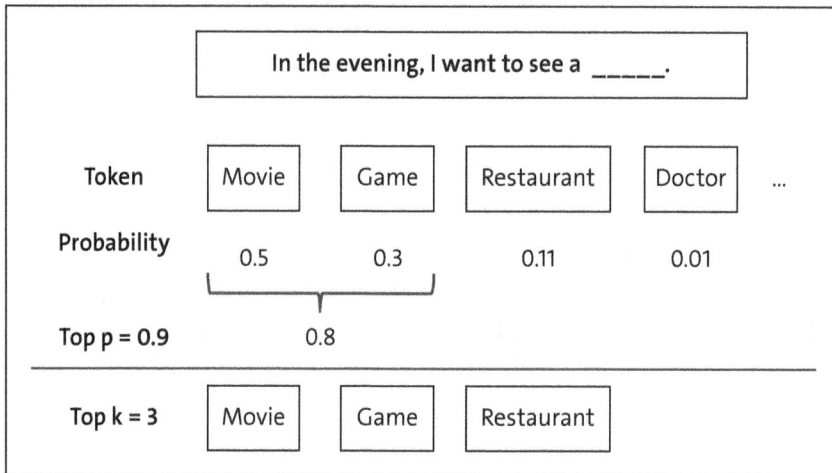

	In the evening, I want to see a _____.				
Token	Movie	Game	Restaurant	Doctor	...
Probability	0.5	0.3	0.11	0.01	
Top p = 0.9		0.8			
Top k = 3	Movie	Game	Restaurant		

Figure 9.7 Model Parameter: Top-p and Top-k Parameter

The example user prompt is, "In the evening, I want to see a <MASK>", where the gap (or MASK) is to be filled in. There are again several possible tokens.

We can calculate the probabilities of these tokens based on a specific temperature. The tokens are displayed in descending order, and the model considers all probabilities that add up to less than top-p. In this case, *movie* and *game* have a probability of 80%. If we were to add the next token (*restaurant*), the aggregated probability would be 91% and therefore above the top p value.

Top-k is simple. Here, top-k is set to 3, so the model selects the three most likely tokens and chooses the final prediction from them at random.

9.2.3   Best Practices

The balance between these parameters that you should choose depends on the area of application or the task you're trying to perform.

- For creative writing, you should choose a higher temperature (0.8–1.0) combined with a moderate top-p (0.9–1.0) and top-k (50–100) to explore a variety of creative outputs.

- When generating code, you'll want to obtain more reliable snippets and therefore choose a low temperature (0.1–0.3) with a small top-k (10–20) and top-p (0.7–0.9) to ensure syntactically correct output.

- In customer service or chatbot applications, the model output must be reliable, focused, and consistent, so you'll want to choose a low temperature of 0.2–0.4 to ensures predictable responses. You'll want to choose a top-p of 0.7–0.9 to allow the model to select highly probable tokens but retain some flexibility to make the interaction natural and thus avoid robotic responses. You'll also want to choose a top-k range of 20–50 to keep the model focused on relevant responses.

We've seen which parameters we can use to influence the model response, so now, let's look at which parameters we should pay attention to in order to select the right model.

9.3   Model Selection

In the previous section, we had our first interactions with different LLMs. We presented a selection of models, but you may have wondered about how to choose the right model for your task. Depending on your project, there are hard and soft criteria that you need to consider, as follows:

- If you want to process long input prompts, the *context window* (Section 9.1.2) is a crucial factor.

- If you want to interact with a model that takes current developments and trends into account, the model *cutoff date* could be extremely important.

- Other important parameters are *cost*, *latency*, and *performance*.

In the following sections, we'll look at each of these topics in addition to considerations about on-premise versus cloud hosting and open-source, open-weight, and propriety models. Let's first look at the performance of the model.

9.3.1   Performance

You can check the performance of different models on the *LMArena* leaderboard (*https://lmarena.ai/leaderboard*) and get a result that looks like Figure 9.8.

Leaderboard Overview

See how leading models stack up across text, image, vision, and beyond. This page gives you a snapshot of each Arena, you can explore deeper insights in their dedicated tabs. Learn more about it here.

📖 Text				🖥 WebDev			
Rank (UB)	Model	Score ↓	Votes	Rank (UB)	Model	Score ↓	Votes
1	G gemini-2.5-pro	1452	61,890	1	GPT-5 (high)	1473	8,004
1	A\ claude-sonnet-4-5-20250929-t...	1448	12,986	1	A\ Claude Opus 4.1 thinking-16k...	1458	8,726
1	A\ claude-opus-4-1-20250805-thi...	1448	28,595	2	A\ Claude Opus 4.1 (20250805)	1451	8,986
2	gpt-4.5-preview-2025-02-27	1442	14,644	4	A\ Claude Sonnet 4.5 (thinking ...	1420	4,863
2	A\ claude-opus-4-1-20250805	1439	41,049	4	MiniMax-M2	1405	3,515
4	chatgpt-4o-latest-20250326	1438	47,608	5	G Gemini-2.5-Pro	1399	14,628
4	gpt-5-high	1437	30,138	5	✗ GLM-4.6	1395	7,563
2	A\ claude-sonnet-4-5-20250929	1436	5,484	5	DeepSeek-R1-0528	1393	4,800
4	o3-2025-04-16	1433	58,575	6	A\ Claude Sonnet 4.5	1387	7,855
4	qwen3-max-preview	1433	25,067	7	A\ Claude Opus 4 (20250514)	1383	9,238
	View all				View all		

Figure 9.8 LMArena Leaderboard, Snapshot of 2025-11-09 (Source: https: //lmarena.ai/ leaderboard)

The models are sorted according to their arena score, but how do we evaluate this arena score? It's not called an arena (*https://lmarena.ai/*) for nothing. In the arena, the user interacts with two models: Model A and Model B.

The user can define a prompt and receives the answers from the two models, which the user must then evaluate to find out which performs better. Thus, we have a double-blind test setting, which is considered the gold standard in the evaluation of test results. In the screenshot, you can see that several models share the same rank, and that's because the model considers the 95% confidence interval. The ranks change often, so your ranking will probably look increasingly different as time passes.

The main **Text** and **WebDev** categories are displayed at the top, but you can select other categories such as **Vision**, **Text-to-Image**, **CoPilot**, or **Search** and check the ranking.

However, performance is not the only relevant factor.

9.3.2 Cutoff Date

Each model has a *knowledge cutoff date*. This means that the training data was finalized on a specific date, the model is trained on this data, and no more recent data can be represented in the model weights. It's therefore important to know the cutoff date, because if you ask a model for information, such as an event or other fact, the model can't know if it happened after the cutoff date.

For chatbots, this parameter is becoming less relevant as these models increasingly have the ability to search the internet for up-to-date information. But for you as a developer of AI systems, this could be an important factor that you'll want to consider.

9.3.3 On-Premise Versus Cloud Hosting

Another important aspect when choosing a model is data protection. If you work with confidential information, you or your customers may not want the data to leave the company network. It's also important to know what data the model has been trained on and whether it's GDPR compliant.

If you've selected a local model taking these parameters into account, you can use it in your own network without risk and assume that your data won't leave your network.

9.3.4 Open-Source, Open-Weight, and Proprietary Models

There are proprietary models that are made available to users via web applications or APIs. A well-known representative of this class is Anthropic, which usually offers its models in this way.

Google and OpenAI handle this differently. Their models are provided as proprietary models via APIs (e.g., Gemini, GPT 5), but other model classes such as Gemma or GPT-OSS are offered as open-weight models.

To be completely correct, we should distinguish between open-source and open-weight. Real open-source models include all details such as model architecture and training data used, but most open-source models don't include these details. The provider publishes the trained model with its weights for the public, but specific details about the underlying data and training details remain secret. A well-known example from this group is Meta with its Llama model family. The public can use these models free of charge, but the company keeps the details of the training data secret.

9.3.5 Cost

The cost of using an LLM service can be a decisive factor for you when choosing models. Typically, proprietary models are billed on a token basis. To be precise, a distinction is made between input tokens and output tokens: input tokens are usually cheaper than output tokens. The current prices for OpenAI models can be found at *https://openai.com/api/pricing/*, and the current prices for Anthropic models can be found at *https://www.anthropic.com/pricing#anthropic-api*.

You should estimate how many API requests and how many tokens will be processed, and based on that, you can create an estimate of your total costs.

9.3.6   Context Window

Your project may involve processing very long documents, and you may need to pass as much information as possible to the model. Therefore, the context window is a decisive factor for the best choice of model.

If you look at the models on Groq (*https://console.groq.com/docs/models*), for example, you'll find models with rather small context windows, such as LlaVa 1.5 7B (with a context window of 4,096 tokens) and Kimi K2 0905 (with an extremely large context window of 250,000 tokens).

9.3.7   Latency

Some use cases require very fast model responses, and there's a dependency on the provision of a model or how long it takes for the response to be provided (the time to first token). If latency is not an issue, you could even run an open-source model on a CPU, but in other cases, latency could be the most important factor.

For example, if you want to pair an LLM with voice generation to enable real-time chats, then an LLM can easily become a bottleneck and affect the user experience because there's no "natural" conversation if the other party has long response times.

9.4   Message Types

In Section 9.1, we took our first steps with LLMs. We called the model objects, sent a simple message, and received a reply. But in a more realistic chat, there are different types of messages, and each message has a specific role and content. In the following sections, we'll look at the most common message types: user/human messages (in Section 9.4.1), system messages (in Section 9.4.2), and assistant messages (in Section 9.4.3).

9.4.1   User/Human Messages

This message type refers to the human message and represents the user's input. The effectiveness of an LLM response depends on the clarity of the user message, and an entire field called *prompt engineering* is basically concerned with optimizing the user message. The user message defines the task that the LLM is expected to perform, such as answering a question or summarizing a text. Alongside the task, you can add contextual information here to provide guidance to the model. The structure of a user/human message can range from a single word or sentence to a multiparagraph document.

9.4.2   System Messages

In addition to the user input, we can define a system message to define how the model should behave and work, like in a role-playing game. For example, if you want to set up a general assistant, a typical system message could look like this: "You are a helpful AI assistant designed to provide accurate, concise, and polite responses. Always ensure that your answers are clear and informative."

Imagine you want your model to behave like a technical support assistant. You could instruct the model with the following system message: "You are a technical support AI assistant specializing in troubleshooting and explaining software-related issues. Respond with clear, step-by-step instructions, avoiding technical jargon whenever possible."

With the system message, you define the initial instructions for the model. You define its role, tone, and specific goals before the interaction with the user begins. The system message is crucial for setting the boundaries and expectations of the model, and it helps to guide the model so that it behaves in accordance with the user's requirements.

However, system messages have their limitations. While they can shape initial model behavior, they can't enforce strict compliance throughout the conversation. This means that models can "drift" in their tone or behavior when confronted with unexpected input from the user.

Moreover, system messages alone can't enforce fine-grained control over content accuracy or ethical considerations without complementary guardrails or moderation.

9.4.3   Assistant Messages

The message type for the wizard corresponds to the model's response, and the main property is the content that contains the output of the model. There's also a property called `response_metadata`, which contains some model-specific output: typically, the token usage and the duration of the query.

Now that we understand the available message types, let's find out how we can use them in prompts. LangChain offers a very flexible interface to set up prompts: prompt templates.

9.5   Prompt Templates

Before we call the LLM and send a request, we set up a prompt in a consistent and structured way by using LangChain's prompt templates. This allows us to instruct the model how to act. Also, with the help of the prompt templates, the model can better understand the user and their intentions. Let's see this in action.

9.5.1   Coding: ChatPromptTemplates

The most flexible prompt template option is to use ChatPromptTemplates, which allows you to pass a list of messages. In Listing 9.13, you can see an example of the code to set up a prompt template.

Let's start with a simple example that illustrates this idea. First, we need to import the ChatPromptTemplate class, and then, we create an instance of this class by calling the from_messages() method. These messages are a list of tuples, and each tuple has the form ("message type", "content"). So, we define a system message that tells the model how to behave and a human message that contains the actual user request. What is important here is how we define variables that are set up as placeholders and filled later. In the example, the variables in the user or human messages are enclosed in curly brackets and we set up input and target_language as variables.

Although the message looks a bit like a Python f-string, it's not the same. In an f-string, predefined variables are passed and replaced by the string representation of the variables. But here, we haven't predefined any variable input or target_language in advance.

In the last step, we call the prompt template and replace the variables with the actual content. To do this, we simply need to call the invoke() method of the prompt_template object. We pass a dictionary as a parameter, which uses keys that correspond to the variables, and the values correspond to the content that we will use.

Finally, since we don't store the invoke prompt in a new variable, the output will be simply displayed in the terminal. The prompt template has been converted into a ChatPromptValue object with a SystemMessage and HumanMessage.

```
#%% packages
from langchain_core.prompts import ChatPromptTemplate


#%% set up prompt template
prompt_template = ChatPromptTemplate.from_messages([
    ("system", "You are an AI assistant that translates English into another
language."),
    ("user", "Translate this sentence: '{input}' into {target_language}"),
])


#%% invoke prompt template
prompt_template.invoke({"input": "I love programming.", "target_language":
"German"})
ChatPromptValue(messages=[
SystemMessage(content='You are an AI assistant that translates English into an-
other language.'),
HumanMessage(content="Translate this sentence: 'I love programming.' into
German")])
```

Listing 9.13 Prompt Template: Setup of Prompt Template

What is the purpose of this approach? With the prompt template, we have a flexible first component that we can pass to a model to get an answer. The "Prompt to LLM" approach is a simple sequence of steps.

We'll look at long-chain chains in Section 9.6. But before that, let's use the wisdom of the crowd to develop a good prompt. LangChain has created an ecosystem that allows users to share and explore prompts with LangChain Hub.

9.5.2 Coding: Improving a Prompt with LangChain Hub

You can find the LangChain Hub at *https://smith.langchain.com/hub*, where you can view prompts that have been created by others for various purposes. From there, you can get help with the creation of a prompt. If you search for "prompt maker," you'll find the hardkothari/prompt-maker prompt, which was created to generate a more detailed prompt. In our example, we'll find out how this works. The code in Listing 9.14 corresponds to the *100_LLM\60_prompt_hub.py* file. We need to load the required packages, and the newcomer here is hub from the langchain package.

```
from langchain import hub
from langchain_openai import ChatOpenAI
from langchain_core.output_parsers import StrOutputParser
from dotenv import load_dotenv
load_dotenv('.env')
from pprint import pprint
```

Listing 9.14 Prompt Hub: Required Packages

To use prompt creation, we need to call the pull method of hub, as follows:

```
#%% fetch prompt
prompt = hub.pull("hardkothari/prompt-maker")
```

There are some input variables that we can access via the input_variables property, as follows:

```
#%% get input variables
prompt.input_variables
```

['lazy_prompt', 'task']

Next, in Listing 9.15, we create an improved prompt. We need to set up a model and run it in a chain, and even though that runs over into the subject matter area of the next section, bear with us. We'll look at chains in more detail there.

```
# %% model instance
model = ChatOpenAI(model="gpt-4o-mini",
                   temperature=0)
# %% chain
chain = prompt | model | StrOutputParser()
```

Listing 9.15 Prompt Hub: Chain Setup

We call the chain in Listing 9.16 and pass the relevant parameters: lazy_prompt and task to get an improved prompt.

```
# %% invoke chain
lazy_prompt = "summer, vacation, beach"
task = "Shakespeare poem"
improved_prompt = chain.invoke({"lazy_prompt": lazy_prompt, "task": task})
print(improved_prompt)
```

As a skilled poet in the style of William Shakespeare, compose a sonnet that captures the essence of summer, vacation, and the beach. ### Instructions: Your poem should reflect the beauty and joy of a summer getaway, using rich imagery and evocative language typical of Shakespearean verse. Aim for 14 lines, adhering to the traditional iambic pentameter and ABABCDCDEFEFGG rhyme scheme. ### Context: Incorporate themes of nature, leisure, and the fleeting nature of time, while evoking a sense of nostalgia and warmth. Use metaphors and similes to enhance the emotional depth of the poem and consider including references to the sun, the sea, and the carefree spirit of summer. ### Desired Outcome: The final piece should resonate with readers, transporting them to a sun-drenched beach, filled with laughter and the gentle sound of waves, while also reflecting on the transitory beauty of such moments.

Listing 9.16 Prompt Hub: Chain Invocation

This improved prompt describes the role of the model in much more detail and gives comprehensive instructions for the task as well as a description of the desired result. Let's run the model with the improved prompt in Listing 9.17.

```
res = model.invoke(improved_prompt)
print(res.content)
```

Upon the golden sands where sunbeams play,
The azure waves do kiss the shore with glee,
In summer's arms, we cast our cares away,
As laughter dances on the breeze so free.

The sun, a sovereign in the heavens high,
Doth paint the sky with hues of rose and gold,

While seagulls wheel and cry, as if to vie
For joy that in this fleeting time we hold.

O, let us linger in this warm embrace,
Where time, like grains of sand, slips through our hands,
Each moment savored, sweet as summer's grace,
A tapestry of dreams upon the strands.

Yet, as the twilight calls the day to rest,
We cherish memories, our hearts possessed.

Listing 9.17 Prompt Hub: Chain Output for Improved Prompt

I leave it to you to run the model based only on the lazy_prompt and the task and compare the two results. You can find my example solution in the script.

At this point, we've seen a small example of how a chain combines different building blocks. This concept is very powerful due to its structure, as you'll see in the next section.

9.6 Chains

Chains are such an important concept that they are part of the package called Lang-Chain. A *chain* is a sequence of process steps that are linked together to accomplish a task, and typically, they consist of several components. We'll start with the smallest and simplest chain.

9.6.1 A Simple Sequential Chain

The simplest chain could be a "Prompt to LLM" chain, as shown in Figure 9.9. In it, a user input is passed to a prompt template, the prompt template itself passes its output to an LLM step, and the LLM step generates a model output.

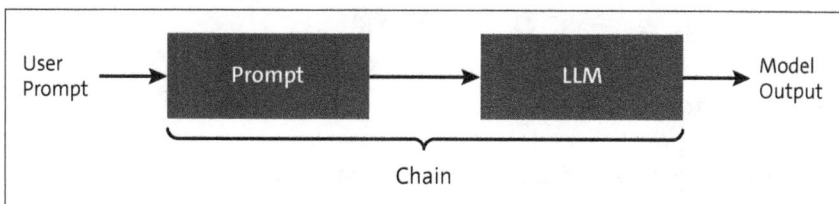

Figure 9.9 Simple Sequential Chain

But you're not limited to using only sequential chains. Figure 9.10 shows how you can also use more complex structures such as parallel chains (on the left) and router chains (on the right).

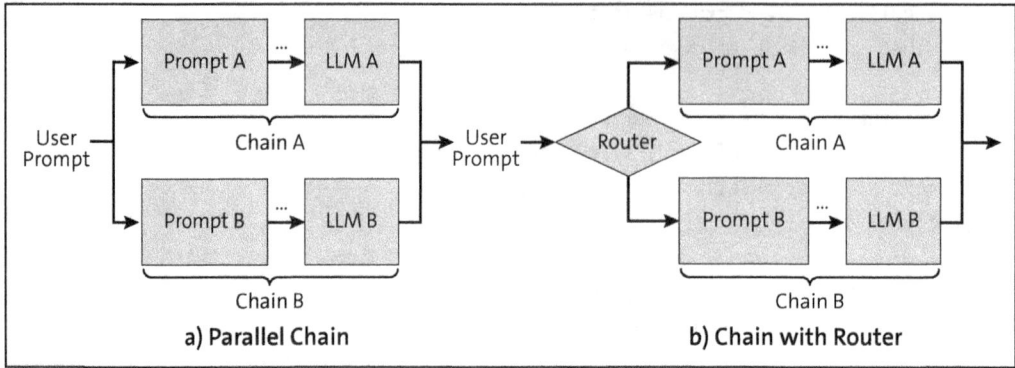

Figure 9.10 More Complex Chain Setups: Parallel (on the Left) and Router (on the Right)

9.6.2 Coding: A Simple Sequential Chain

We'll set up a chain that consists of a prompt template. The prompt template is passed to an LLM, and after the model has created an output, the output is passed to a StrOutputParser. The corresponding code can be found at *100_LLM/70_simple_chain.py*.

In Listing 9.18, you'll start by importing relevant packages and API keys. You use OpenAI as the LLM provider, and you load the API key via dotenv. Please make sure to save an *.env* file containing an entry for OPENAI_API_KEY in the working folder.

```
from langchain_openai import ChatOpenAI
from langchain_core.prompts import ChatPromptTemplate
from dotenv import load_dotenv
from langchain_core.output_parsers import StrOutputParser
load_dotenv('.env')
```

Listing 9.18 Sequential Chain: Required Packages

In Listing 9.19, we extend the prompt template from Chapter 1.6 and create a ChatPromptTemplate based on a system and user message. The task is to translate an input text into a target language (target_language).

```
#%% set up prompt template
prompt_template = ChatPromptTemplate.from_messages([
    ("system", "You are an AI assistant that translates English into another
language."),
    ("user", "Translate this sentence: '{input}' into {target_language}"),
])
```

Listing 9.19 Sequential Chain: Prompt Template Creation

The next component we use is an LLM. We use the GPT-4o-mini model and create the model instance model with ChatOpenAI, as follows:

```
model = ChatOpenAI(model="gpt-4o-mini",
                   temperature=0)
```

Connecting the chain elements couldn't be easier. We only need to use a pipe operator (|) that separates the components. In our example, the prompt is the first component in the chain, the model comes next, and the model output is then passed to StrOutputParser, which parses the model output into the most likely string, as follows:

```
# %% chain
chain = prompt_template | model | StrOutputParser()
```

Everything is prepared so that we can call the chain with input parameters. This gives us the following results:

```
# %% invoke chain
res = chain.invoke({"input": "I love programming.", "target_language": "Ger-
man"})
res
```

The translation of "I love programming" into German is "Ich liebe Programmieren."

Now that we've covered a simple sequential chain, we could aim higher and build much more complex constructs like routers or parallel chains. I'll leave that to you for self-study because next, I want to focus on a capability of language models that is extremely valuable: structured output.

9.7   Structured Outputs

In Section 9.7.1, we'll gain an understanding what structured outputs are, and in Section 9.7.2, we'll implement them in some coding examples.

9.7.1   What Are Structured Outputs?

Language models are very good at generating free text, and that's great for creating stories, sending emails, or even generating code. However, sometimes, a specific format is required—for example, for data capture, process automation, or integrations with other systems—whenever the language model is not at the end of the process but the output of the language model is to be used by other tools or systems. This is exactly where structured outputs come into play.

A *structured output* is a model response that is output not as free text but in a specific, machine-readable format. The most common formats are JSON and XML, and others are CSV and simple lists.

The major advantages of structured outputs are as follows:

- There's seamless communication between language models and other software systems.

- Structured output can be used to automate workflows. For example, an LLM could extract product details from free text and then feed them into an e-commerce system.

- This avoids ambiguities that could arise with free text. Since the model is forced to adhere to a clearly defined schema, the results become more consistent.

Let's take a look at how we can implement structured outputs.

9.7.2   Coding: Structured Outputs

Have you ever been in a situation where you can remember the plot of a movie but you can't remember the title or the actors? You can solve that problem by building a chain as shown in Figure 9.11.

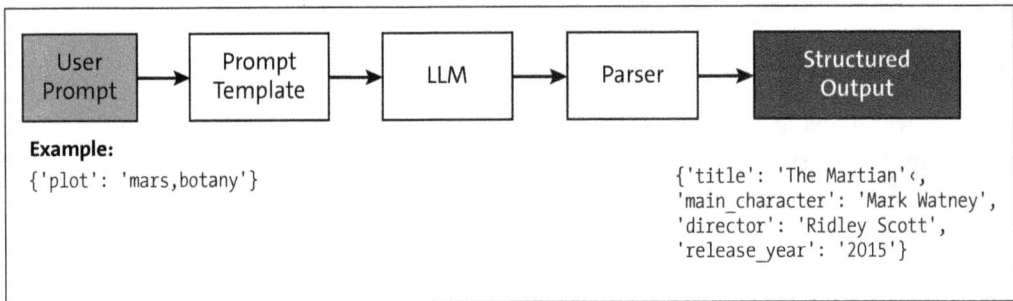

Figure 9.11 Chain for Structured Output

In this chain, the user passes the framework action to the chain, which consists of a prompt template, a model, and an output parser. As a result, we receive the structured output in JSON format. Such a workflow is easy to implement, and we use the code at *100_LLM\80_structured_outputs.py* for it.

We first load the required packages in Listing 9.20. The decisive new functionality comes via the parser, which postprocesses the result of the model. We use PydanticOutputParser for this, and it's connected to the pydantic package, from which we load the BaseModel class.

```
from langchain_core.prompts import ChatPromptTemplate
from langchain_groq import ChatGroq
from dotenv import load_dotenv, find_dotenv
load_dotenv(find_dotenv(usecwd=True))
```

```
from langchain_core.output_parsers import PydanticOutputParser
from pydantic import BaseModel
```

Listing 9.20 Structured Outputs: Required Packages

We expect a very specific format for the output, which we define using our own `MyMovieOutput` class. This class inherits from `BaseModel`, and in it, we define the keys and the data type of the values of the JSON object that is to be returned to us. For example, we can define that the `title` is expected as a string, as follows:

```
#%% pydantic model
class MyMovieOutput(BaseModel):
    title: str
    main_character: str
    director: str
    release_year: str
```

We then pass this output format to the `PydanticOutputParser`, which allows the model to be given clear instructions regarding the output, as follows:

```
# %% prompt
parser = PydanticOutputParser(pydantic_object=MyMovieOutput)
```

We define the model behavior in the messages via the system prompt, and we also transfer the format instructions here. We then store the actual user request in the user prompt, as follows:

```
messages = [
    ("system", "Du bist ein Filmexperte. {format_instructions}"),
    ("user", "Handlung: {plot}")
]
```

Now, we come to the first module of our chain: the `prompt_template`, which we create based on the messages. A new feature here is the `partial()` method, which has the purpose of passing the `format_instructions` parameter to the prompt template. We don't have to write the format instructions ourselves but can fall back directly on the instructions of the parser, as follows:

```
prompt_template = ChatPromptTemplate.from_messages(messages).partial(
    format_instructions=parser.get_format_instructions()
)
```

We create the model instance in the classic way with `ChatGroq`. It makes sense to give the model a low temperature, as follows:

```
MODEL_NAME = "meta-llama/llama-4-scout-17b-16e-instruct"
model = ChatGroq(model=MODEL_NAME, temperature=0.2)
```

Temperature for Structured Outputs

An important note at this point is that for structured outputs, the model should be less creative and work in a more deterministic manner. Therefore, we should set the temperature low (for example, between 0 and 0.3).

Now, we can create the chain, which consists of a sequence of prompt_template, model, and parser, as follows:

```
chain = prompt_template | model | parser
```

At this point, everything is ready for our first test. We define the input as a dictionary in the chain_inputs object and pass it to the chain by using invoke, as follows

```
chain_inputs = {"plot": "mars, botanik"}
res = chain.invoke(chain_inputs)
```

We can retrieve the result in Listing 9.21 by using the model_dump() method.

```
res.model_dump()
```

```
{'title': 'The Martian',
 'main_character': 'Mark Watney',
 'director': 'Ridley Scott',
 'release_year': '2015'}
```

Listing 9.21 Structured Outputs: Model Output

As desired, the result is a JSON object with the specified keys and the correct values.

There's so much more to say about language models and generative AI in general, so at this point, I'd like to refer you to my book *Generative AI with Python* (*https://www.sap-press.com/generative-ai-with-python_6057/*), which is dedicated to these topics in great detail.

The topic of LLMs is technologically inseparable from transformer architecture, so in the next section, we'll take a technical deep dive into this technology.

9.8   Deep Dive: How Do Transformers Work?

Figure 9.12 illustrates the basic structure of a transformer model, which consists of some fundamental building blocks that we need to understand well.

At the lowest level is the text that is passed to the model, which in our example is the sentence "PyTorch is fun." First, the text is first tokenized (i.e., broken down into individual parts). More on that in Section 9.8.1.

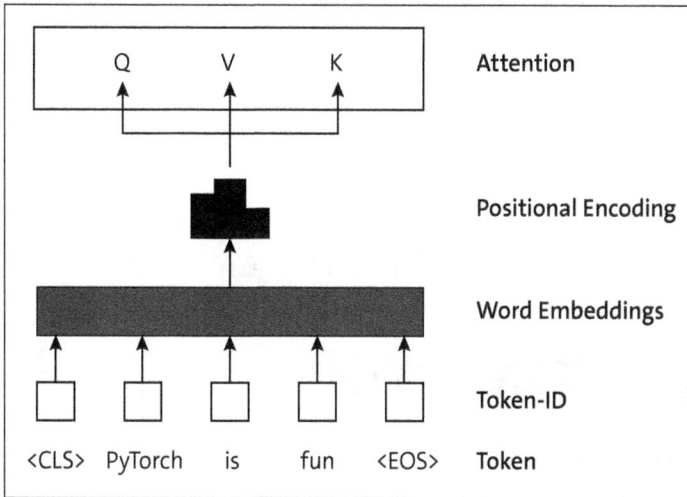

Figure 9.12 Structure of Transformer Model

For the sake of simplicity, we'll assume at this point that the text has been broken into individual words. The words are passed through an initial network module that determines word embeddings. These word embeddings are vectors that reflect the meaning of each word. Further details on this can be found in Section 9.8.2.

The next step is positional encoding. Imagine the sentence "The ball hits the boy," and then imagine the sentence "The boy hits the ball." This shows that the order of the words plays a role: the two sentences use the same words, but the sentences' meanings are completely different due to the different positions of the words. With positional encoding, the model receives information about which word can be found in which position. More on that in Section 9.8.3.

In the next step, the data is fed into a module called `attention`. This is the actual core of the model because it's where the model learns the inner structure of the sentence and which word is related to which other word. We'll look at this concept in more detail in Section 9.8.4.

9.8.1   Tokenization

Tokenization simply means the breaking down of text into smaller units called tokens. The *tokens* can be individual words, parts of sentences, or even individual characters, depending on the tokenization method used. In today's common language models, we use subword tokenization.

Each token is then assigned a unique, numerical value: the token ID, which is a kind of dictionary entry whereby the dictionary enables a mapping between human words and assigned numerical values.

Figure 9.13 shows how the tokenizer works. It processes text input from the user, extracts the individual tokens, and returns the tokens with their corresponding token IDs.

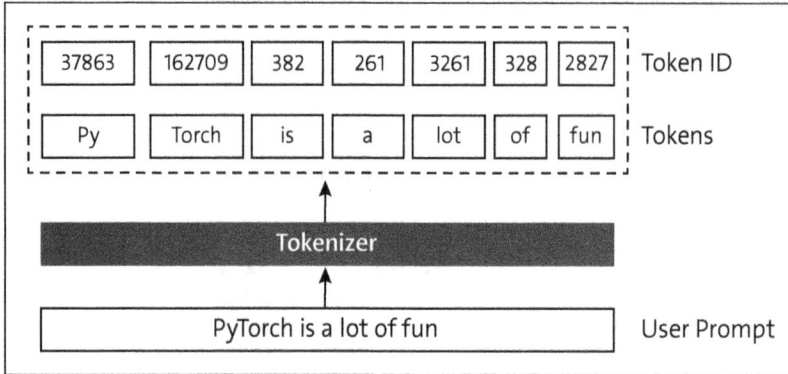

Figure 9.13 Tokenization

In the example, we used the tokenizer from GPT-4o and GPT-4o mini. You can test this yourself with other inputs if you use the OpenAI tokenizer at *https://platform.openai.com/tokenizer*.

This tokenizer is based on subword tokenization, which you can recognize by the fact that the word *PyTorch* has been split into the two tokens "*Py*" and "*Torch.*" It's also important to know that separate tokens are created for special characters. In our example, the dot was coded as a separate token.

In practice, we always talk about tokens, although the concept of words is more understandable for us. It's easy for me to count the words in this chapter, but how many tokens do they correspond to? We deal with this in the info box.

Converting Tokens into Words

Converting tokens into words is not an exact science because it depends on several factors. However, there are good heuristics that we can use, and they vary from language to language:

- In English, the common rule of thumb is that 1 token corresponds to approximately 3/4 of a word. Roughly speaking, this means that 100 tokens can describe 75 words.

- In German, the ratio is less favorable. A rule of thumb is that 1 word corresponds to an average of 2.1 tokens. This is due to the peculiarities of the German language (for example, the fact that many words are compound words), but it's also because most language models are based on English vocabulary, which means that many common English words are treated as individual tokens.

There's not just one "dictionary" that we use in tokenization. Instead, as a developer, you need to make sure to use the tokenizer that matches the model. This step is often encapsulated by the user so that they don't have to worry about it, but not always.

As deep learning models can only work with tensors and more generally with numerical values, these steps take place for all language models. The user input is tokenized, it's processed by the model, and the model output is converted back into human language.

Now, we come to the next important aspect of the transformer model: word embeddings.

9.8.2 Word Embeddings

If you think back to your math classes, you may remember the definition of a *vector*: a line with an arrow that you had to draw in a coordinate system. Back then, you had to draw these lines, which could be represented by two numbers.

A very simple example is shown in Figure 9.14, which shows living things and their position in a 2D diagram in which the two dimensions are the number of legs and the body size. By representing the information in this way, we learn about the world and the semantic meaning of words (for example, that cats and dogs are quite similar in terms of these two characteristics).

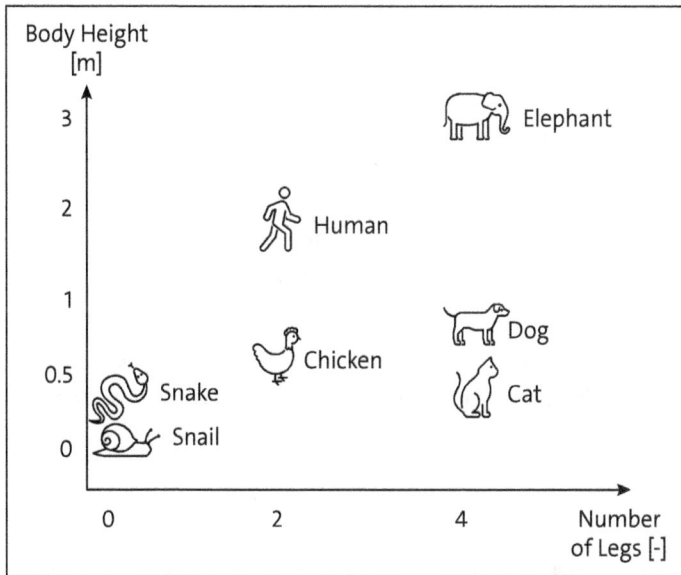

Figure 9.14 Simplified Example of Vector Space

We humans can imagine a point in a 2D space, as shown in the figure, or in a 3D space. We can imagine adding a third dimension like intelligence to the figure.

However, we can't imagine a 1536- or 3072-dimensional space. But computers can do that easily, and that's helpful because the semantic meaning of words, images, sounds, or any other kind of information can be represented in a higher dimension. The crucial aspect here is that similar concepts are closer to each other than concepts that are very different. As in Figure 9.14, we can see that cats and dogs or snakes and snails are comparable— depending on the properties we've selected. With each additional dimension, the computer algorithm gains a better understanding of the meaning of a word.

We humans use languages like English and German to communicate and clarify a concept. However, computers don't work directly with our languages but with their equivalent: numerical, so-called embedding vectors.

In our example in Table 9.1, a dog is defined by [4, 1], a cat is defined by [4, 0.5], and a human is defined by [2, 2] for the computer algorithm. The table shows an example assignment.

Human Concept	Embedding Vector
Dog	[4, 1]
Cat	[4, 0.5]
Human	[4, 2]
Elephant	[4, 3]
Snake	[0, 0.5]
Snail	[0, 0.1]

Table 9.1 Human Concepts and Their Computer Equivalents

The process of translating human texts into vectors is called *embedding*. There are different types of embeddings. We can represent individual words as well as entire text passages with an embedding vector.

At this stage, the transformer model already understands individual words, but what happens when the position of the words changes? This is where *positional encoding* comes into play.

9.8.3   Positional Encoding

Positional encoding is a crucial technique for preserving the order of words. In practice, we add a vector to the word embeddings that contains information about the absolute or relative position of the token in the input sequence. Mathematical implementation uses trigonometric functions such as sine and cosine to enable the model to generalize and learn the positions of words over the entire length of the sequence.

9.8.4 Attention

We explain attention here by using the example of *self-attention*, which calculates the similarity of word embeddings between all words and themselves.

Figure 9.15 shows the process by using the following sentence as an example: "The man ate the pizza because it smelled delicious." If we focus only on the word *it* here, what does *it* refer to? With some basic understanding of language, you could teach an algorithm that *it* refers to a noun—But here, *it* could refer to *man* or *pizza*. This is where the concept of self-attention comes into play. It allows a model to "understand" the relationship between words, so if it has been trained on enough data, it should become clear that *it* is much more likely to refer to *pizza*.

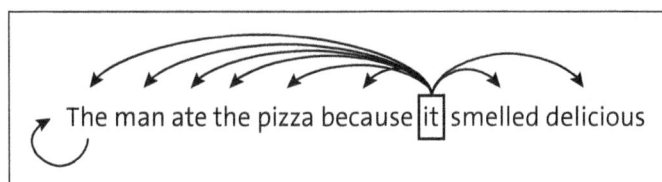

Figure 9.15 Self-Attention Process

Let's look at this in a practical example and check it using the script at *100_LLM\self_attention.py*. At the beginning of the script in Listing 9.22, we loaded the required packages for processing the models: `torch`, `AutoTokenizer`, and `AutoModel`. We'll visualize the results with `matplotlib` and `seaborn`.

```
#%% packages
import torch
from transformers import AutoTokenizer, AutoModel
import matplotlib.pyplot as plt
import seaborn as sns
import numpy as np
```

Listing 9.22 Self-Attention: Load Packages

In Listing 9.23, we pass the `sample_sentence` sample text to the model after we've created the tokenizer. As already mentioned, the tokenizer must match the model, and here, we use the `model` (and the `tokenizer`) `bert-base-uncased`. The inputs are the token IDs that were created with the tokenizer, and the word embeddings result from the last hidden output.`last_hidden_state` layer.

```
#%% test
sample_sentence = "the man ate the pizza because it smelled delicious"

#%% Get word encodings and attention weights from BERT
tokenizer = AutoTokenizer.from_pretrained("bert-base-uncased")
```

```
model = AutoModel.from_pretrained("bert-base-uncased", output_attentions=True)
inputs = tokenizer(sample_sentence, return_tensors="pt", padding=True, trunca-
tion=True)

with torch.no_grad():
    outputs = model(**inputs)
word_encodings = outputs.last_hidden_state
tokens = tokenizer.convert_ids_to_tokens(inputs['input_ids'][0])
```

Listing 9.23 Self-Attention: Creation of Word Embeddings and Tokens

We can now also extract the attention weights from the model results. We determine the average attention (avg_attention) as follows:

```
#%% Get attention weights from all layers and heads
attention_weights = outputs.attentions

last_layer_attention = attention_weights[-1][0]

avg_attention = last_layer_attention.mean(dim=0)
```

Figure 9.16 shows the results of the model. The numbers represent the attention weights based on the word *it*.

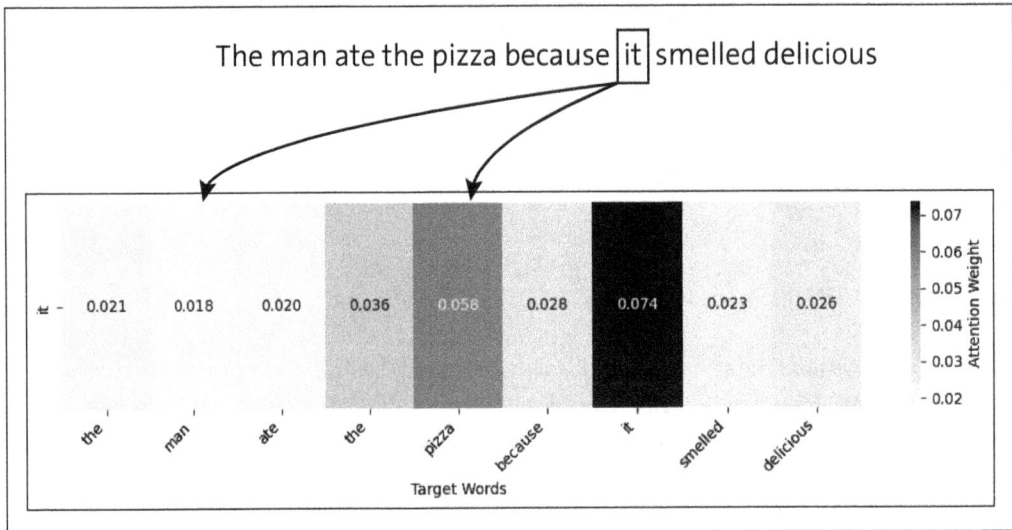

Figure 9.16 Self-Attention of Word it in Relation to Words Man and Pizza

Here, it becomes clear that *it* most likely refers to *pizza*.

In a real transformer, there's not just one *attention module* (which is usually referred to as an *attention head*). There's *multihead attention*, in which each head focuses on different aspects of connections between words. For example, one head may focus on the relationships between verbs and nouns and another may focus on the relationships between adjectives and nouns.

9.9 Summary

In this chapter, you've learned the basics of language models, from their simple use with Python to the meaning of model parameters and various criteria for model selection—including performance, cutoff date, hosting, and deployment options.

We've looked at message types, prompt templates, and chains to design effective interactions. We placed a particular focus on structured outputs that enable us to achieve specific and predictable results, and finally, we provided deeper insight into transformer architecture to give you a better understanding of how these powerful models work. With this knowledge, you're now well equipped to use language models in your own projects.

9

Chapter 10

Pretrained Networks and Fine-Tuning

"Only those who know the past can understand the present and shape the future."
—*August Bebel*

Bebel's quote refers to a historical principle, but it also applies very well to the core of machine learning. In the field of deep learning, the past represents all the knowledge that a model has gained through the training process, and the future is the specific new task for which we're trying to adapt the pretrained model. For this purpose, we can use *transfer learning*, which is a technique for taking knowledge that a pretrained model has learned in one task and transferring it to a new task. In this process, most of the pretrained network is frozen and only the last layers are trained with new data. *Fine-Tuning* is a special form of transfer learning in which we train some or all layers of the pretrained model for a new task. And since transfer learning involves transferring an existing model to new tasks, it uses the extensive knowledge of the past to master the challenges of the present and future.

In Section 10.1, you'll dive into the world of pretrained networks and learn why they've become a foundation of machine learning. We'll start with an introduction to Hugging Face, which is the leading platform for publicly available models. You'll also learn how to use the huge library of models for your own projects.

In Section 10.2, we'll explain the basic concepts of fine-tuning, and you'll learn how to adapt a pretrained model to a specific task. In Section 10.3 and Section 10.4, I will show you how to fine-tune a computer vision model and a language model, using concrete code examples.

Now, let's start with neural networks that have already been trained.

10.1 Pretrained Networks with Hugging Face

Once a network has been trained and is ready for use, we refer to it as a *pretrained network*. Thankfully, there's a lively community that shares such trained networks so that we don't have to reinvent the wheel all the time. The best-known platform on which we can find such models is *Hugging Face*, which offers a huge collection of models for a wide range of applications. Figure 10.1 shows Hugging Face's classification of tasks into different categories, such as computer vision and natural language processing.

Multimodal		
Audio-Text-to-Text	Image-Text-to-Text	
Visual Question Answering		
Document Question Answering	Video-Text-to-Text	
Visual Document Retrieval	Any-to-Any	

Computer Vision		
Depth Estimation	Image Classification	
Object Detection	Image Segmentation	
Text-to-Image	Image-to-Text	Image-to-Image
Image-to-Video	Unconditional Image Generation	
Video Classification	Text-to-Video	
Zero-Shot Image Classification	Mask Generation	
Zero-Shot Object Detection	Text-to-3D	
Image-to-3D	Image Feature Extraction	
Keypoint Detection	Video-to-Video	

Natural Language Processing		
Text Classification	Token Classification	
Table Question Answering	Question Answering	
Zero-Shot Classification	Translation	
Summarization	Feature Extraction	
Text Generation	Fill-Mask	Sentence Similarity
Text Ranking		

Audio		
Text-to-Speech	Text-to-Audio	
Automatic Speech Recognition	Audio-to-Audio	
Audio Classification	Voice Activity Detection	

Tabular		
Tabular Classification	Tabular Regression	
Time Series Forecasting		

Reinforcement Learning		
Reinforcement Learning	Robotics	

Other	
Graph Machine Learning	

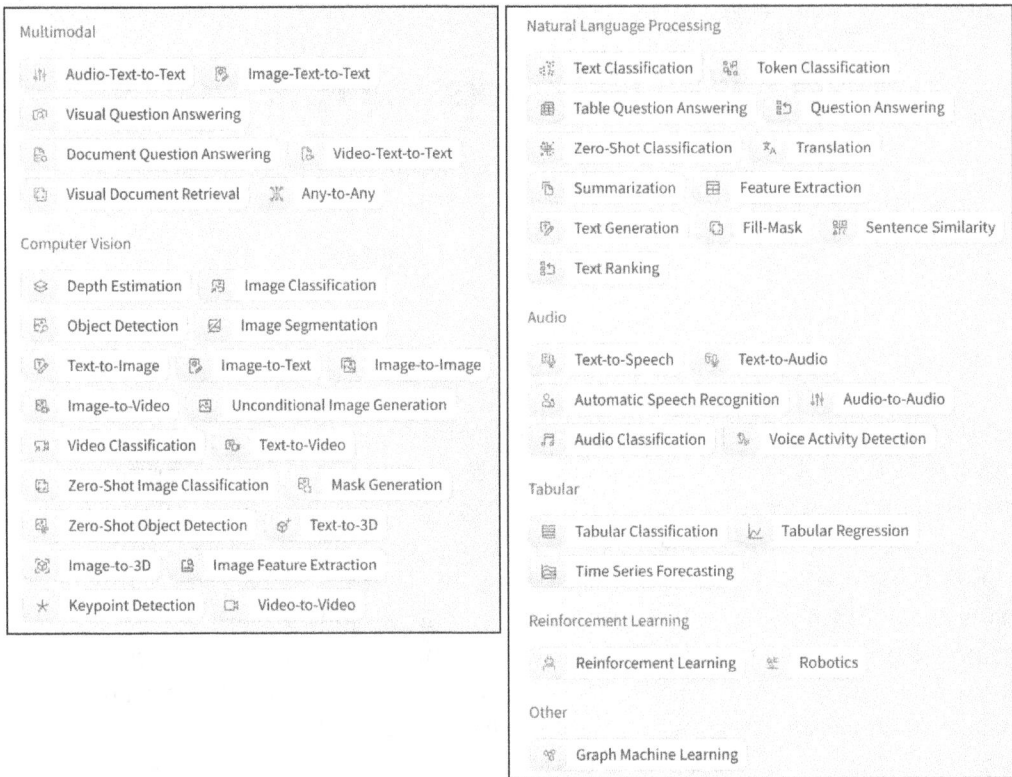

Figure 10.1 Hugging Face: Tasks (Source: https://huggingface.co/models)

I recommend that you invest the time to get an overview of the areas of responsibility for which models are available on Hugging Face. This is important because of an old saying: "It is tempting, if the only tool you have is a hammer, to treat everything as if it were a nail." You need to have a toolbox that contains more than just one tool—and that's what familiarizing yourself with all the tasks will get you.

How do you go about selecting the right model for your application from the almost infinite number of models that are available? You can narrow down the model class by limiting it according to the super category that describes the problem you want to solve.

Imagine you want to mark objects in images (including their contours), in order to color them, replace them with other objects, or do something similar. Figure 10.2 shows the initial image ❶ and the image with superimposed model output ❷.

Once you familiarize yourself with the task types on Hugging Face, you'll know that the problem belongs to the **Computer Vision** category and the **Mask Generation** task class. You'll have come closer to a good solution—you'll know that there's not just one but many possible and good solutions. But there will still be too many models left, so you can limit the number of parameters because you may not have an extremely powerful GPU at your disposal.

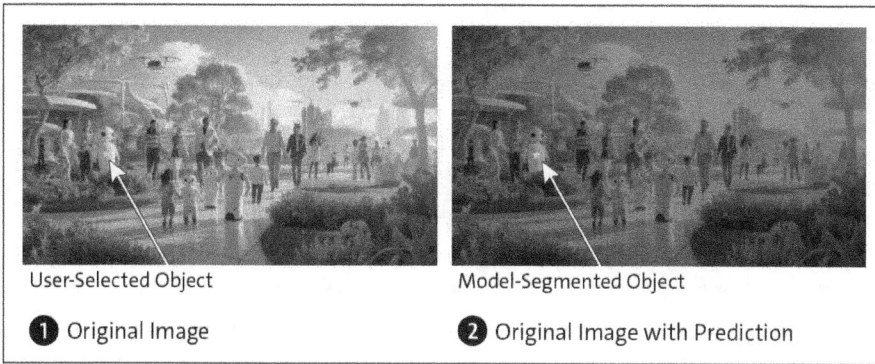

Figure 10.2 Initial Image and Model Result

Another important filter is the license. Not all models listed on Hugging Face are licensed for all use cases. For example, if you want to use the model for commercial purposes, the license under which the model is provided must also allow this. For example, you can select Apache 2.0 and see whether the license for the model allows you to use it for your use case.

You'll also notice that certain models—such as facebook/sam2-hiera-tiny, facebook/sam2-hiera-small, facebook/sam2-hiera-large, and facebook/sam2-hiera-huge—are available in different sizes. Larger models usually perform better than smaller models, but they also require significantly more resources.

Once you get the models that are right for your use case down to a manageable number, you can start taking a closer look at the model descriptions. For example, Figure 10.3 shows the model card of the facebook/sam2-hiera-tiny model.

The model card is the starting point for using a model, and it contains the following information:

- How well the model performs
- How it should be implemented
- Which scientific paper it's based on

Models often also have a direct inference, meaning you can enter text for language models or an image for image models, execute the inference, and get a result. This can help you evaluate the operation of the model and its output. You can also use the **Files and Versions** tab to check the size of the model you've selected for download.

Marking Objects in Images

The actual implementation that leads to the display in Figure 10.2 can be found at *110_PreTrained_FineTuning\sam_example.py*.

```
facebook / sam2-hiera-tiny    ♡ like  23     Follow  AI at Meta  7.45k

   Mask Generation     Transformers   8 Safetensors    sam2_video    feature-extraction    arxiv:2408.00714


   Model card     Files and versions   xet     Community  2
```

Repository for SAM 2: Segment Anything in Images and Videos, a foundation model towards solving promptable visual segmentation in images and videos from FAIR. See the SAM 2 paper for more information.

The official code is publicly release in this repo.

Usage

For image prediction:

```
import torch
from sam2.sam2_image_predictor import SAM2ImagePredictor

predictor = SAM2ImagePredictor.from_pretrained("facebook/sam2-hiera-tiny")

with torch.inference_mode(), torch.autocast("cuda", dtype=torch.bfloat16):
    predictor.set_image(<your_image>)
    masks, _, _ = predictor.predict(<input_prompts>)
```

Figure 10.3 Model Card of Model on Hugging Face

10.2 Transfer Learning

Transfer learning is the process in which a neural network that has already been pre-trained on a large amount of data is further trained for a specific new task with a smaller, task-related dataset.

We discuss advantages of transfer learning in Section 10.2.1 and learn about different approaches in Section 10.2.2.

10.2.1 Advantages of Transfer Learning

Why should you perform transfer learning when you could simply train a network from scratch? The answer lies in efficiency and is illustrated in Figure 10.4.

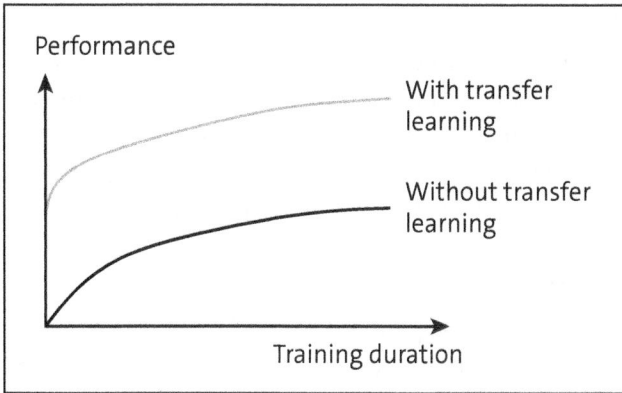

Figure 10.4 Influence of Transfer Learning on Model Performance

While a model that is trained from scratch (depicted by the solid "Without Transfer Learning" line in the figure) initially starts without knowledge and therefore can't show any performance at the beginning, a model that already starts with prior knowledge (depicted by the dashed "with Transfer Learning line in the figure) starts from a higher level. The latter model requires significantly less computing power, as enormous amounts of computing power have already gone into training the original model.

Transfer learning therefore requires much less computing power and energy than training a model from scratch. This means that the model converges faster, as it already starts from a well-trained state and doesn't start from random weights. You can therefore use transfer learning to complete model training in fewer epochs.

Another important point is that fine-tuning requires less data. While model training from zero requires large datasets, fine-tuning usually manages with much smaller, specific datasets. That's because the pretrained network has already learned to recognize many general features from the original data and can reuse them. In image processing, for example, this could be the recognition of edges or textures, or in language processing, it could be the recognition of grammatical structures. The network doesn't have to relearn this knowledge but can transfer it to a new task and thus achieve higher performance right from the start.

There are also very practical considerations. Without transfer learning, most companies and researchers would not be able to use the best models because training them is already prohibitively expensive, whereas transfer learning is significantly cheaper due to the lower resource requirements.

Now, we want to gain a better understanding of how we can adapt a model to our task based on an already trained network.

10.2.2   Transfer Learning Approaches

There are different approaches to transfer learning. These range from training the complete model and all its parameters based on a pretrained network to only training a small subset of the parameters and using many model features based on frozen weights. If the latter approach doesn't provide sufficient results, you can progressively unfreeze layers. All these concepts are explained in the following paragraphs.

Transfer Learning with the Complete Model

In this method, we train all parameters of the pretrained model for the new specific problem. As a rule, we use a very low learning rate (e.g., 1E-5 or smaller), which means that we only minimally adjust the originally learned weights to keep the model from "forgetting" the general knowledge.

This approach is suitable if your dataset is large and diverse enough to justify a meaningful adjustment of all layers without the model losing its initial capabilities.

Feature Extraction

Figure 10.5 shows the procedure for fine-tuning with feature extraction, using the example of a model that has been trained to classify cats and that we now want to adapt so that our new model can recognize dogs in future.

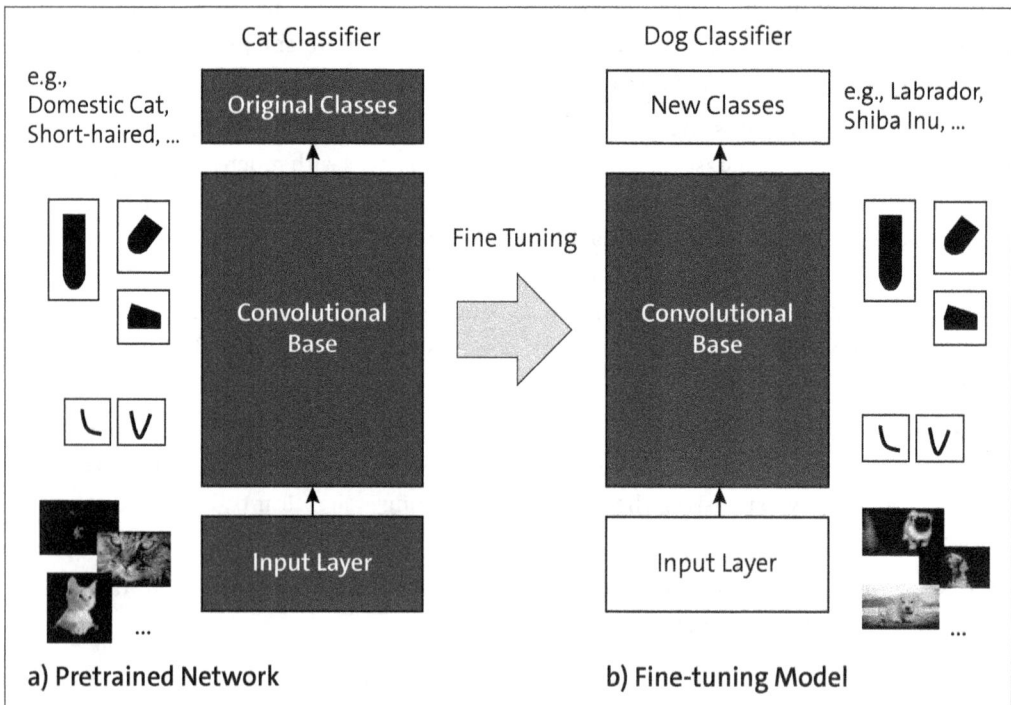

Figure 10.5 Fine-Tuning Procedure

Let's start with the left-hand side of the figure: the pretrained model. It was trained with many cat pictures, of which there are quite a few on the internet, and the target variable in the model is the name of the cat species. The trained model in the example is a CNN, but the approach also works with any other model class.

Our task is to use this model to train a dog classification model based on it. Why can this work at all? Well, the actual network learns different information from the training images. It starts with simple textures, and the deeper you go into the network, the more complex structures (such as ears, noses, etc.) it recognizes. Although dogs and cats undoubtedly have some differences, they share many more similarities: fur, four legs, one mouth, two eyes, and one nose. The fine-tuned model doesn't have to learn these similarities again from scratch but can simply adopt them from the pretrained model.

In CNNs, these frozen network weights and layers are called *convolutional bases*. What changes in the fine-tuned network are the images that are fed into the input layer and the target sizes of the dog breeds.

If the result is still not satisfactory, you can retrain further layers from the end of the network—but you should definitely leave most of the network unchanged. The pretrained model can be highly complex, yet you can simply use it, retrain a few layers, provide your data, and start the training.

We'll put this into practice in the following sections, and we'll start by fine-tuning an image classification model.

Progressive Unfreezing

This is a step-by-step method that combines freezing and full fine-tuning. First, as with the feature extraction approach, we train only the last part of the model (the classification layers) while the rest remains frozen.

We then thaw the next layers closest to the output layer, layer by layer, and continue the training process. We keep doing this step by step until we've released the number of shifts we want for training. This approach lets the model learn the specific features of the new dataset without running the risk of destroying the deep, generally learned features of the first layers.

10.3 Coding: Fine-Tuning a Computer Vision Model

In this example, we'll use a pretrained network as a starting point and train it with a special dataset. The model we use will be *Densenet-121*, which is a very deep CNN that connects each layer directly to the next layer through its dense connections to maximize the flow of information. This network was trained with the *ImageNet* dataset, which is one of the best-known bases for training image classification models. It consists of over 1 million images divided into 1,000 different object classes.

Figure 10.6 shows the basic network architecture with the different network layers that we all know from Chapter 4. We'll freeze all weights and only retrain some dense layers and the output layer at the end of the network.

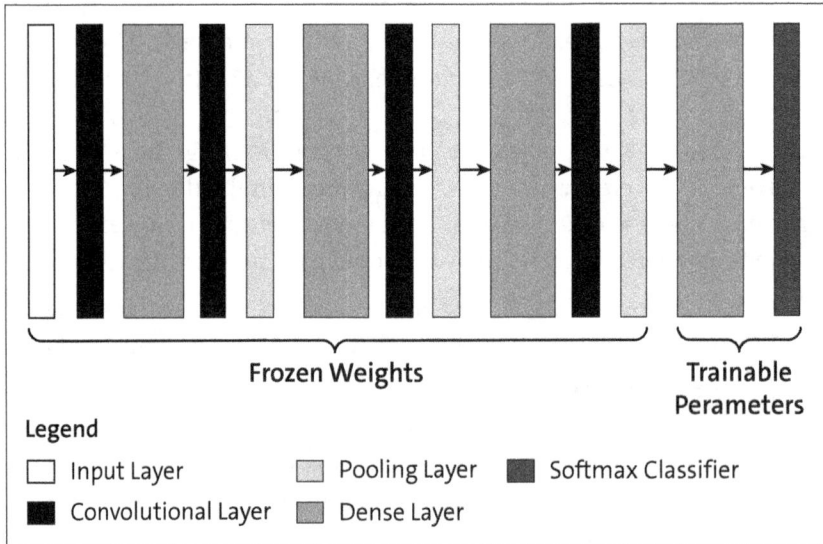

Figure 10.6 Densenet-121 Structure and Parameters

We'll use this model as a starting point and train it to recognize different classes of garbage. The corresponding dataset can be found at Kaggle, and it contains 2,300 to 2,500 images from six different classes: plastic, metal, glass, cardboard, paper, and trash. We'll load the packages and data in Section 10.3.1, fine-tune the model in Section 10.3.2, and check the model results in Section 10.3.3.

The entire script can be found at *110_PreTrained_FineTuning\cv_transfer_learning.py*.

10.3.1   Data Preparation

We load the required packages in Listing 10.1, which has the following elements:

- **Data preparation**
  We use `kagglehub` for data import, `numpy` for numerical calculations, and `random_split` and `DataLoader` for the creation of the dataset.

- **Model training**
  We use multiple `torch` modules, `OrderedDict` for model setup, and some modules from `torchvision` for model loading and preprocessing of images.

- **Model evaluation**
  We use `seaborn` and `matplotlib` for visualization and `sklearn` for determination of metrics.

```
# data handling
from torch.utils.data import random_split, DataLoader
import os
import numpy as np
import kagglehub
# modeling
from collections import OrderedDict
import torch
import torch.nn as nn
import torchvision
from torchvision import transforms, models
from sklearn.dummy import DummyClassifier
# evaluation
import seaborn as sns
from sklearn.metrics import confusion_matrix, accuracy_score
import matplotlib.pyplot as plt
```

Listing 10.1 Computer Vision Fine-Tuning: Loading Packages

We then download the data is Listing 10.2 from Kaggle to the local computer with data-set_download.

```
#%% data import
# Download latest version
path = kagglehub.dataset_download("zlatan599/garbage-dataset-classification")

print("Path to dataset files:", path)
folder_path = os.path.join(path, "Garbage_Dataset_Classification", "images")
```

Listing 10.2 Computer Vision Fine-Tuning: Dataset Download

We control the model training via the hyperparameters defined in Listing 10.3.

```
BATCH_SIZE = 32
LEARNING_RATE = 0.01
EPOCHS = 10
DEVICE = torch.device("cuda" if torch.cuda.is_available() else "cpu")
```

Listing 10.3 Computer Vision Fine-Tuning: Defining Hyperparameters

Now, let's move on to preparing the dataset. This is where we employ the transformations we defined in Listing 10.4, which we use to create the complete dataset (full_dataset). This uses the ImageFolder method from the torchvision package, and we store the images in various subfolders in the folder_path folder, according to their class. The method takes care of the structure of the dataset object.

```
transform = transforms.Compose([
    transforms.Resize((224, 224)),
    transforms.ToTensor()
])

# Load the full dataset
full_dataset = torchvision.datasets.ImageFolder(
    root=folder_path,
    transform=transform
)
```

Listing 10.4 Computer Vision Fine-Tuning: Creating Dataset

We carve the three buckets (train_dataset, val_dataset, and test_dataset) out of the overall full_dataset object in Listing 10.5. Here, we use 60% for training and 20% each for validation and testing.

```
train_size = int(0.6 * len(full_dataset))
val_size = int(0.2 * len(full_dataset))
test_size = len(full_dataset) - train_size - val_size

train_dataset, val_dataset, test_dataset = random_split(
    full_dataset,
    [train_size, val_size, test_size]
)
```

Listing 10.5 Computer Vision Fine-Tuning: Creating Dataset Instances

In Listing 10.6, we create the DataLoader instances from the datasets.

```
train_loader = DataLoader(dataset=train_dataset, batch_size=BATCH_SIZE, shuf-
fle=True)
val_loader = DataLoader(dataset=val_dataset, batch_size=BATCH_SIZE, shuffle=
False)
test_loader = DataLoader(dataset=test_dataset, batch_size=BATCH_SIZE, shuffle=
False)
```

Listing 10.6 Computer Vision Fine-Tuning: Creating DataLoader Instances

Now, let's look at Listing 10.7 to see which classes are present in the dataset. We can determine the number of output nodes of the OUTPUT_FEATURES network directly from the number of image classes.

```
#%% get the class labels
class_labels = full_dataset.classes
OUTPUT_FEATURES = len(class_labels)
```

```
class_labels
```

```
['cardboard', 'glass', 'metal', 'paper', 'plastic', 'trash']
```

Listing 10.7 Computer Vision Fine-Tuning: Image Classes

10.3.2   Model Training

We can load the pretrained network via the models module and its densenet121 function, and we can also load many more models via this module. We also need to pass the pretrained = True parameter so that the model weights are loaded.

Then, we need to "freeze" all model parameters by setting the requires_grad property to False for each parameter, as shown in Listing 10.8.

```
model = models.densenet121(pretrained = True)

#%% freeze ALL model layers
for params in model.parameters():
    params.requires_grad = False
```

Listing 10.8 Computer Vision Fine-Tuning: Creating Model Instance and Freezing Parameters

Of course, the model can't learn anything if all parameters are frozen. The model maps the network layers via nn.Sequential, and it's set up in such a way that the layers are subdivided into features and classifiers. So, because we only want to adapt the last layers, we can simply overwrite the model.classifier in Listing 10.9. Note that OrderedDict means that the order in the dictionary must be adhered to.

The fully connected fc1 layer is adjusted because the number of classes to be recognized has changed. The original model predicted 1,000 classes, but in our case, there are only 6 classes, which we pass generically via the OUTPUT_FEATURES variable. We also should not activate the output with nn.Softmax to obtain the probability of the individual classes, because nn.CrossEntropyLoss internally performs this step.

```
model.classifier = nn.Sequential(OrderedDict([
    ('fc1', nn.Linear(in_features=1024, out_features=OUTPUT_FEATURES))]))
model = model.to(DEVICE)
```

Listing 10.9 Computer Vision Fine-Tuning: Defining Trainable Parameters

Now, we're back in familiar territory. There are no surprises with the loss function and the optimizer shown in Listing 10.10: we use CrossEntropyLoss and Adam.

```
loss_function = nn.CrossEntropyLoss()
optimizer = torch.optim.Adam(model.parameters(), lr=LEARNING_RATE)
```

Listing 10.10 Computer Vision Fine-Tuning: Defining Loss Function and Optimizer

The training in Listing 10.11 also holds no surprises. We train the model using the training data, and we use the validation data to check how the training is progressing.

```
losses_train, losses_val = [], []
for epoch in range(EPOCHS):
    loss_epoch_train, loss_epoch_val = 0, 0

    for _, (X_batch, y_batch) in enumerate(train_loader):
        # zero gradients
        optimizer.zero_grad()
        X_batch = X_batch.to(DEVICE)
        y_batch = y_batch.to(DEVICE)
        # forward pass
        y_batch_pred = model(X_batch)


        # loss calculation
        loss = loss_function(y_batch_pred, y_batch)
        loss_epoch_train += loss.item()
        # backward pass
        loss.backward()

        # update weights
        optimizer.step()
    losses_train.append(loss_epoch_train)
    # validation loop
    model.eval()
    val_loss_epoch = 0
    with torch.no_grad():
        for _, (X_batch, y_batch) in enumerate(val_loader):
            X_batch = X_batch.to(DEVICE)
            y_batch = y_batch.to(DEVICE)
            y_batch_pred = model(X_batch)
            loss = loss_function(y_batch_pred, y_batch)
            val_loss_epoch += loss.item()
    losses_val.append(val_loss_epoch)
    print(f"Epoch: {epoch}, Loss: {loss_epoch_train}, Val Loss: {val_loss_ep-
och}")

Epoch: 0, Loss: 385.32334792613983, Val Loss: 125.89920091629028
...
Epoch: 9, Loss: 328.9187126159668, Val Loss: 112.78194999694824
```

Listing 10.11 Computer Vision Fine-Tuning: Performing Training

This training reduced the training and validation losses, but longer training could improve the result even more. Nevertheless, we're optimistic and will assess the model.

10.3.3   Model Evaluation

For the evaluation, we need predictions for the test data (y_test_pred), which we can then compare with the actual data (y_test_true). We can determine this data via the code in Listing 10.12. The models provides logits of which we need to find the index with the maximum value with np.argmax to extract the predicted class.

```
model.eval()
y_test_true = []
y_test_pred = []
for _, (X_batch, y_batch) in enumerate(test_loader):
    # forward pass
    with torch.no_grad():
        X_batch = X_batch.to(DEVICE)
        y_test_pred_batch = model(X_batch).detach().cpu().numpy()
    y_test_pred_class = np.argmax(y_test_pred_batch, axis=1).tolist()
    y_test_pred.extend(y_test_pred_class)
    y_test_true.extend(y_batch.numpy().tolist())
```

Listing 10.12 Computer Vision Fine-Tuning: Creating Model Predictions

Then, we can compare the predictions to the real values in a confusion matrix to find out which classes we can predict well, and we can use the code in Listing 10.13 for this purpose.

We use confusion_matrix to determine the data basis for the graphic we'll create later. A special feature here is the color scaling, which we carry out by using the cm_normalized object. The color scale should represent the value range from 0% to 100% for the correct predictions on the main diagonal and from –100% to 0% for incorrect predictions outside the main diagonal. High values represent dark colors, and low values represent light colors to no color at all.

```
cm = confusion_matrix(y_test_true, y_test_pred)
cm_normalized = confusion_matrix(y_test_true, y_test_pred, normalize=
'true')*100
cm_normalized = cm_normalized - 2*np.triu(cm_normalized, 1) - 2*np.tril(cm_nor-
malized, -1)
plt.title("Konfusionsmatrix")
sns.heatmap(cm_normalized, xticklabels=class_labels, yticklabels=class_labels,
annot=cm, fmt='.0f', vmin=-100, vmax=100, cmap='PiYG', cbar_kws=
{'format':'%d%%'})
```

Listing 10.13 Computer Vision Fine-Tuning: Creating Confusion Matrix

The confusion matrix is shown in Figure 10.7, where you can immediately see that most classes are predicted very well because the fields appear in dark colors. However, the model still has some weaknesses when differentiating between glass and plastic.

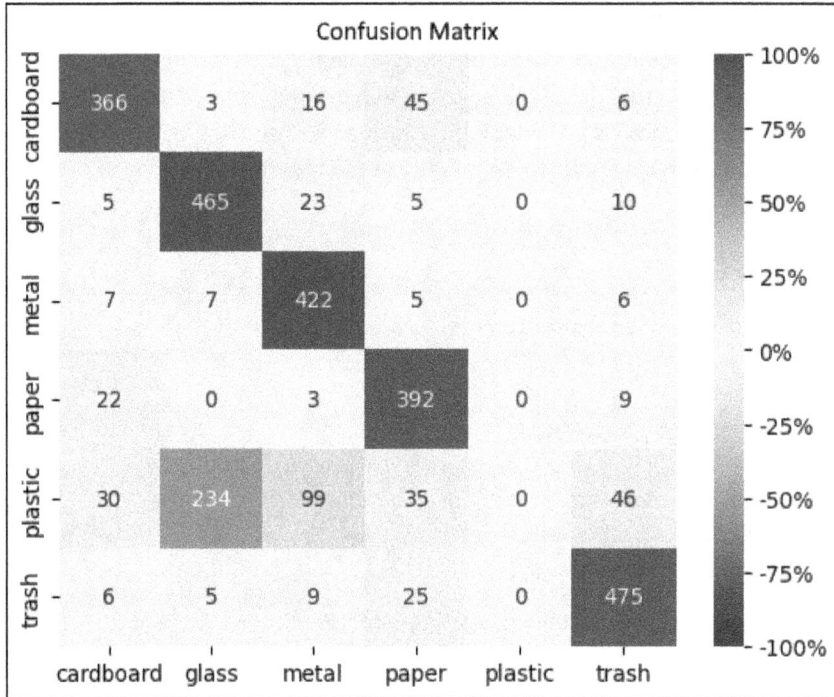

Figure 10.7 Computer Vision Fine-Tuning: Confusion Matrix

However, the overall picture looks good. We can see this in the accuracy (i.e., the percentage of correctly predicted classes), which we can determine by using the accuracy_ score. We can calculate it as 76%, as follows:

```
accuracy_score(y_test_true, y_test_pred)
```

0.76

However, we can only classify this if we compare the value with a dummy classifier that represents the results of pure guessing. We create this model in Listing 10.14, and it determines the accuracy to be 27%.

```
dummy_clf = DummyClassifier(strategy="most_frequent")
y_pred_dummy_all = []
y_true_dummy_all = []
for _, (X_batch, y_batch) in enumerate(test_loader):
    dummy_clf.fit(X_batch, y_batch)
    y_pred_dummy = dummy_clf.predict(X_batch)
```

```
y_pred_dummy_all.extend(y_pred_dummy.tolist())
y_true_dummy_all.extend(y_batch.numpy().tolist())

accuracy_score(y_true_dummy_all, y_pred_dummy_all)
```

`0.27`

Listing 10.14 Computer Vision Fine-Tuning: Dummy Classifier

We can therefore say that we've trained a usable model in a very short time, as we've started with a very good model and adapted it to the new task by fine-tuning it. Of course, fine-tuning is not only possible in the field of computer vision but with any type of model. To show that this is possible with language models, I will guide you through the fine-tuning of a language model in the following section.

10.4   Coding: Fine-Tuning a Language Model

In this section, we'll use a small language model and train it for a specific task. In this context, we speak of *small language models* (SLMs) as models that are usually derived from a large language model and are created through a process called *distillation*.

The model we use here is Gemma-3, which is characterized by the fact that it has very few parameters. In its smallest version, it only requires 270 million parameters, which is a factor of 1,000 less than the heavyweights such as GPT-5 and Gemini-2.5 Pro.

Now, we'll train the SLM to distinguish spam from nonspam. You can find the corresponding dataset on Kaggle, and it contains almost 8,200 text examples in the training dataset and 2,700 examples in the test data. In each case, there's a text and the corresponding label. Table 10.1 shows two sample entries from the dataset.

Text	Label
hey I am looking for Xray baggage datasets can you provide me with the same	not_spam
"Get rich quick! Make millions in just days with our new and revolutionary system! Don't miss out on this amazing opportunity!"	spam

Table 10.1 Spam Detection Dataset: Example

We'll now adapt the language model we've selected to this dataset so that it delivers the corresponding assessment for a given text as to whether it's spam or nonspam. The corresponding script can be found at *110_PreTrained_FineTuning\llm_finetuning.py*. First, we prepare the data, and then, we train and evaluate the model.

Note

Training language models is far more computationally intensive than any other training we've covered in this book, and it only makes sense for you to perform the training we describe here if you have a GPU with at least 8 GB. If your local computer doesn't have this equipment, then I suggest that you perform the training in the cloud (e.g., using Google Colab [*https://colab.research.google.com/*]), where you can use GPU capabilities free of charge.

10.4.1   Data Preparation

We load the packages with the code in Listing 10.15. Two new classes at this point are GemmaForSequenceClassification and GemmaTokenizer, which correspond directly to the model we're using. All other packages and classes are familiar from previous chapters.

```
# datapreparation
import numpy as np
from collections import Counter
from datasets import load_dataset
import huggingface_hub
# modeling
from transformers import GemmaForSequenceClassification, GemmaTokenizer
import torch
from torch.utils.data import DataLoader
# evaluation
from sklearn.metrics import accuracy_score
```

Listing 10.15 SLM Fine-Tuning: Loading Packages

You should make sure that the GPU memory is cleared by using the following command:

```
torch.cuda.empty_cache()
```

Now, we can define the hyperparameters for the later training in Listing 10.16. If there's an error message later due to insufficient GPU memory, you can reduce the BATCH_SIZE here to transfer less data in each run.

```
EPOCHS = 2
DEVICE = torch.device("cuda" if torch.cuda.is_available() else "cpu")
BATCH_SIZE = 8
LR = 0.001
```

Listing 10.16 SLM Fine-Tuning: Defining Hyperparameters

We load the model with the code in Listing 10.17 via Hugging Face. To do this, we can use the GemmaForSequenceClassification model class, pass the model_name to load the model we select, and load the model onto the GPU with model.to(DEVICE).

In this case, we must create the corresponding tokenizer by using GemmaTokenizer.

```
model_name = "google/gemma-3-270m-it"
model = GemmaForSequenceClassification.from_pretrained(model_name)
model.to(DEVICE)
tokenizer = GemmaTokenizer.from_pretrained(model_name)
```

Listing 10.17 SLM Fine-Tuning: Loading Model and Tokenizer

At that point, the model will be available and we'll be able to turn to the data. We can load the dataset in Listing 10.18 with load_dataset, and we can code via label2id that not_spam corresponds to a value of 0 and spam corresponds to a value of 1. As the labels are still available in text form in the dataset, we convert them into the corresponding numerical values by using the lambda function.

```
dataset = load_dataset("Deysi/spam-detection-dataset")

# Map string labels to integer class ids
label2id = {"not_spam": 0, "spam": 1}
id2label = {v: k for k, v in label2id.items()}
dataset = dataset.map(
    lambda batch: {"label": [label2id[label] for label in batch["label"]]},
    batched=True,
)
```

Listing 10.18 SLM Fine-Tuning: Loading Dataset and Encoding Target Variable

Given a dataset on spam, you could assume that the data is very unbalanced, meaning that there's significantly more non-spam than spam. At least, that's what my inbox reflects. We can check this by determining the number of 0 and 1 labels via the counter class, which shows that 50.5% of the data is labeled as spam.

```
print(Counter(dataset["train"]["label"]))
```

Counter({1: 4125, 0: 4050})

So, in fact, the dataset is almost perfectly balanced.

Now, we must convert the dataset into tokens by passing the text features to the tokenizer instance in Listing 10.19. We must also pad the text sequences to the same length with padding="max_length". If the texts are longer than 512 tokens, they are truncated (truncation=True).

```
def tokenize_function(examples):
    return tokenizer(examples["text"], padding="max_length", truncation=True,
max_length=512)

tokenized_datasets = dataset.map(tokenize_function, batched=True)
```

Listing 10.19 SLM Fine-Tuning: Tokenizing Dataset

In Listing 10.20, we delete the original text column and rename the label column labels. Later, it's important for use to use the set_format, which converts the object into torch tensors.

```
tokenized_datasets = tokenized_datasets.remove_columns(["text"])
tokenized_datasets = tokenized_datasets.rename_column("label", "labels")
tokenized_datasets.set_format(
    type="torch",
    columns=["input_ids", "attention_mask", "labels"],
)
```

Listing 10.20 SLM Fine-Tuning: Processing of Tokenized Dataset

In Listing 10.21, we ensure that the label2id and id2label mappings are configured correctly in the model.

```
model.config.label2id = label2id
model.config.id2label = id2label
```

Listing 10.21 SLM Fine-Tuning: Label Mappings

We create the DataLoader instances for training (train_loader) and validation (val_loader) in Listing 10.22.

```
train_loader = DataLoader(tokenized_datasets["train"], batch_size=BATCH_SIZE,
shuffle=True)
val_loader = DataLoader(tokenized_datasets["test"], batch_size=BATCH_SIZE, shuffle=False)
```

Listing 10.22 SLM Fine-Tuning: Creating DataLoader Instances

The data will then be available to us in the correct form, so we can continue on to training the model.

10.4.2   Model Training

To train the model, we need an optimizer and a loss function. We create both in Listing 10.23.

```
loss_fn = torch.nn.CrossEntropyLoss()
optimizer = torch.optim.AdamW(model.parameters(), lr=LR)
```

Listing 10.23 SLM Fine-Tuning: Defining Loss Function and Optimizer

We perform the model training as shown in Listing 10.24. The train_loader operates on the CPU, so we must copy the batch elements to the GPU. The result is a new dictionary on the GPU.

We can then pass the current batch to the model in the forward pass. We use the **batch construct, which leads to the dictionary being unpacked. The prediction outputs contain the logits, which are compared with the actual batch["labels"] values in the loss_ fn loss function. The other instructions ensure that the losses are stored in the train_ losses list. The training is very computationally intensive; just two epochs take a relatively long time.

```
model.train()
train_losses = []
for epoch in range(EPOCHS):
    train_loss_epoch = 0
    for batch_idx, batch in enumerate(train_loader):
        optimizer.zero_grad()
        batch = {k: (v.to(DEVICE) if isinstance(v, torch.Tensor) else v) for
                k, v in batch.items()}
        outputs = model(**batch)

        loss = loss_fn(outputs.logits, batch["labels"])

        loss.backward()
        optimizer.step()
        train_loss_epoch += loss.item()
        print(f"Epoch {epoch+1}/{EPOCHS}:
                Training batch {batch_idx + 1}/{len(train_loader)+1}:
                Loss: {loss.item()}")
    train_losses.append(train_loss_epoch/len(train_loader))
    print(f"Epoch {epoch+1}/{EPOCHS}: Loss: {train_losses[-1]}")
```

Epoch 1/2: Training batch 1/1023: Loss: 0.875882625579834
...
Epoch 2/2: Training batch 350/1023: Loss: 0.003161453874781728

Listing 10.24 SLM Fine-Tuning: Performing Model Training

At this point, we've trained the model, so in the next section, we'll see how well we've adapted it to the dataset.

10.4.3   Model Evaluation

First, we create the predictions batch by batch in Listing 10.25 and save them in the predicted_labels_all list.

```
model.eval()
predicted_labels_all, true_labels_all = [], []

with torch.no_grad():
    for batch in val_loader:
        input_ids = batch["input_ids"].to(DEVICE)
        attention_mask = batch["attention_mask"].to(DEVICE)
        true_labels = batch["labels"].numpy().tolist()

        pred_labels = model(input_ids, attention_mask=attention_mask)
        logits = pred_labels.logits
        predicted_labels = torch.argmax(logits, dim=1).cpu().numpy().tolist()

        predicted_labels_all.extend(predicted_labels)
        true_labels_all.extend(true_labels)
```

Listing 10.25 SLM Fine-Tuning: Creating Predictions for Test Dataset

Finally, we want to check how well the model can distinguish spam from nonspam, and we determine the accuracy for this in Listing 10.26.

```
accuracy = accuracy_score(true_labels_all, predicted_labels_all)
print(f "Validation accuracy: {(accuracy*100):.1f}%")
```

Validation accuracy: 95.6%

Listing 10.26 SLM Fine-Tuning: Determining Model Accuracy

The accuracy is 95.6%, and as a reminder, the dominant class in the dataset comprises just over 50% of the data. We can therefore be satisfied with our model, which has proven to be both reliable and efficient.

10.5   Summary

In Section 10.1 of this chapter, you first learned what pretrained networks are. You got to know Hugging Face as a central point of contact for such networks and familiarized yourself with how to find a suitable model. In Section 10.2 on fine-tuning, we highlighted the advantages of this approach over training from scratch, and we examined various approaches to fine-tuning theoretically.

In Section 10.3, we moved on to the practical side, where we first adapted a model for sorting waste into different classes. Finally, in Section 10.4 we further trained a language model for classifying text into spam and nonspam.

10

Chapter 11

PyTorch Lightning

"Order is the highest form of power. It allows energy to be channeled rather than wasted."
—Unknown

When working with frameworks like PyTorch, building a training workflow can become a complex endeavor. While PyTorch offers tremendous flexibility, manually managing training loops, validation loops, checkpoints, and the logic for different hardware configurations requires a significant amount of *boilerplate code*—which is program code that we can use in different contexts with little or no modification. We've already seen some of this up to this point. Just think of the forward pass, backward pass, and parameter updates in the training loop.

This is where *PyTorch Lightning* comes into play. It's not a new programming language and not a replacement for PyTorch—it's a layer of abstraction on top of it, and it can help you organize the models and the model training. This framework introduces a clear structure that allows us to focus more on the essentials—the logic of the model itself. With its clear and consistent structure, PyTorch Lightning channels the development process and ultimately makes us more efficient and productive.

In this chapter, we'll learn how PyTorch Lightning helps us use the "power of order" for our purposes and to speed up and simplify our projects. We'll start with model training and learn what model training with PyTorch looks like and what changes when we use PyTorch Lightning Finally, we'll gain an understanding of what callbacks are and how we can utilize them with PyTorch Lightning.

11.1 PyTorch Versus PyTorch Lightning

Figure 11.1 shows the general structure of model training with PyTorch on the left side and with PyTorch Lightning on the right side.

There are some differences, which are indicated by the callout numbers:

❶ When we create our own model class, it inherits from `torch.nn.Module` in PyTorch and from `pl.LightningModule` in PyTorch Lightning.

❷ We use the optimizer at various points in the script in PyTorch, and after we initialize it, we use it to reset the parameters with `optimizer.zero_grad()` and update the parameters with `optimizer.step()`. In contrast, in PyTorch Lightning, we only define the `configure_optimizers()` method in a single location.

❸ In classic PyTorch, we define a training loop in which we process the data, optimizer, loss function, and losses. Here, too, the elements "fly" around freely. However, in PyTorch Lightning, the code is much tidier, and in the model class, we only define the `training_step` and (optionally) `validation_step` methods, which take care of training and validation.

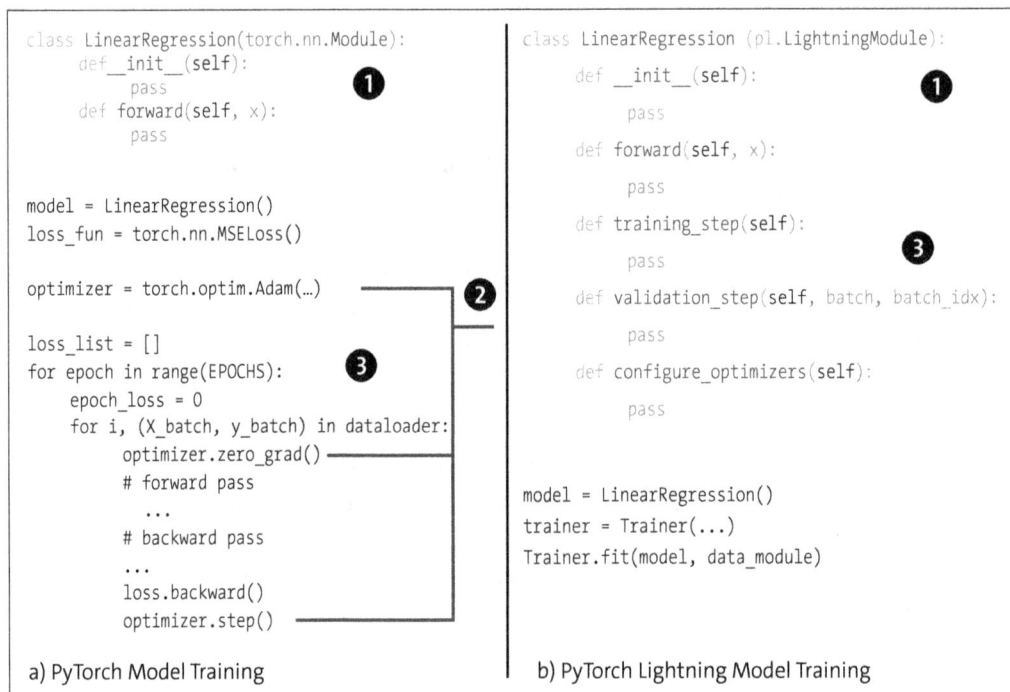

```
class LinearRegression(torch.nn.Module):           class LinearRegression (pl.LightningModule):
    def __init__(self):                                def __init__(self):              ❶
        pass                         ❶                     pass
    def forward(self, x):
        pass                                           def forward(self, x):
                                                           pass
                                                       def training_step(self):
model = LinearRegression()                                 pass                          ❸
loss_fun = torch.nn.MSELoss()
                                                       def validation_step(self, batch, batch_idx):
optimizer = torch.optim.Adam(…)  ──────── ❷              pass

loss_list = []                                         def configure_optimizers(self):
for epoch in range(EPOCHS):      ❸                          pass
    epoch_loss = 0
    for i, (X_batch, y_batch) in dataloader:
        optimizer.zero_grad() ─────                  model = LinearRegression()
        # forward pass                               trainer = Trainer(...)
        ...                                          Trainer.fit(model, data_module)
        # backward pass
        ...
        loss.backward()
        optimizer.step() ─────────

a) PyTorch Model Training                            b) PyTorch Lightning Model Training
```

Figure 11.1 Comparison of PyTorch Model Training with PyTorch Lightning Model Training

In general, we can say that the code in PyTorch Lightning is much tidier. As with classic PyTorch, we must define a model class and create a model instance, but what's new here is that we create a trainer-instance. We then start the training by calling the `fit()` method, as you might be used to from scikit-learn.

11.2   Coding: Model Training

The entire script for this example can be found at *12O_PyTorch_Lightning\pytorch_lightning_approach.py*. We'll train a very simple model from an earlier chapter, but this time, we'll base it on PyTorch Lightning. In Listing 11.1, we load the required packages as usual.

```
import numpy as np
import torch
from torch.utils.data import Dataset, DataLoader
from DataPrep import X, y
import seaborn as sns
import matplotlib.pyplot as plt
from torch.utils.data import random_split
import pytorch_lightning as pl
from pytorch_lightning import Trainer
from pytorch_lightning.callbacks import ModelCheckpoint, EarlyStopping
```

Listing 11.1 PyTorch Lightning Training: Loading Packages

The pytorch_lightning package, which is usually used with the pl alias, is new here. We also load the trainer class from the package, and we load the ModelCheckpoint and Early-Stopping functionalities from the callbacks submodule. We also define the EPOCHS, LEARNING_RATE and BATCH_SIZE hyperparameters, as follows:

```
EPOCHS = 10
LEARNING_RATE = 0.1
BATCH_SIZE = 512
```

We'll again work with the Anxiety dataset from Chapter 2. Listing 11.2 shows the corresponding AnxietyDatamodule, which inherits from the pl.LightningDataModule class.

```
class AnxietyDataModule(pl.LightningDataModule):
    def __init__(self, X, y, batch_size=512, val_split=0.2):
        super().__init__()
        self.X = torch.tensor(X, dtype=torch.float32)
        self.y = torch.tensor(y, dtype=torch.float32)
        self.batch_size = batch_size
        self.val_split = val_split

    def setup(self, stage=None):
        full_dataset = torch.utils.data.TensorDataset(self.X, self.y)

        val_size = int(len(full_dataset) * self.val_split)
        train_size = len(full_dataset) - val_size

        self.train_dataset, self.val_dataset = random_split(
            full_dataset, [train_size, val_size])

        print(f"Training samples: {len(self.train_dataset)}")
        print(f"Validation samples: {len(self.val_dataset)}")
```

```
def train_dataloader(self):
    return DataLoader(self.train_dataset, batch_size=self.batch_size,
        shuffle=True)

def val_dataloader(self):
    return DataLoader(self.val_dataset, batch_size=self.batch_size,
        shuffle=False)
```

Listing 11.2 PyTorch Lightning Training: Creating Dataset and DataLoader

The class has the following methods:

- **__init__**
  In this method, the independent features (X) and the dependent features (y) are assigned to internal objects.

- **setup**
  This method creates training, validation, and test datasets. If necessary, we'll perform appropriate transformations of the data here. Later, the trainer calls this method automatically with `trainer.fit()`.

- **train_dataloader and val_dataloader**
  These two methods return the training data and validation data, respectively. As we discussed in previous chapters, we can use the batch size and the shuffle parameter here for random sampling.

Listing 11.3 shows the `LinearRegression` model class, which inherits from the `pl.LightningModule` class. Here, we define the model architecture, optimizer, and actual model training in a cleanly structured manner in one place.

```
class LinearRegression(pl.LightningModule):
    def __init__(self, input_size, output_size, learning_rate=0.1):
        super().__init__()
        self.linear = torch.nn.Linear(input_size, output_size)
        self.learning_rate = learning_rate
        self.loss_fun = torch.nn.MSELoss()

        # Store training and validation losses for plotting
        self.training_losses = []
        self.validation_losses = []

    def forward(self, x):
        return self.linear(x)

    def training_step(self, batch, batch_idx):
        X_batch, y_batch = batch
```

```
        y_predict = self(X_batch)
        loss = self.loss_fun(y_predict, y_batch.reshape(-1, 1))

        # Log loss for monitoring
        self.log('train_loss', loss, on_step=True, on_epoch=True, prog_bar=True)

        # Store loss for plotting
        self.training_losses.append(loss.item())

        return loss

    def validation_step(self, batch, batch_idx):
        X_batch, y_batch = batch
        y_predict = self(X_batch)
        loss = self.loss_fun(y_predict, y_batch.reshape(-1, 1))

        # Log validation loss
        self.log('val_loss', loss, on_epoch=True, prog_bar=True)

        # Store validation loss for plotting (only at epoch end)
        if batch_idx == 0:   # Only store once per epoch
            self.validation_losses.append(loss.item())

        return loss

    def configure_optimizers(self):
        return torch.optim.Adam(self.parameters(), lr=self.learning_rate)
```

Listing 11.3 PyTorch Lightning Training: Model Class

This model class has the following important methods that we need to define:

- **__init__**
  In this method, we initialize the network layers, the learning rate (self.learning_rate), and the lists for storing the losses (self.training_losses and self.validation_losses).

- **forward**
  We're already familiar with this method from previous models. It's where the network is built.

- **training_step**
  This method is where the actual model training takes place. It has two parameters: the current batch and the index of the current batch (batch_idx). This is where the network training is performed, and a new addition is self.log, a method that's

responsible for monitoring and logging metrics during training. In addition to losses, we can log metrics such as accuracy and model parameters such as the learning rate. We can also specify when the values should be logged, and we can do this at the end of each batch (on_step) or at the end of each epoch (on_epoch).

- **validation_step**
  This function, which performs validation in each epoch, behaves very much like training_step. It's optional.

- **configure_optimizers**
  We use this method to specify the optimizer and hyperparameters such as the learning rate. Lightning takes care of everything else: we no longer need to reset the optimizer to zero or update the model parameters.

Now, we can create an instance of the dataset class (data_module) and an instance of the model (model), as follows:

```
data_module = AnxietyDataModule(X, y, batch_size=BATCH_SIZE)
model = LinearRegression(X.shape[1], 1, learning_rate=LEARNING_RATE)
```

Next, we come to the trainer class, which acts like the conductor who ensures that many different musicians play in unison. Let's take a closer look at some of the skills of this orchestrator:

- It manages the training loop, which means we no longer have to manually set up and manage the epoch loop and batch loop.

- It automatically detects and uses hardware such as GPUs and CPUs. This means we no longer have to manually copy the data and model to the GPU in order to run the model training there.

- It executes callbacks such as checkpoints or early stopping are executed. More on this in the following section.

- For very large model training, it may be necessary and advantageous to perform the training on multiple GPUs simultaneously. The trainer class takes care of this complex task if necessary.

In Listing 11.4, we create a trainer instance and pass parameters for the use of the features we've mentioned. The max_epochs parameter defines the maximum number of epochs, and we can use log_every_n_steps to specify how often logging should take place. We can either set the hardware accelerator (accelerator) manually (e.g., cpu, gpu) or leave the selection to Lightning with auto. We can specify the number of hardware devices via devices, so we can easily control training on multiple GPUs via a parameter.

```
trainer = Trainer(
    max_epochs=EPOCHS,
    log_every_n_steps=10,
    accelerator='auto',
```

```
    devices='auto'
)
```

```
GPU available: True (cuda), used: True
TPU available: False, using: 0 TPU cores
HPU available: False, using: 0 HPUs
```

Listing 11.4 PyTorch Lightning Training: Creating Trainer Instance

On the console, we receive information about the available hardware and which one is being used. In our case, a GPU has been detected and is being used.

We start the training in Listing 11.5 by calling the fit method. This one line does a lot automatically: it executes the training and validation loop, logs parameters, and executes callbacks.

```
trainer.fit(model, data_module)
```

```
You are using a CUDA device ('NVIDIA RTX A2000 8GB Laptop GPU') that has Tensor
Cores.
...
Training samples: 8800
Validation samples: 2200


  | Name     | Type    | Params | Mode
-----------------------------------------------
0 | linear   | Linear  | 31     | train
1 | loss_fun | MSELoss | 0      | train
-----------------------------------------------
31        Trainable params
0         Non-trainable params
31        Total params
0.000     Total estimated model params size (MB)
2         Modules in train mode
0         Modules in eval mode
...
Epoch 9: 100%|??????????| 18/18 [00:00<00:00, 93.87it/s, v_num=4, train_loss_
step=1.200, val_loss=1.330, train_loss_epoch=1.290]
`Trainer.fit` stopped: `max_epochs=10` reached.
Epoch 9: 100%|??????????| 18/18 [00:00<00:00, 91.44it/s, v_num=4, train_loss_
step=1.200, val_loss=1.330, train_loss_epoch=1.290]
```

Listing 11.5 PyTorch Lightning Training: Performing Model Training

Helpful information is displayed on the console during training. In addition to the hardware used, the number of data and the trainable model parameters are displayed. At the end, the respective progress in the epoch is displayed so that you can determine whether the training is going in the right direction or whether you should stop it if necessary.

Now, we want to check whether the training has been successful. To do this, we use seaborn and `matplotlib` in Listing 11.6 to display the training losses that we've saved per batch and the validation losses that we've logged per epoch.

```
#%% Plot training and validation loss
plt.figure(figsize=(10, 6))

sns.lineplot(x=range(len(model.training_losses)), y=model.training_losses,
    label='Training Loss (per Batch)')

if len(model.validation_losses) > 0:
    val_x = np.linspace(0, len(model.training_losses)-1, len(model.validation_
losses))
    sns.lineplot(x=val_x, y=model.validation_losses,
    label='Validation Loss (per Epoch)', marker='o')

plt.title('Training and Validation Loss Over Time')
plt.xlabel('Training Steps')
plt.ylabel('Loss')
```

Listing 11.6 PyTorch Lightning Training: Visualizing Training Losses and Validation Losses

The losses over the epochs are shown in Figure 11.2.

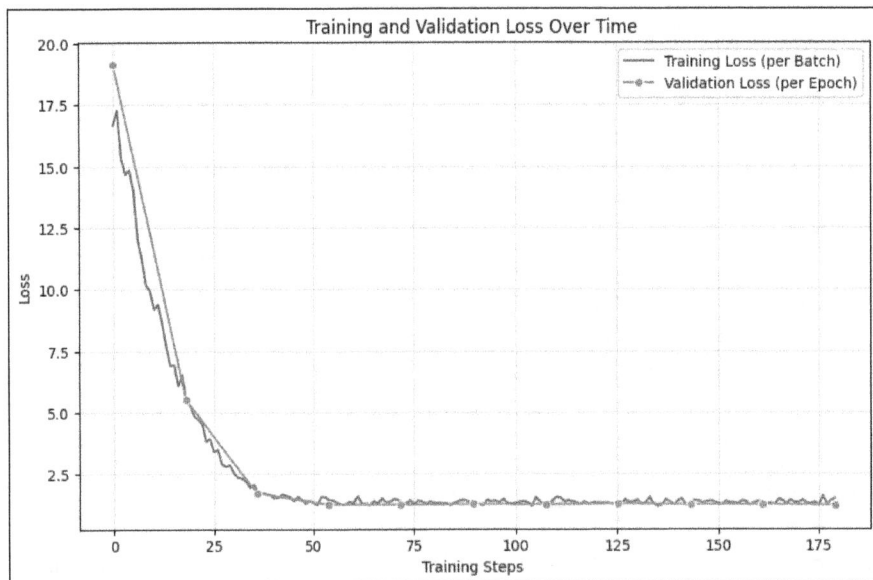

Figure 11.2 PyTorch Lightning Training: Training Loss and Validation Loss

The progression looks good—the losses start high, quickly decrease, and asymptotically converge to one value, which indicates that the model hardly learns any more thereafter. As expected, the training losses are "snappier" as they are based on the batches. Also, the validation losses are slightly higher than the training losses, which reflects normal and desired behavior.

We've already learned about some of the advantages of Lightning. So far, we've only mentioned callbacks briefly, but since it's a very powerful concept, we'll take a closer look at it in the following section.

11.3   Callbacks

Callbacks are classes that provide us with methods that are called at certain points in the training and validation process. For example, we can call callbacks at these times:

- At the beginning of training
- At the end of an epoch
- At the beginning of a validation run
- At the end of the entire training

Callbacks have several advantages:

- They are reusable, and we can use them in different projects.
- Above all, they promote code modularity. We define the logic for training, such as calculating loss and updating weights, in the model class, but we can encapsulate additional tasks (such as saving checkpoints) in separate callback classes. This makes the code cleaner and easier to maintain.
- PyTorch Lightning offers a variety of readymade callbacks that you can easily embed in your training.
- Integrating callbacks is very easy, and we only need to pass them to the trainer as parameters.

As examples, we'll take a closer look at two callbacks: model checkpoints and early stopping.

11.3.1   Model Checkpoints

The `ModelCheckpoint` callback enables us to save the training progress very easily. It monitors a defined metric, such as the validation loss (`val_loss`) or the validation accuracy (`val_accuracy`), and it automatically saves the model weights at the point in time when this metric has reached the best value. If the validation result deteriorates again briefly during model training, which is quite normal, the model is not saved, but only if it represents an improvement. We can restore this model at the end, even if the training continues afterward.

You can also use this callback to check how many checkpoints should be saved, and you can define the save_top_k parameter for this purpose. In our case, we always save a maximum of the best three models. We can use the `filename` parameter to define the naming convention by making parameters such as epoch and val_loss part of the naming scheme. We specify the storage location via dirpath, and we store all checkpoints in this folder. Listing 11.7 shows our example implementation of the ModelCheckpoint.

```
checkpoint_callback = ModelCheckpoint(
    monitor='val_loss',
    dirpath='./checkpoints',
    filename='anxiety_model-{epoch:02d}-{val_loss:.2f}',
    save_top_k=3,
    mode='min',
)
```

Listing 11.7 PyTorch Lightning Callbacks: Creating ModelCheckpoint

The PyTorch Lightning developers make it very easy for us to use this callback. All we have to do is pass all callbacks in a list in the callback parameter of the trainer class. Listing 11.8 shows how to transfer a callback to the trainer class.

```
trainer = Trainer(
    ...
    callbacks=[checkpoint_callback],
    ...
)
```

Listing 11.8 PyTorch Lightning Callbacks: Passing ModelCheckpoint

We save the models in the checkpoints subfolder and follow the naming convention we've defined. We obtained the output by calling tree /f under Windows (or tree -f under Linux) on the command line.

```
├──────checkpoints
|        anxiety_model-epoch=02-val_loss=1.32.ckpt
|        anxiety_model-epoch=04-val_loss=1.32.ckpt
|        anxiety_model-epoch=07-val_loss=1.33.ckpt
```

At a later point in time, we may not want to train the model again but directly load the best model instead. To do this, in Listing 11.9, we only need to call our LinearRegression model class with the load_from_checkpoint method. We must transfer the checkpoint path here, and we could pass it hardcoded or referenced via the checkpoint_callback.best_model_path object. The other parameters (input_size and output_size) result from our model class.

```
best_model = LinearRegression.load_from_checkpoint(
    checkpoint_callback.best_model_path,
    input_size=X.shape[1],
    output_size=1
)
```

Listing 11.9 PyTorch Lightning Callbacks: Loading Best Model

Now, we come to another important callback: early stopping.

11.3.2   Early Stopping

Early stopping is a callback strategy that automatically ends training when the performance of the model no longer improves. This can prevent overfitting and save computing resources during training, and the advantage of this is that the training is terminated precisely when the model begins to adapt too strongly to the training data and the generalization capability decreases.

Figure 11.3 shows the error for the training and validation data over the epochs. It shows a classic progression, in which the training data initially decreases sharply but the reduction in the error becomes smaller over time. The validation loss initially behaves similarly and decreases sharply, but there comes a point at which the validation loss increases again. This is exactly the point at which the model starts to overfit the training data, and the idea of early stopping is to stop the model training exactly at the point of minimum validation loss.

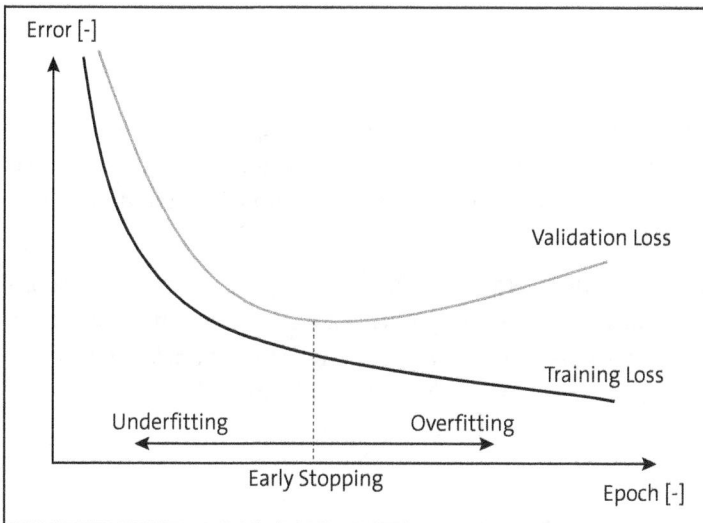

Figure 11.3 Early Stopping

This callback usually monitors the validation loss (val_loss), and if this loss value shows no improvement, it stops the training. It waits for a certain number of epochs, which you can define via the patience parameter. Listing 11.10 shows how to set up the early stopping callback.

```
early_stopping_callback = EarlyStopping(
    monitor='val_loss',
    patience=3,
    mode='min',
    verbose=True
)
```

Listing 11.10 PyTorch Lightning Callbacks: Defining EarlyStopping

Listing 11.11 shows how the earlystopping callback is passed to the callbacks parameter.

```
trainer = Trainer(
    max_epochs=EPOCHS,
    callbacks=[checkpoint_callback, earlystopping],
    ...
)
```

Listing 11.11 PyTorch Lightning Callbacks: Passing EarlyStopping Callback

When the training starts via trainer.fit, we see that the training stops after four epochs, although the maximum number of epochs is ten.

11.4 Summary

In Section 11.1 of this chapter, we learned about the direct comparison between classic PyTorch and model training using PyTorch Lightning. In Section 11.2, we implemented a model with Lightning and discussed the differences in the model class and the actual training using the trainer class.

In Section 11.3, we learned what callbacks are and what advantages they bring. For example, we can define ready-made capabilities in separate callback classes that we can reuse, and we can fall back on many callbacks that are already available. We also got to know model checkpoints and early stopping as extremely helpful modules.

Chapter 12
Model Evaluation, Logging, and Monitoring

"Violence is the last resort of the incompetent."
—Isaac Asimov, Foundation

Training models is often like fighting while blindfolded. You change hyperparameters, adjust the model architecture, or simply try something out on spec because you don't know what else to do. This is the "violence" of which Asimov speaks—an act of desperation that becomes necessary because we lack the right tools. Competence in training deep learning models isn't demonstrated by blind trial and error. Systematic observation, comparison, and evaluation are better and more effective.

Frameworks such as TensorBoard, MLflow, and Weights & Biases (WandB) were developed for this purpose. They help to bring structure to the otherwise chaotic model training process and allow you to track results later. These tools also complement each other, and their uses range from direct observation during training with TensorBoard, to documentation of all decisions and results in the MLflow logbook, to mission control with WandB, which maintains an overview from the outside and enables collaboration.

This chapter will focus on how to evaluate, document and monitor models to transform training from an uncontrolled experiment into a traceable, reproducible process. In Section 12.1, you'll get to know the *TensorBoard* framework, which is the standard visualization tool in the deep learning environment that can help you interactively get to know training metrics, model graphs and distributions. In Section 12.2, we deal with *MLflow*, which is a platform-independent solution for tracking experiments, managing models, and for reproducible deployments. Finally, in Section 12.3, we explore WandB.

Let's start with what's probably the best-known framework: TensorBoard.

12.1 TensorBoard

At its core, TensorBoard is a web interface that clearly summarizes relevant information on model training. Although it was developed by the team for TensorFlow, the other major framework for training deep learning models, you can easily use TensorBoard for PyTorch models as well. It is completely open source and has established itself among users who want to visualize and evaluate training processes. For example, you can use it to create interactive visualizations of data and results.

We start by helping you gain a basic understanding of how TensorBoard works and how you can use it, and we'll focus on how to use the dashboard.

12.1.1   How It Works

Figure 12.1 illustrates how TensorBoard works via the following features:

- **Logging**
  During model training, TensorBoard logs important values (i.e., it records them for later analysis and stores them in a file [the event file]). These can be model metrics (such as accuracy) or hyperparameters (such as the learning rate).

- **Reading event files**
  TensorBoard reads these log files to visualize them in an attractive dashboard.

- **Providing dashboard access**
  When TensorBoard is running, a small web server starts up and gives you access to the dashboard.

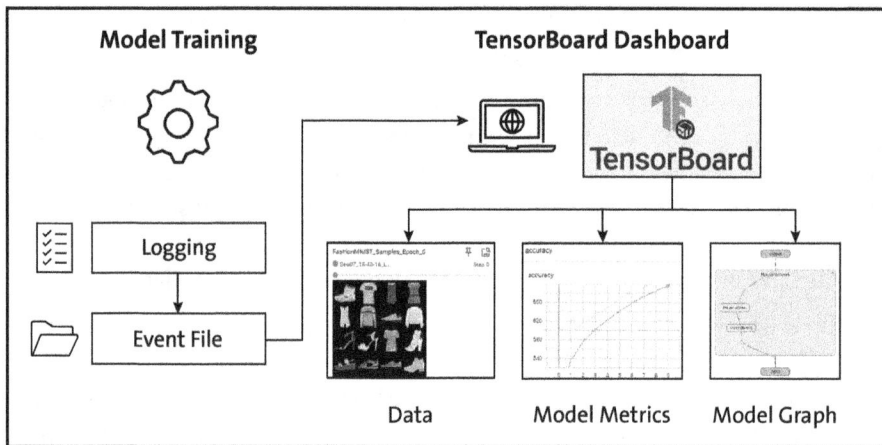

Figure 12.1 TensorBoard Architecture

You can use TensorBoard to display images, audio, text, and model graphs in addition to the classic metrics.

12.1.2   Using TensorBoard

In this section, you'll see how you can practically integrate TensorBoard into model training, and you can use the script at *130_Evaluation\tensorboard_intro.py* as a guide. The required packages are listed in Listing 12.1, and of note here is the SummaryWriter class from the torch.utils.tensorboard module, which is the interface you can use to write log files for TensorBoard.

```
import torch
from torch import nn
from torch.utils.data import DataLoader
from torch.utils.tensorboard import SummaryWriter
from torchvision import datasets
from torchvision.transforms import ToTensor
from torchvision.utils import make_grid
from torchmetrics.classification import Accuracy
```

Listing 12.1 TensorBoard: Loading Required Packages

You can make direct use of the SummaryWriter and create the writer instance, and you can also initialize this class without parameters. You can then store the log files in the *runs* directory with a timestamp, but you can also pass a specific directory via the log_dir parameter. We'll access the instance later to log the desired information.

```
writer = SummaryWriter(log_dir="runs/tensorboard_intro")
```

Next, we can pass the hyperparameters in Listing 12.2.

```
EPOCHS = 10
BATCH_SIZE = 64
LEARNING_RATE = 0.001
```

Listing 12.2 TensorBoard: Defining Hyperparameters

At this point, our focus isn't on a specific dataset, but since we obviously need data for the training, we'll choose the FashionMNIST dataset for the creation of the training and validation data. This is a dataset with 70,000 grayscale images of items of clothing. The images are available with a resolution of 28 × 28 pixels and belong to one of 10 classes, which are balanced and generally more difficult to train than the even more famous MNIST dataset. We create the datasets and DataLoader with the code in Listing 12.3.

```
training_data = datasets.FashionMNIST(
    root="data",
    train=True,
    download=True,
    transform=ToTensor(),
)
validation_data = datasets.FashionMNIST(
    root="data",
    train=False,
    download=True,
    transform=ToTensor(),
)
```

```
train_dataloader = DataLoader(training_data, batch_size=BATCH_SIZE)
validation_dataloader = DataLoader(validation_data, batch_size=BATCH_SIZE)
```

Listing 12.3 TensorBoard: Creating Training and Validation Dataset

In the FashionMnistCnn model class from Listing 12.4, we use a CNN. In previous chapters, we learned about the model structure using nn.Sequential, where we describe the individual layers one by one. But in most cases, we created a model class using a separate class definition—while here, we combine both approaches and use nn.Sequential to give the network more structure.

This divides the network into features and classifiers at the main level. In the features layers, we use the Conv2d and MaxPool2d layers to extract the features from the data, and in the classifier layers, we use the extracted features to derive the class scores.

```
device = "cuda" if torch.cuda.is_available() else "cpu"


class FashionMnistCnn(nn.Module):
    def __init__(self):
        super().__init__()
        self.features = nn.Sequential(
            nn.Conv2d(in_channels=1, out_channels=32, kernel_size=3,
                    padding=1),
            nn.ReLU(inplace=True),
            nn.MaxPool2d(kernel_size=2, stride=2),
            nn.Conv2d(in_channels=32, out_channels=64, kernel_size=3,
                    padding=1),
            nn.ReLU(inplace=True),
            nn.MaxPool2d(kernel_size=2, stride=2),
        )
        self.classifier = nn.Sequential(
            nn.Flatten(),
            nn.Linear(64 * 7 * 7, 128),
            nn.ReLU(inplace=True),
            nn.Linear(128, 10),
        )

    def forward(self, x):
        x = self.features(x)
        logits = self.classifier(x)
        return logits
```

Listing 12.4 TensorBoard: Model Class

In Listing 12.5, we write a function for logging images in TensorBoard. First, we extract the images and the corresponding labels from the dataloader, and then, with make_grid, we create a grid of 4 × 4 images which we log with writer.add_image.

In addition to the grid, we can log individual images with their associated labels. We do this by once again calling writer.add_image, in which individual images are transferred via the img_tensor parameter and the labels are transferred via the tag parameter. We must also specify the data format as CHW to ensure that the images are displayed correctly.

```python
def log_images_to_tensorboard(dataloader, epoch, num_images=16):
    """Log a grid of FashionMNIST images to TensorBoard"""
    # Get a batch of images
    dataiter = iter(dataloader)
    images, labels = next(dataiter)

    # Take only the first num_images
    images = images[:num_images]
    labels = labels[:num_images]

    # Create a grid of images
    img_grid = make_grid(images, nrow=4, normalize=True, scale_each=True)

    # Log to TensorBoard
    writer.add_image(
        tag=f'FashionMNIST_Samples_Epoch_{epoch}',
        img_tensor=img_grid,
        global_step=epoch,
        dataformats="CHW",
    )

    # Also log individual images with their labels
    class_names = ['T-shirt/top', 'Trouser', 'Pullover', 'Dress', 'Coat',
                   'Sandal', 'Shirt', 'Sneaker', 'Bag', 'Ankle boot']

    for i in range(min(4, len(images))):  # Log first 4 individual images
        writer.add_image(
            tag=f'Individual_Images/{class_names[labels[i]]}',
            img_tensor=images[i],
            global_step=epoch,
            dataformats="CHW",
        )
```

Listing 12.5 TensorBoard: Logging of Images

We now come to the creation of the model instance, the loss function, and the optimizer in Listing 12.6. In addition to these old acquaintances, the training and validation metrics (metric_train and metric_val) are listed in this block. This can be helpful as it defines the specific metric outside the training loop.

```
loss_fn = nn.CrossEntropyLoss()
metric_train = Accuracy(task="multiclass", num_classes=10).to(device)
metric_val = Accuracy(task="multiclass", num_classes=10).to(device)
model = FashionMnistCnn().to(device)
optimizer = torch.optim. Adam(model.parameters(), lr= LEARNING_RATE)
```

Listing 12.6 TensorBoard: Loss Function, Model, and Optimizer

It's also very helpful to log the entire model graph, and Listing 12.7 shows exactly how to do this. First, we retrieve a specific image (sample_images) from the DataLoader and log it by using writer.add_graph, whereby we must pass the model in addition to the image.

```
sample_batch = next(iter(train_dataloader))
sample_images = sample_batch[0][:1].to(device)
writer.add_graph(model, sample_images)
```

Listing 12.7 TensorBoard: Logging Graph

The actual model training is shown in Listing 12.8. The most important parts are as follows, and they're marked in bold in the listing:

- **log_images_to_tensorboard**
  We call this function at the beginning of each epoch to log the images.

- **metric_train** and **metric_val**
  We update these metrics during the batch loops to pass the predictions (y_pred) and the actual values (y_train). After we run the last batch loop, metric_train.compute() calculates the aggregated epoch_accuracy_train metric from all previous batches since the last reset(). At the end of the batch loop, we must reset the metrics by using their reset method.

- **writer.add_scalar**
  We call this method to log individual metrics or hyperparameters. A label tag, the scalar value, and a global_step (in our case, the epoch) are passed.

- **writer.flush()**
  This method ensures that all buffered logs are written to the event files. This is helpful when we use TensorBoard interactively, as it ensures that we can view the current model results "live" at the end of each epoch.

- **writer.close()**
  We ultimately apply this outside the training loop to save the latest data and close the writer at the same time. We ignore other add_* methods.

```
train_loss_list, val_loss_list = [], []
train_accuracy_list, val_accuracy_list = [], []
for epoch in range(EPOCHS):
    print(f"Epoch {epoch+1}\n")
    epoch_loss_train = 0
    epoch_loss_val = 0

    log_images_to_tensorboard(train_dataloader, epoch)

    model.train()
    for batch, (X_train, y_train) in enumerate(train_dataloader):
        X_train, y_train = X_train.to(device), y_train.to(device)

        y_pred = model(X_train)
        loss = loss_fn(y_pred, y_train)
        metric_train.update(y_pred, y_train)

        # Backpropagation.
        loss.backward()
        optimizer.step()
        optimizer.zero_grad()

        # store loss and accuracy
        epoch_loss_train += loss.item()
    # compute & reset train accuracy per epoch
    epoch_accuracy_train = metric_train.compute().item()
    metric_train.reset()
    train_loss_list.append(epoch_loss_train)
    train_accuracy_list.append(epoch_accuracy_train)
    writer.add_scalar("train_loss", epoch_loss_train, global_step=epoch)
    writer.add_scalar("train_accuracy", epoch_accuracy_train,
                    global_step= epoch)

    # Validation loop
    model.eval()
    with torch.no_grad():
        for batch, (X_val, y_val) in enumerate(validation_dataloader):
            X_val, y_val = X_val.to(device), y_val.to(device)
            y_pred = model(X_val)
            loss = loss_fn(y_pred, y_val)
            metric_val.update(y_pred, y_val)
            epoch_loss_val += loss.item()
    # compute & reset val accuracy per epoch
    epoch_accuracy_val = metric_val.compute().item()
```

12

```
metric_val.reset()
val_loss_list.append(epoch_loss_val)
val_accuracy_list.append(epoch_accuracy_val)
writer.add_scalar("val_loss", epoch_loss_val, global_step=epoch)
writer.add_scalar("val_accuracy", epoch_accuracy_val, global_step=epoch)

# Ensure events are written to disk for TensorBoard to pick up
writer.flush()
```

```
writer.close()
```

Listing 12.8 TensorBoard: Training Model

After or during the model training, which also includes logging, we can check the results by opening the TensorBoard dashboard. All we have to do is execute the following on the command line:

```
tensorboard --logdir=runs
```

TensorBoard 2.19.0 at http://localhost:6006/ (Press CTRL+C to quit)

We must also pass the logdir parameter and make sure it refers to the logging directory runs. The URL via which we can reach the server is specified in the output, and we can access this URL by pressing CTRL while clicking our mouse. Alternatively, we can access the browser that's integrated in VSC (or Cursor) by clicking **File • View • Command Palette** or pressing CTRL + SHIFT + P.

That will open a menu where you can search for "Browser" and find **Simple Browser: Display**. In the next section, we'll find out how to use the dashboard.

12.1.3   TensorBoard Dashboard

The dashboard offers various tabs in the menu bar:

- **Time Series**
  Here, you can view metrics such as losses or accuracy over time. It's particularly helpful to overlay multiple training runs to reveal trends and differences.

- **Scalars**
  This area provides a quick overview of individual key figures, allowing you to grasp the development of key metrics at a glance.

- **Images**
  This area visualizes image data that was logged during training, like input images for computer vision tasks, reconstructed images for autoencoders, or feature maps. This area allows you to check what the input data looks like and whether the model is actually learning the desired structures.

■ **Graphs**

The model is displayed as a graph that makes it very easy for you to understand how the individual layers are connected to each other.

Figure 12.2 shows an example of the `train_accuracy` for scalars over the number of epochs. The graphs in TensorBoard are usually interactive, so you can freely zoom in and out.

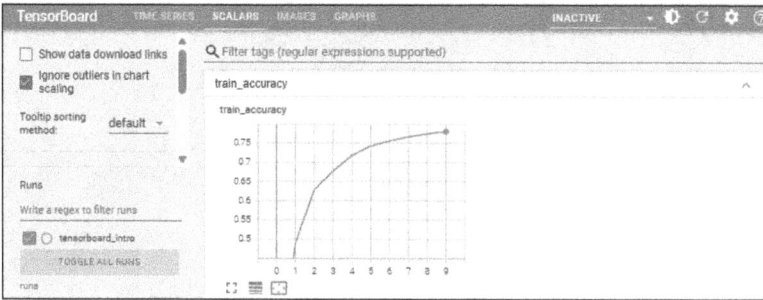

Figure 12.2 TensorBoard: Scalars

Figure 12.3 illustrates the **Images** tab, where you can see all the images in a 4 × 4 grid or, as we've specified, individual images with their corresponding categories.

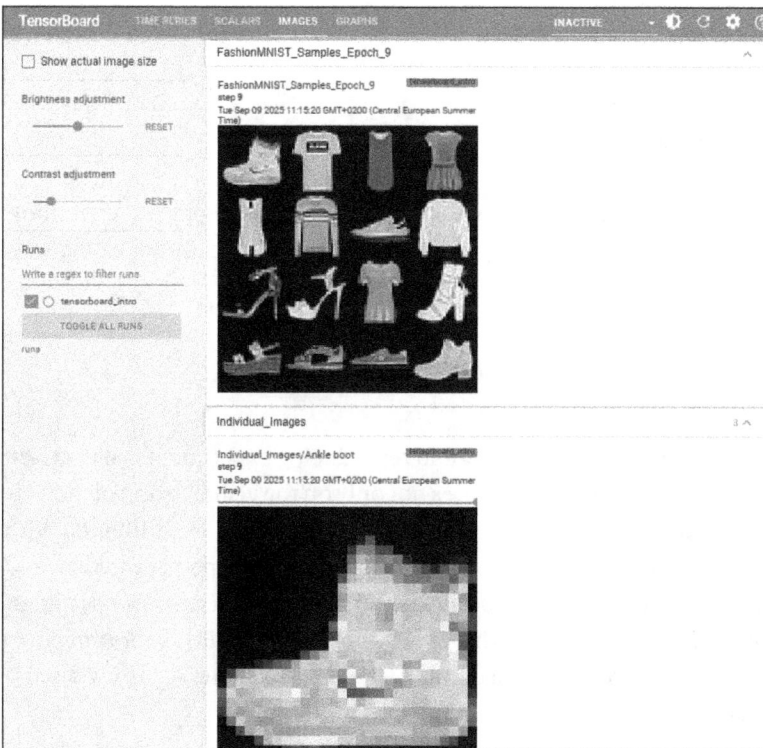

Figure 12.3 TensorBoard: Images

12

371

Figure 12.4 shows the model graph, which shows the advantages of the approach with module blocks. We can expand the individual blocks (such as classifier or features) to display the network layers defined in them.

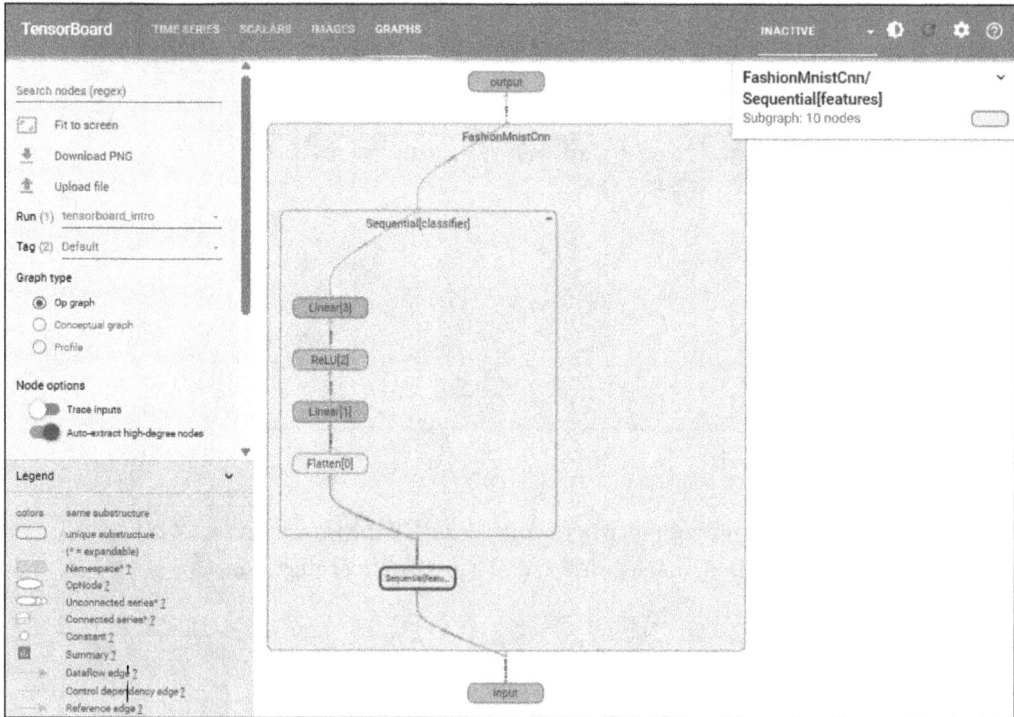

Figure 12.4 TensorBoard: Graph

In this section, you've gotten to know TensorBoard as a tool for logging relevant information about the model training locally and presenting it in a visually appealing way. This will be extremely helpful for reproducing the training later.

12.2   MLflow

MLflow is a platform for tracking, comparing and reproducing experiments. In contrast, to TensorBoard, the MLflow framework focuses not only on the visualization of individual training sessions but also on the systematic management of experiments. With MLflow, you can store and compare different model versions, hyperparameters, and results in a central location. MLflow also ensures that research and development remain traceable, and the complete script can be found at*130_Evaluation\mlflow_ intro.py*. As a large part of the code from Section 12.1.2 is repeated here, we'll only discuss the relevant changes.

We'll log the data, find out how to save and load models, and get acquainted with the MLflow dashboard.

12.2.1 Data Logging

Before we can log data, we must initialize MLflow with the `mlflow.set_tracking_uri` method as in Listing 12.9. This URI can be a local file, a local server, a remote server, or a database. In our case, we're working with the local *mlruns* folder, where we'll save all experiments, runs, and metrics for later traceability. With the `set_experiment` method, for example, we can assign a unique and comprehensible name to the experiment: in this case, `Fashion MNIST`.

```
import mlflow
mlflow.set_tracking_uri(uri="file:./mlruns")
mlflow.set_experiment(experiment_name="Fashion MNIST")
```

Listing 12.9 MLflow: Initialization

The folder structure shown in Listing 12.10 looks like this:

- All information about the experiment with ID 323771178417362930 is stored in the folder.
- There, you'll find a corresponding entry for each run: in our case, 62CE5EB539404B30947B9A5782DFE438.
- All artifacts, metrics, data, parameters, and models are stored in the run.

```
tree mlruns
```

```
Volumeseries number : 6EE5-4D85
C:\PROJECTS\BOOKSARTICLES\PYTORCHBOOK\CODE\130_EVALUATION\MLRUNS
├───.trash
├───323771178417362930          # Experiment ID
│   └───62ce5eb539404b30947b9a5782dfe438          # RUN-ID
│       ├───artifacts
│       │   └───fashion_mnist_model
│       │       └───data
│       ├───metrics
│       ├───params
│       └───tags
└───models
    └───fashion_mnist_model
    ├───version-1
    └───version-2
```

Listing 12.10 MLflow: Folder Structure

We carry out the model training within the scope of `mlflow.start_run()`, and we use the following logging options in the code shown in Listing 12.11:

- `mlflow.log_params()`
  We can log parameters with this function.

- `mlflow.log_metric()`
  We can save loss information (`train_loss` and `val_loss`) or accuracies with this function, which requires a name and the value of the metric. Since these metrics change during model training with each batch or epoch, we must pass the current step via the parameter step.

- `mlflow.log_artifact()`
  MLflow can use this function to save additional files that are created during the experiment. These can be configuration files, trained models, visualizations, or reports. We log the model summary that was saved in the *model_summary.txt* file.

```python
with mlflow.start_run():
    # Log training parameters
    params = {
        "epochs": EPOCHS,
        "learning_rate": LEARNING_RATE,
        "batch_size": BATCH_SIZE,
        "loss_function": loss_fn.__class__.__name__,
        "metric_function": metric_fn.__class__.__name__,
        "optimizer": "Adam",
    }
    mlflow.log_params(params)

    train_loss_list, val_loss_list = [], []
    train_accuracy_list, val_accuracy_list = [], []
    for epoch in range(EPOCHS):
        ...
        mlflow.log_metric("train_loss", epoch_loss_train, step=epoch)
        mlflow.log_metric("train_accuracy", epoch_accuracy_train, step=epoch)

        # Evaluation
        mlflow.log_metric("val_loss", epoch_loss_val, step=epoch)
        mlflow.log_metric("val_accuracy", epoch_accuracy_val, step=epoch)

    with open("model_summary.txt", "w", encoding="utf-8") as f:
        f.write(str(summary(model, input_size=(1, 1, 28, 28))))
    mlflow.log_artifact("model_summary.txt")
```

Listing 12.11 MLflow: Logging During Training

We've seen various ways to log information during model training, and now, we come to the question of how to save and load the trained models.

12.2.2 Saving and Loading the Model

We can use the approaches shown in Listing 12.12 to save the models, as follows:

- `log_model()`
  You use this to save the model of the current run. The model is tied to this run, and this approach is more suitable for small experiments and teams.

- `register_model()`
  Alternatively, you can use this to register the model in the model registry. Multiple versions of the same model are possible, and this approach is suitable for production systems.

```
model_info = mlflow.pytorch.log_model(model, "fashion_mnist_model")

# model registration
mlflow.register_model(
    model_uri=model_info.model_uri,
    name="fashion_mnist_model"
)
```

Listing 12.12 MLflow: Model Saving

You can determine which models have been registered with `mlflow.search_registered_models()`, as shown in Listing 12.13.

```
#%% get all registered models
registered_models = mlflow.search_registered_models()
print(registered_models)
```

```
[<RegisteredModel: aliases={}, creation_timestamp=1757431063233, description=
None, last_updated_timestamp=1757431286263, latest_versions=[<ModelVersion:
aliases=[], creation_timestamp=1757431286263, current_stage='None', description=
None, last_updated_timestamp=1757431286263, name='fashion_mnist_model', run_id=
'62ce5eb539404b30947b9a5782dfe438', run_link=None, source='file:c:/Projects/
BooksArticles/PyTorchBook/Code/130_Evaluation/mlruns/323771178417362930/
62ce5eb539404b30947b9a5782dfe438/artifacts/fashion_mnist_model', status='READY',
status_message=None, tags={}, user_id=None, version=2>], name='fashion_mnist_
model', tags={}>]
```

Listing 12.13 MLflow: Determination of Registered Models

You can store the model on your local system (or on a server if you set up the system in that way). To load the model, you need to know its location, and for this, you use the model URI, which is similar to a file name. You use the model_uri parameter to load the model.

```
model_uri = registered_models[0].latest_versions[0].source
print(model_uri)
```

file:c:/Projects/BooksArticles/PyTorchBook/Code/130_Evaluation/mlruns/ 323771178417362930/62ce5eb539404b30947b9a5782dfe438/artifacts/fashion_mnist_ model

You load the models with the load_model() method, to which you transfer the model_ uri. The model will then be ready for use, and you'll be able to make predictions directly with predict(), as follows:

```
#%% Load model from local storage
loaded_model = mlflow.pyfunc.load_model(model_uri=model_uri)
loaded_model.predict(np.random.randn(1, 1, 28, 28).astype(np.float32))
```

Now, you know how to save and load models, so in the next section, we'll look at the MLflow dashboard.

12.2.3   MLflow Dashboard

Before you can use the dashboard, you must start the internal MLflow server locally by executing the following command on the command line. You can then start the server and access it via the specified URL:

```
mlflow ui --host 127.0.0.1 --port 8080
```

INFO:waitress:Serving on http://127.0.0.1:8080

Figure 12.5 shows an overview of the experiments listed at the top left. The experiments include one or more completed runs, and the logged metrics are presented for the selected run. If there's a very large number of runs, the filters can help you to find the specific run again.

In addition to the experiments, you can register the models with MLflow. Figure 12.6 shows the overview of the model registry for the fashion_mnist_model class, and several versions of the model have already been created.

In this section, you've become familiar with MLflow, which is a central platform for making machine learning experiments comprehensible, comparable, and reproducible. MLflow takes care of tracking runs, storing parameters and metrics, and managing artifacts and models. This allows you to provide models in various serving environments, such as locally or in the cloud. This turns every model training session into a documented experiment, the results of which you can evaluate, compare, and repeat later.

Figure 12.5 MLflow: Experiments

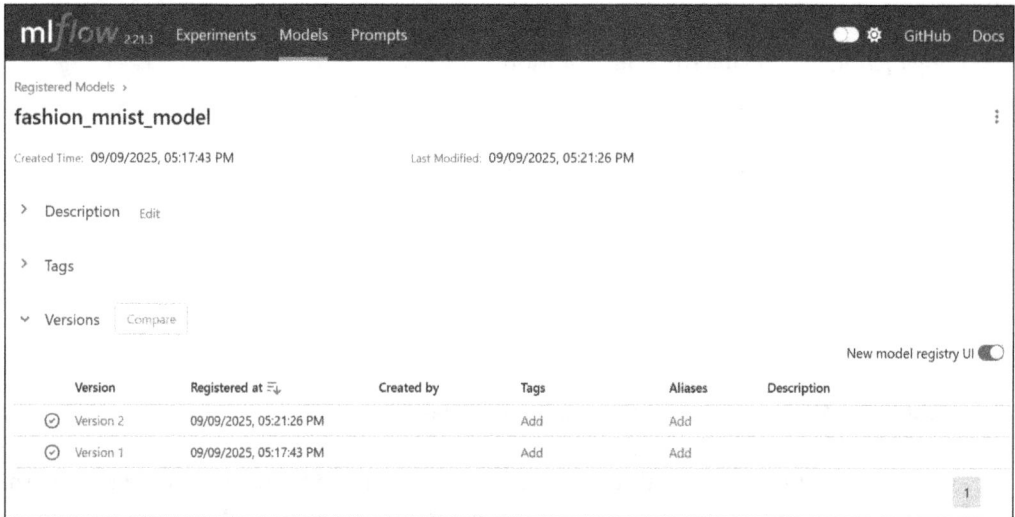

Figure 12.6 MLflow: Registered Models

12.3   Weights & Biases: WandB

When your experiments become complex, it's no longer enough to just visualize them locally or write a single logbook. Development teams need tools that centralize the results and make them usable together, and that's exactly where *Weights & Biases* (WandB) comes in. It's a cloud-based platform that not only stores metrics and artifacts but also focuses on collaborative work.

If you think of TensorBoard as a radar and MLflow as a logbook, WandB is mission control. All data streams can converge there, and the entire team can maintain an overview.

To use WandB, you need to set up an account at *wandb.ai*. The various possibilities offered by WandB are described in the script at *130_Evaluation\wandb_intro.py*.

Most of the code in this section is based on the model training in the previous sections of this chapter. Therefore, I will only go into the relevant parts that are required for use. We'll initialize the framework, use it to log metrics and artifacts, and then work with sweeps, which can be helpful in the performance of parameter studies.

12.3.1   Initialization

You must load the wandb package and authenticate yourself via wandb.login(), as follows:

```
import wandb
wandb.login()
```

Then, you create the hyperparameters with the code in Listing 12.14, add them to a config dictionary, and log this configuration.

```
EPOCHS = 10
BATCH_SIZE = 32
LEARNING_RATE = 0.001
config = {
    "epochs": EPOCHS,
    "batch_size": BATCH_SIZE,
    "learning_rate": LEARNING_RATE,
}
```

Listing 12.14 WandB: Hyperparameters and Configuration

The current model run is always part of a project, and you must first initialize it with wandb.init(). You need to give it a model name, and it makes sense to pass the hyperparameters to the config parameter here, as follows:

```
run = wandb.init(project="fashion_mnist_model",
                notes="CNN for Fashion MNIST, reduced batch size",
                config=config)
```

12.3.2   Logging Metrics

You can log metrics within the training loop via wandb.log(), as follows:

```
for epoch in range(EPOCHS):
    ...
    for batch, (X_train, y_train) in enumerate(train_dataloader):
        ...
        wandb.log({
            "train_loss": epoch_loss_train,
            "train_accuracy": epoch_accuracy_train,
        })
```

You can view this information live online during model training. The overview in Figure 12.7 shows all runs within the project, along with the hyperparameters and model results.

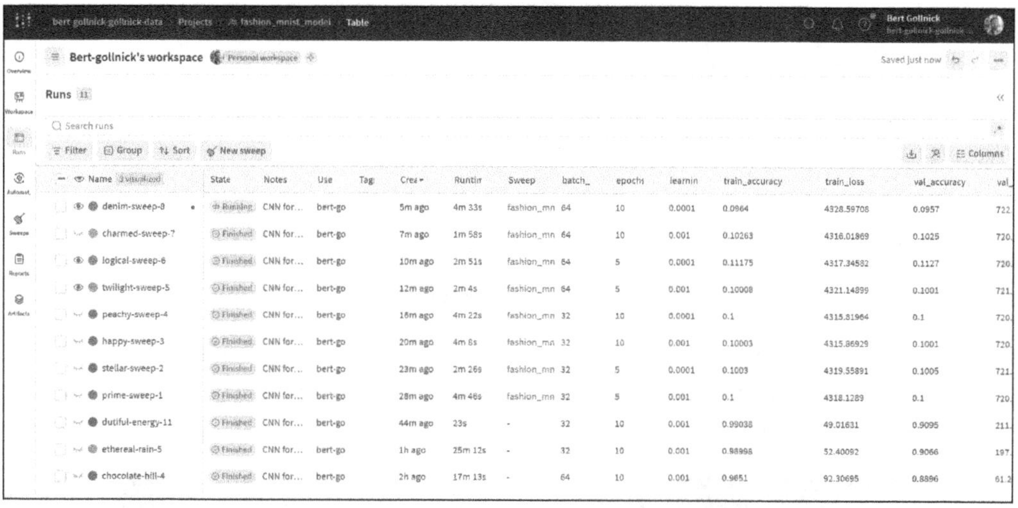

Figure 12.7 WandB: Runs

If you click on a specific run, you'll see the metrics shown in Figure 12.8. The individual graphics are linked to each other so that you can view the information on a selected epoch at a glance.

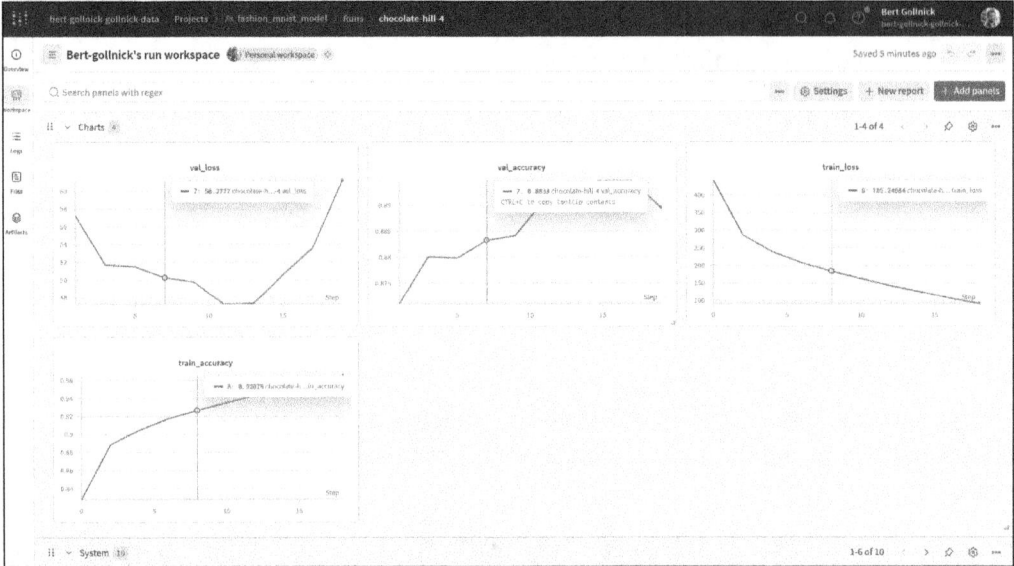

Figure 12.8 WandB: Metrics of Run

12.3.3   Logging Artifacts

In addition to classic metrics, you can use WandB to save files with the `wandb.save()` method, as follows:

```
torch.save(train_loss_list, "train_loss_list.pth")
wandb.save("train_loss_list.pth")
```

Figure 12.9 shows files that are assigned to a given run, and you can find these by clicking **Files** in the menu bar on the left.

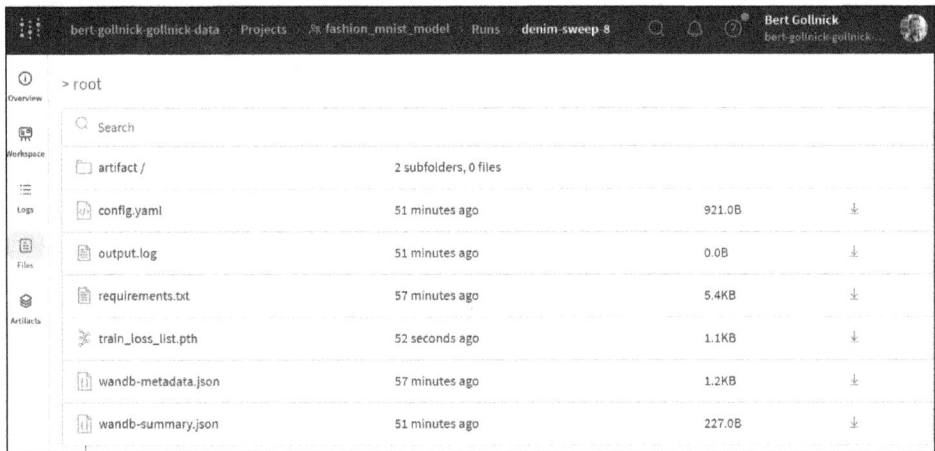

Figure 12.9 WandB: Files

You can save artifacts (i.e., files that are used in the context of an experiment) in a slightly different way. WandB distinguishes between file and model artifacts, and Listing 12.15 shows the procedure for logging artifacts (which must be available as files):

1. Save the Python object.
2. Initialize the `model_artifact`.
3. Add the file to this model instance by using `add_file()`.
4. Log the artifact by using `wandb.log_artifact()`.

```
torch.save(model.state_dict(), "model.pth")
model_artifact = wandb.Artifact("model_artifact", type="model")
model_artifact.add_file("model.pth")
wandb.log_artifact(model_artifact)
```

Listing 12.15 WandB: Saving Model Artifacts

You can then find the artifact as shown in Figure 12.10 by clicking **Artifacts** in the menu bar on the left.

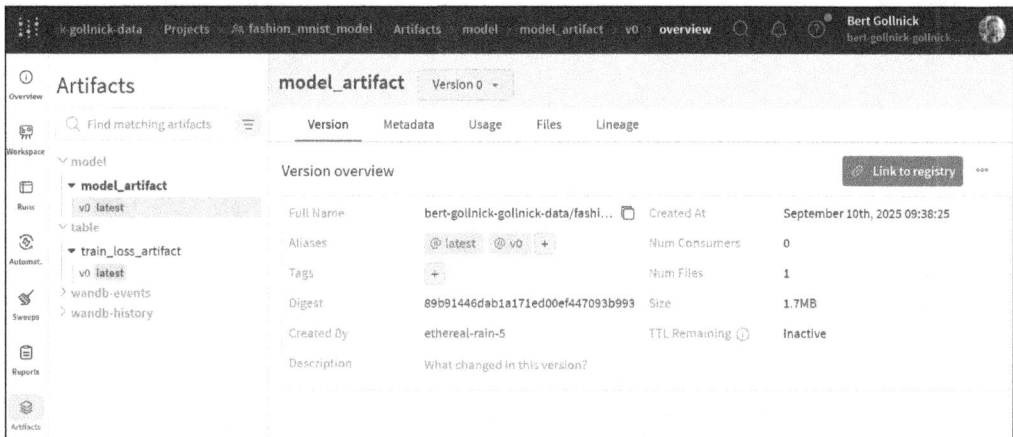

Figure 12.10 WandB: Model Artifact

You can also save tables as artifacts, as follows:

1. Create the table by using `wandb.Table()`.
2. Create an artifact by using `wandb.Artifact()` and the `table` type.
3. Add the `train_loss_table` table object to the `train_loss_artifact` by using `add()`.
4. Log the artifact by using `log_artifact()`.

Listing 12.16 shows the code for saving a table artifact.

```
train_loss_table = wandb.Table(dataframe=pd.DataFrame(
    {"train_loss": train_loss_list,
     "train_accuracy": train_accuracy_list}))
train_loss_artifact = wandb.Artifact("train_loss_artifact", type="table")
train_loss_artifact.add(train_loss_table, "train_loss_table")
wandb.log_artifact(train_loss_artifact)
```

Listing 12.16 WandB: Saving Table Artifacts

Figure 12.11 shows the table artifact in the frontend. You can also find it again by clicking **Artifacts** in the menu bar on the left and selecting **Table** from the dropdown list.

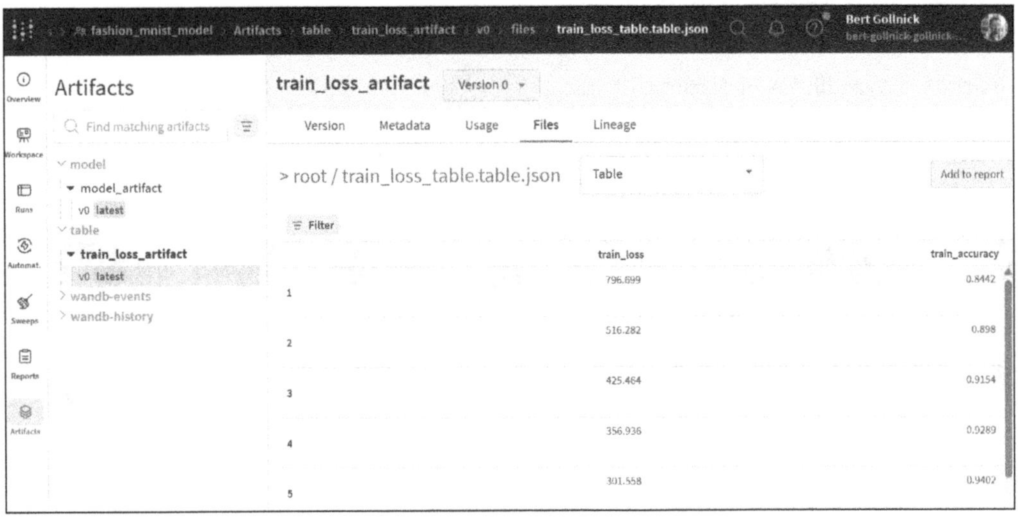

Figure 12.11 WandB: Table Artifact

Now, we have come to some very powerful tools in WandB: sweeps, which we can use to automatically optimize hyperparameters.

12.3.4   Sweeps

Up to now, you've adjusted the model parameters manually and incorporated your know-how. But how can you carry out complex optimization and automatically test parameter combinations? This is where sweeps come into play, and we can easily illustrate this idea with an example.

In Listing 12.17, the parameters for a grid search are defined and possible values are defined for various hyperparameters (e.g., 5 and 10 for the epoch). The grid search then creates all possible combinations of the parameters (i.e., a grid in the hyperparameter space), and models are then trained and validated for all these combinations.

```
sweep_config = {
    "name": "fashion_mnist_model",
    "method": "grid",
    "parameters": {
        "epochs": {"values": [5, 10]},
        "batch_size": {"values": [32, 64]},
        "learning_rate": {"values": [0.001, 0.0001]},
    },
}
```

Listing 12.17 WandB: Sweep Configuration

Listing 12.18 shows the `sweep_run()` function for running and logging a single training session. The run is initialized, the known model training is carried out, and at the end of the epoch, the losses (`train_loss`) and the accuracy (`train_accuracy`) are logged with `wandb.log()`.

```
def sweep_run():
    run = wandb.init(project="fashion_mnist_model",
                     notes="CNN for Fashion MNIST, reduced batch size",
                     config=config)
    model = FashionMnistCnn().to(device)
    train_loss_list, val_loss_list = [], []
    train_accuracy_list, val_accuracy_list = [], []
    for epoch in range(EPOCHS):
        ...
        wandb.log({
            "train_loss": epoch_loss_train,
            "train_accuracy": epoch_accuracy_train,
        })

    wandb.finish()
```

Listing 12.18 WandB: Function for Conducting Training

Now, as shown in Listing 12.19, we can initialize the grid search with `wandb.sweep()` and then start it with `wandb.agent()`.

```
sweep_id = wandb.sweep(sweep_config, project="fashion_mnist_model")
wandb.agent(sweep_id, function=sweep_run)
```

Listing 12.19 WandB: Performing Sweep Simulations

We can then carry out simulations for the combination of all hyperparameters, and we can visualize this as shown in Figure 12.12.

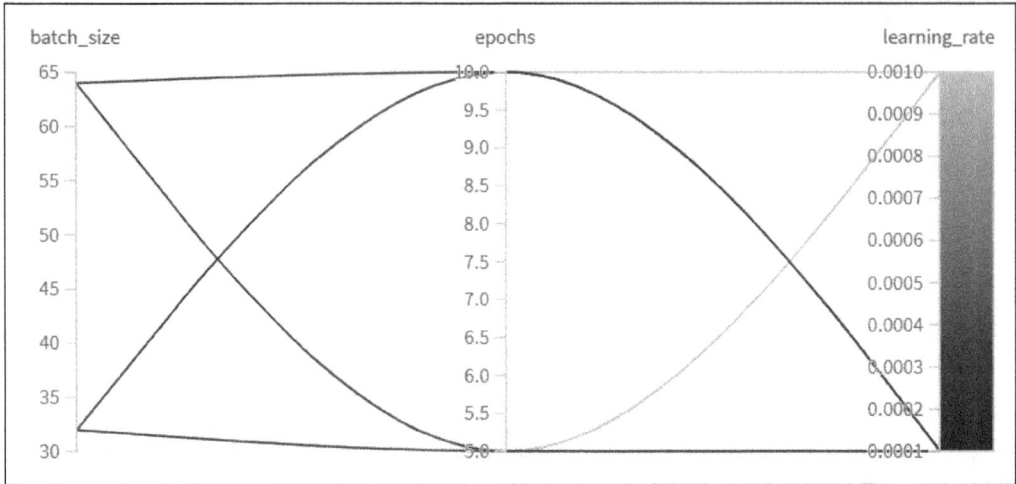

Figure 12.12 WandB: Sweep

Using this approach can save us a lot of manual effort, and we can filter out the most promising parameter combination from the runs we've carried out in the dashboard with little effort.

12.4   Summary

In this chapter, you've learned about three central tools for model evaluation, logging, and monitoring:

- **TensorBoard**
  This enables the visualization of training metrics, model graphs, and input data in real time. It provides a direct view of the current state of training.

- **MLflow**
  This extends the approach to include the systematic management of experiments. In addition to hyperparameters and metrics, it stores models and artifacts to ensure that experiments remain reproducible and comparable.

- **WandB**
  This offers a platform that enables real-time tracking, hyperparameter sweeps, artifact management, and real-time teamwork. It ensures that teams can coordinate complex projects internally.

Together, these frameworks make it clear that modern deep learning development isn't just about training itself but also about visualizing, documenting, and coordinating the results. This creates a comprehensible, structured development process from individual experiments.

Chapter 13
Deployment

"There will be no deployment on Friday."
—*DevOps wisdom*

This simple sentence reveals a lot about the risks of deployments, because nobody wants to spend the weekend searching for errors. But the core of the statement is also that deployment means responsibility. Your colleagues and customers are relying on you. Deployment is about integrating the development—for example, of a new deep learning model—into real business processes and making the models available for others to use.

In Section 13.1, we'll look at different deployment strategies and how they can look. In Section 13.2, we'll learn how to provide a model locally, which is usually the first step in testing whether everything works as desired. Only after that can we access resources, usually in the cloud, to provide the model there. We'll get to know two representatives of this: Heroku and Microsoft Azure.

In Section 13.3, we'll cover deployment with Heroku, which is practical for small projects and start-ups because it's straightforward and very easy to implement. In Section 13.4, we'll cover how to deploy our model using Microsoft Azure, which may be appropriate if you work in a large company or want to map a very complex workflow.

13.1 Deployment Strategies

The specific strategy or architecture you'll want to use to deploy an application depends heavily on the application itself. We can only go into simple approaches here. The topic of deployment is an integral part of the more comprehensive process of *continuous integration/continuous delivery* (CI/CD), which is a central practice in modern software development. It's a series of automated processes that aim to bring software changes quickly and reliably from the developer to the end user.

In this context, we often refer to a *CI/CD pipeline*, which is an automated chain of steps that is run through every time the code is changed. CI focuses on integrating code from multiple developers into a shared repository (usually a Git repository) in an automated manner. An automatic process is triggered each time code is uploaded. This process can also include automatic tests (*unit tests* and *integration tests*) or automatic builds (combining the code into an executable file or a Docker container). The aim is to identify problems as quickly as possible so that the process stops and no faulty code is deployed.

CD can have two meanings: *continuous delivery* or *continuous deployment*. This depends on whether the changes are deployed directly (continuous deployment) or are initially only made available for manual release (continuous delivery).

A simple diagram showing the data flow of an internet service deployment is shown in Figure 13.1. Here, the end user accesses the service via their web browser. In doing so, they communicate implicitly with the frontend server, which displays the website correctly and communicates with the user's browser by using a communication protocol.

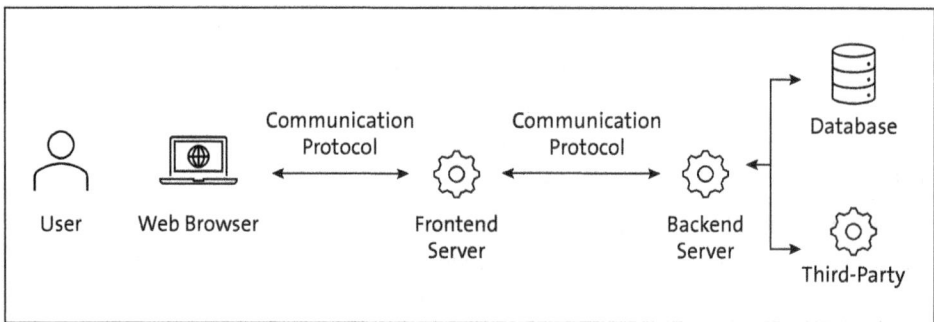

Figure 13.1 Deployment Architecture with Frontend and Backend

The frontend server is generally not static but reacts to user input. Let's look at the example of a chatbot, in which the user can enter text that is then processed. The frontend server communicates with one or more backend servers, and various services (called *microservices*) run on the backend server. The entered text is then sent to a backend server, which processes the data. In the example, the text would be forwarded to an LLM provided by a third-party provider (such as OpenAI) and data may be retrieved from databases. The decisive factor here is that the backend servers don't communicate directly with the end user. Instead, communication only takes place via the middleman of the frontend server.

How is the data exchanged? There are several different communication protocols used for communication between different services, and the most common are REST API, gRPC, and GraphQL:

- **REST API**
  This is first and foremost among communication protocols, and it's one of the most important pillars of the internet because almost all communication between services is based on it. The REST API is based on the HTTP protocol, and its actions are defined by HTTP methods such as GET, POST, PUT, and DELETE. REST is simple and widely used, and it uses the standardized JSON format for data exchange. This makes interaction with clients such as browsers very straightforward.

- **gRPC**

  This is a modern framework developed by Google. Unlike REST, which exchanges text-based data, this protocol uses special gRPC buffers for binary communication. This makes this protocol faster and more efficient because it has to transfer less data. When high speed and low latency are important, this is the protocol of choice.

- **GraphQL**

  This is a query language for APIs developed by Facebook (now Meta). It has no rigid endpoints (as does REST), so clients can request exactly the data that they need. They can retrieve data from multiple resources in a single request, which would require multiple requests with REST. This makes GraphQL suitable for apps that use complex data structures.

In the following sections, we'll use a trained model and initially make it available on the local computer using the REST API.

13.2   Local Deployment

In this section, we'll create a REST API service that responds to requests. A REST API service listens to requests at different endpoints, as shown in Figure 13.2. An *endpoint* is the specific URL to which a client sends a request to interact with the REST API, and it's also the access point for using the server.

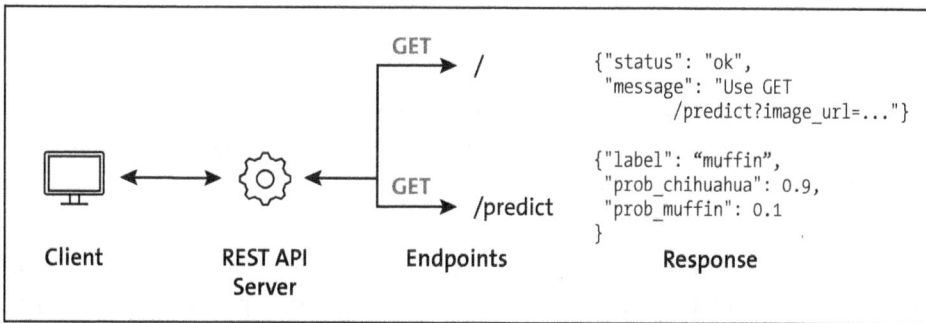

```
GET
      ──▶   /          {"status": "ok",
                         "message": "Use GET
                                  /predict?image_url=..."}

                        {"label": "muffin",
GET                      "prob_chihuahua": 0.9,
      ──▶   /predict     "prob_muffin": 0.1
                        }

Client   REST API    Endpoints        Response
         Server
```

Figure 13.2 REST API Server

If you call up my website in the browser via *https://gollnickdata.de*, it will implicitly call up the default "/" endpoint. Then, if you call up the information on my book *Generative AI with Python* on the website, for example, you'll see in the address bar of the browser that the endpoint has changed, as in *https://gollnickdata.de/#/genai_book_de*. In this way, we'll create a REST API with two endpoints: "/" and "/predict".

The following section is dedicated to the implementation of this REST API, which is the system that we'll subsequently deploy and test.

13.2.1 API Development

We'll deploy the binary classification model from Chapter 4, which sorts images into the categories of "Muffin" and "Chihuahua." To do this, we must export the trained model. In the script at *050_ComputerVision\010_BinaryImageClassification\binary_img_classification.py*, the model is first trained and then exported as shown in Listing 13.1, and the ONNX format is also applied here.

> **ONNX**
>
> *Open Neural Network Exchange* (ONNX) is an open format that improves interoperability among different deep learning frameworks such as PyTorch, TensorFlow, and Keras. With ONNX, you can train a model with PyTorch and then export it to another runtime environment. This makes it particularly useful for deployment, as many production environments use optimized runtimes that may not be compatible with the training framework.

```
dummy_input = torch.randn(1, 1, 32, 32)
torch.onnx.export(model=model.to("cpu"),
                  args=dummy_input,
                  f="bin_class_model.onnx",
                  verbose=True,
                  opset_version=12)
```

Listing 13.1 Model Export via ONNX

This model file is then copied to the working folder for this section (*140_Deployment\heroku*). Now, we must develop the REST API server that will answer the requests in file *140_Deployment\heroku\app.py*. The core of the REST API will be the /predict endpoint, which receives an image URL as a parameter and returns a prediction as to whether the image shows a muffin or a Chihuahua.

To do this, we need the packages listed in Listing 13.2. We create the REST API with the fastapi package, and we use pydantic for type safety and PIL to process the image. The REST API starts with uvicorn, where uvicorn is the asynchronous server that executes the FastAPI app.

```
from fastapi import FastAPI, HTTPException, Query
import uvicorn
from pydantic import HttpUrl
import onnxruntime as ort
from PIL import Image
import numpy as np
import io
import requests
```

Listing 13.2 REST API Server: Loading Packages

Next, you must create a FASTAPI app instance. If you want API documentation to be created automatically, you can pass the title parameter, as follows:

```
app = FastAPI(title="Muffin vs Chihuahua Classifier")
```

Now, we must load the previously saved ONNX model in Listing 13.3 and create an inference session for it. WE pass the path to the model file, and with the providers parameter and the CPUExecutionProvider value, we force an execution on the CPU.

```
session = ort.InferenceSession(
    "bin_class_model.onnx",
    providers=["CPUExecutionProvider"],
)
```

Listing 13.3 REST API Server: Loading Model

The default endpoint "/" is defined in Listing 13.4, and we'll use it later to test the accessibility of the server. We use the @app.get("/") decorator to define a GET route to the "/" path. Then, we define the function. The name of the function (root in this case) doesn't matter, so we can use any name. This function doesn't receive any parameters and returns a static dictionary.

```
@app.get("/")
def root():
    return {"status": "ok", "message": "Use GET /predict?image_url=..."}
```

Listing 13.4 REST API Server: Defining Default Endpoint

Now, let's turn to the implementation of the "/predict" endpoint in Listing 13.5. We pass a URL to this endpoint as a parameter, and we can basically implemented this via a GET or POST request. Here, we use the parameter transfer via GET, and we pass the parameter directly as part of the URL. Let's look at the individual steps we must run through:

1. We pass the image URL as an HttpUrl. We must download the image stored under the URL, and since images can often only be downloaded via a browser, we must "emulate" a browser by using the headers object, which is passed when we download the image by using requests.get(). Then, it's converted into a NumPy array with the correct dimensions.

2. We must preprocess the downloaded image by reducing its dimensions to 32 × 32 pixels and converting it to grayscale.

3. We pass the image through the model in a forward pass to make a prediction. Based on the probability value (prob), the model assigns a label (muffin or chihuahua).

4. The results are returned as JSON with the label, prob_chihuahua, and prob_muffin keys.

```
@app.get("/predict")
def predict(image_url: HttpUrl = Query(...)):
    # 1. Image upload to server
    try:
        headers = {
            "User-Agent": "Mozilla/5.0 (Windows NT 10.0; Win64; x64) AppleWeb-
Kit/537.36 (KHTML, like Gecko) Chrome/126.0.0.0 Safari/537.36",
            "Accept": "text/html,application/xhtml+xml,application/xml;q=
0.9,image/avif,image/webp,image/apng,*/*;q=0.8,application/signed-exchange;v=
b3;q=0.7",
            "Accept-Language": "en-US,en;q=0.9",
            "Referer": "https://www.google.com/",
            "Connection": "keep-alive",
        }
        resp = requests.get(str(image_url), timeout=15, headers=headers)
        resp.raise_for_status()
        image_bytes = resp.content
    except Exception as e:
        raise HTTPException(status_code=400, detail=f"Failed to download image:
{e}")

    # 2. Image preprocessing
    try:
        img = Image.open(io.BytesIO(image_bytes)).convert("L")
        img = img.resize((32, 32))
        arr = np.array(img).astype(np.float32) / 255.0
        arr = (arr - 0.5) / 0.5
        arr = np.expand_dims(arr, axis=0)
        arr = np.expand_dims(arr, axis=0)
    except Exception as e:
        raise HTTPException(status_code=400, detail=f"Failed to preprocess im-
age: {e}")

    # 3. Model-inference
    input_name = session.get_inputs()[0].name
    output_name = session.get_outputs()[0].name
    logits = session.run([output_name], {input_name: arr})[0]
    logit = float(logits[0][0])
    prob = 1.0 / (1.0 + np.exp(-logit))
    label = "chihuahua" if prob > 0.5 else "muffin"
    # 4. Return statement
    return {"label": label, "prob_chihuahua": prob, "prob_muffin": 1.0 - prob}
```

Listing 13.5 REST API Server: Predicting Route Implementation

Listing 13.6 shows the code (within if __name__ == "__main__"), which is only executed if we start the file directly. We start the server via uvicorn.run, and we set the host parameter so that it listens to requests on the port with the number 8,000 on all network interfaces.

```
if __name__ == "__main__":
    uvicorn.run(app, host="0.0.0.0", port=8000)
```

Listing 13.6 REST API Server: Server Start

We've created the code for our REST API server, so we can deploy it in the next step.

13.2.2   Deployment

We can test the local deployment by executing app.py. To do this, we execute the following command in the command line, as shown in Listing 13.7.

```
python app.py

INFO:    Started server process [31508]
INFO:    Waiting for application startup.
INFO:    Application startup complete.
INFO:    Uvicorn running on http://0.0.0.0:8000 (Press CTRL+C to quit)
```

Listing 13.7 Local Deployment: App Start

The output on the command line shows us that the server has been started without errors and is listening for requests. We can stop the server by pressing CTRL + C.

Now, we want to find out whether the server can respond correctly to the requests.

13.2.3   Test

Before we move on to more advanced approaches to testing the server, let's check the accessibility of the server itself by navigating to "localhost:8000" in the browser, as shown in Figure 13.3. This request uses the standard endpoint and returns a status message of **ok** as well as a message that we should use the /predict method.

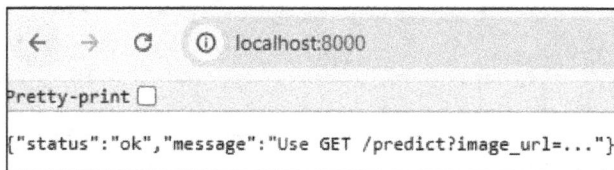

Figure 13.3 Local Deployment: Accessibility of Server

Now that we have ensured that the server is accessible, let's move on to testing the /predict endpoint.

In this case, we can also simply call up the URL (*http://localhost:8000/predict?image_url=https://upload.wikimedia.org/wikipedia/commons/6/65/Chihuahuasmooth-coat.jpg*) in the browser and receive the API response. The URL request contains the endpoint and the image_url parameter, and as this is a GET request, we can pass parameters as part of the URL.

However, this approach reaches its limits as soon as we use other HTTP methods. Therefore, I would like to introduce you to a platform for testing REST API: *Postman*, which is one of the leading and most widely used platforms for developing and testing APIs. It's free and available for various platforms at *https://www.postman.com/*.

Postman's main purpose is to provide developers and testers with a simple graphical interface for sending HTTP requests to APIs and examining the responses. All common HTTP methods (GET, POST, PUT, and DELETE) are supported. You can also organize multiple requests into collections, which is useful when you're using it for testing in multiple projects. Postman has become a comprehensive platform that also facilitates teamwork through shared collections.

Figure 13.4 shows how you need to define a request. In addition to selecting the correct method (which is GET in this case), you must specify the URL (which is *http://127.0.0.1:8000/predict*). Because this endpoint expects the image_url parameter, you must define it via the Params tab. You can use any image URL as the value, so as a test, you can use this image of a Chihuahua from Wikipedia: *https://upload.wikimedia.org/wikipedia/commons/6/65/Chihuahuasmoothcoat.jpg*.

Figure 13.4 Local Deployment: Postman Test

You can see the response from the server in the lower section of the screen: **200 OK** means that the server sent the HTTP code 200, which is delivered in the event of a successful request. The actual response from the server is available as JSON, and the server was able to correctly predict the Chihuahua with a high probability (86%).

In this section, you've successfully created a REST API, deployed it locally, and tested it. Now that we know that the API works, we'll deploy it to the cloud in the next section.

13.3   Heroku

Heroku is a popular service that simplifies the deployment and management of applications. It's a cloud provider known for its user-friendly interface for developers and its easy integration with Git. Heroku allows developers to deploy with minimal operational effort: the idea is that you deploy the code and the service takes care of infrastructure complexities such as server setup and scaling. This allows you to focus on developing applications.

One of the supported languages is Python, so we're ready to get started. The first step is to create an account at *https://signup.heroku.com/*. Each app execution only costs a few cents, but you still need to add your payment details. Once you've done that, you can move on to creating a new app.

It's helpful to use Heroku's command line interface, so we'll find out how to install it and use it for the deployment process. Once we've completed the deployment, we'll test the system, and at the end, we'll clean up by stopping and deleting the app to avoid unnecessary costs.

13.3.1   Command Line Interface and Login

Heroku comes with its own *command line interface* (CLI), and you can download and install it at *https://devcenter.heroku.com/articles/heroku-cli*. You can start the login process via the command line as shown in Figure 13.5:

```
heroku login
```

```
C:\Projects\BooksArticles\PyTorchBook\Code\140_Deployment\heroku>heroku login
heroku: Press any key to open up the browser to login or q to exit:
Opening browser to https://cli-auth.heroku.com/auth/cli/browser/90912a4a-b222-4f42-8c58-cded6f414a55?r
equestor=SFMyNTY.g2gDbQAAAA04OC43NC4yMTAuMTI5bgYAED1mTZkBYgABUYA.XgkcS_zAjUFrHOjbOGS9dEBSmHzYSLcpwwXSm
vbxJ1E
Logging in... done
Logged in as info@gollnickdata.de
```

Figure 13.5 Heroku: Login Successful

That will open the browser so you can log in, and you can then return to the IDE and verify that your login was successful.

13.3.2 Deployment

Deployment is organized in such a way that you first create all the necessary files that Heroku requires for deployment and then create an app and load the data into the app. In this process step, all data is copied to Heroku and the app is deployed.

Let's start with the required files. In addition to the *app.py* we've already discussed, we need further files for deployment with Heroku. Listing 13.8 shows the files required in our heroku working folder, as follows:

- *app.py*
  This contains the REST API that we created in the last section.

- *bin_class_model.onnx*
  This model file is also a must.

- *Procfile*
  This tells Heroku which commands need to be started for the application to run.

- *requirements.txt*
  This summarizes all the packages needed to run the Python script.

- *runtime.txt*
  This is a text file that specifies the exact Python version. It's optional, but it can prevent compatibility issues caused by version differences.

```
heroku
└──app.py
└──bin_class_model.onnx
└──Procfile
└──requirements.txt
└──runtime.txt
```

Listing 13.8 Heroku: Folder Structure and Required Files

The *Procfile* from Listing 13.9 contains the exact command that Heroku needs to start the FastAPI application correctly. The process type web tells Heroku that it's a web server that should respond to HTTP requests.

Also, uvicorn is the *asynchronous server gateway interface* (ASGI), which is required to run an asynchronous Python application such as FastAPI.

The module path (app:app) refers to the file name (app.py), and the second :app is the name of the FastAPI instance within the file. The --host parameter instructs the server to listen to all network interfaces, while the --port parameter refers to the $PORT environment variable. This is set dynamically by Heroku.

```
web: uvicorn app:app --host 0.0.0.0 --port $PORT
```

Listing 13.9 Heroku: Procfile

Now, let's turn to the *requirements.txt* file in Listing 13.10. You must include all relevant packages in it, ideally with the software version, so that you can replicate your local system to Heroku in the best possible way.

```
fastapi==0.115.5
uvicorn[standard]==0.34.0
pydantic==2.9.2
pillow==11.0.0
numpy==2.1.3
onnxruntime==1.19.2
requests==2.32.3
```

Listing 13.10 Heroku: requirements.txt

Finally, we come to the *runtime.txt* file (see Listing 13.11), which contains the Python version to be used on the server. The file is not mandatory, but it can be very helpful for avoiding version conflicts.

```
python-3.12.5
```

Listing 13.11 Heroku: runtime.txt

At this point, we've prepared the files, so we need to create an app at Heroku in which we can use the files. In principle, we could also do this via the command line, but for starters, it's easier to do it directly on the Heroku website.

In the dashboard on Heroku, you can click on **Create New App**. Figure 13.6 shows the options when creating a new app, as follows:

- You must assign a name to the app (e.g., "pytorch-pred").
- You must deploy the app on a server in either the United States or Europe.
- You have the option of making the app part of a larger pipeline (but we won't do that at this point).
- You click **Create app**.

Then, there are the options shown in Figure 13.7 for providing an app by Heroku. We'll use the first option shown: **Heroku Git**. The steps we'll follow in this section are described on the website.

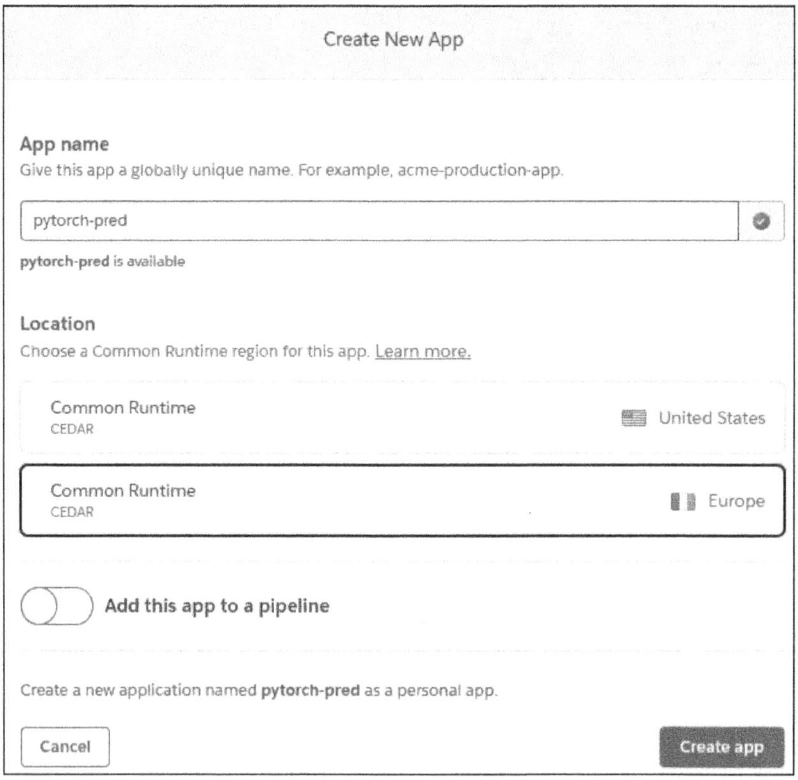

Figure 13.6 Heroku: Creating New App

Figure 13.7 Heroku: Deployment Options

Then, we can switch back to the command line and transfer the files to Heroku using Git. This assumes that you've installed Git on your system. (It's the most popular version control program and the basis of GitHub and GitLab.) If you haven't, you can obtain and install the latest version of Git at *https://git-scm.com/downloads*. Then, you can create an empty Git repository in the current directory as follows:

```
git init
```

You can then use the following command to connect your local Git repository to the special Git repository provided by Heroku. You must enter the app name that was used in the previous step, as follows:

```
heroku git:remote -a deinappname
```

You can add the files that are to be made available via Heroku as follows:

```
git add .
```

This doesn't copy the files to Heroku yet but puts them in a local subdirectory called the *staging area*. Then, with git commit, you can permanently save the files that are in the staging area in the local Git repository. The parameter -m stands for message, as each commit requires a corresponding message, as follows:

```
git commit -m "Initial commit"
```

There are various branches in git, and Heroku expects the files to be in the main branch. However, if your system creates a master branch by default, you must change the name of the default branch from master to main, as follows:

```
git branch -m master main
```

The last step is to push the local changes to Heroku. You do this by transferring the entire local main branch to the Git repository that Heroku hosts for your application, as follows:

```
git push heroku main
```

This starts the provisioning process and copies all files to Heroku. It then sets up the environment based on the *requirements.txt*, and it starts the server as described in the *procfile*. When the process is done and the deployment is complete, you should get the feedback shown in Figure 13.8.

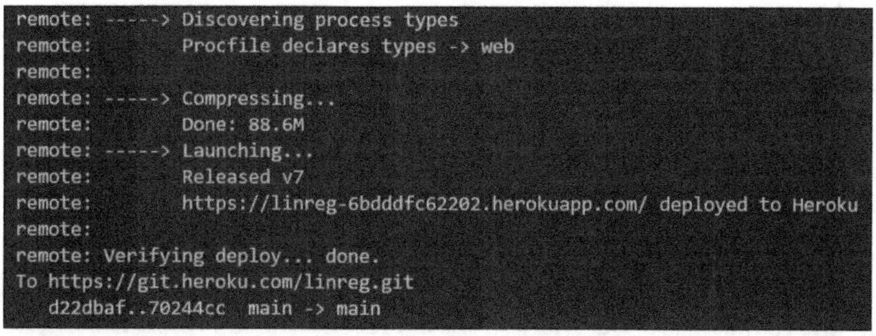

```
remote: -----> Discovering process types
remote:        Procfile declares types -> web
remote:
remote: -----> Compressing...
remote:        Done: 88.6M
remote: -----> Launching...
remote:        Released v7
remote:        https://linreg-6bdddfc62202.herokuapp.com/ deployed to Heroku
remote:
remote: Verifying deploy... done.
To https://git.heroku.com/linreg.git
   d22dbaf..70244cc  main -> main
```

Figure 13.8 Heroku: Deployment Successful

13.3.3 Test

Next, we need to test whether the server is generally accessible, and we use the "/" default route for that. As this is a normal GET request, you can simply open the URL in the browser. Figure 13.9 shows the return of the server: the status is "ok", so the server is accessible and can answer requests.

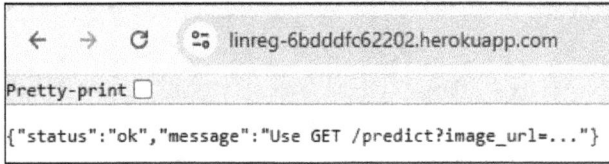

Figure 13.9 Heroku: Server Accessible

Once we've verified this, we'll want to test whether the predict route also works. For this, we use Postman again. We can use the same parameters as in the previous section, and we only need to adjust the URL, as follows:

- HTTP method: GET
- URL: ihrappname.herokuapp.com/predict
- Parameter: pass an image URL to the image_url parameter (e.g., https://upload.wiki-media.org/wikipedia/commons/8/8a/Muffin_NIH.jpg)

Figure 13.10 shows a successful test of the /predict route.

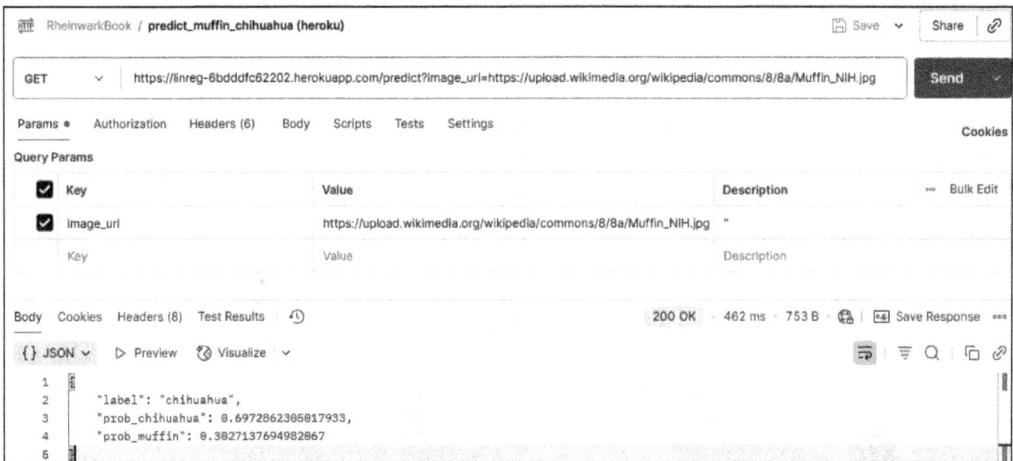

Figure 13.10 Heroku: Test of Predict Route

You can execute the request via **Send**, and you should receive the API response.

13.3.4   Stopping and Deleting the App

To avoid unnecessary costs, you should stop the app when you no longer need it. This option is available in the **Resources** tab. Figure 13.11 shows the relevant area with the hourly costs incurred, and it shows the app has already been stopped. You can switch between start and stop by clicking on the pencil (which is the edit button) to adjust the slider.

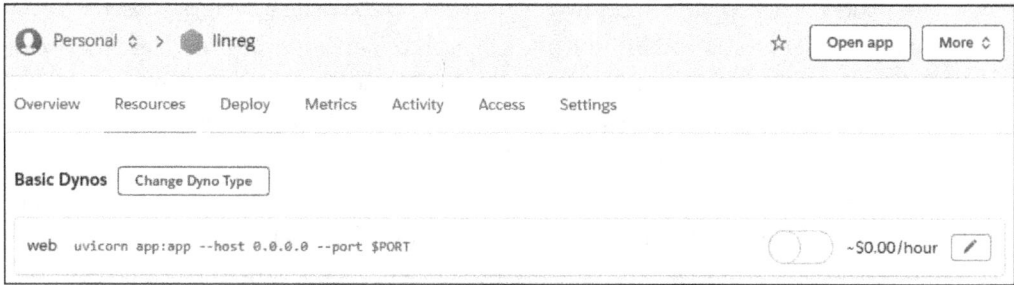

Figure 13.11 Heroku: Stopping App

The option to permanently delete the app can be found in the **Settings** tab, where you can scroll all the way down and find the **Delete app...** button, as depicted in Figure 13.12.

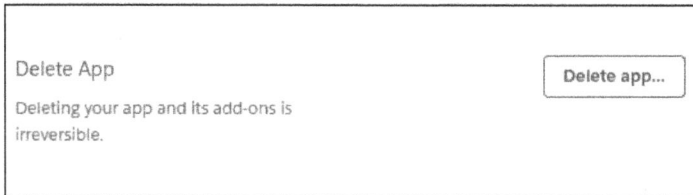

Figure 13.12 Heroku: Deleting App

In this section, you've learned how to deploy a simple REST API in the cloud by using Heroku. In the next section, we'll look at another very popular cloud provider and deploy our REST API by using Microsoft Azure.

13.4   Microsoft Azure

Microsoft Azure is a comprehensive cloud computing platform developed by Microsoft. It offers an extremely wide range of services, and its global infrastructure and scalability make it a popular choice for companies of all sizes. Since it's a gigantic framework, we can only focus here on the aspect of providing the REST API.

To get started, we'll learn how to set up an Azure subscription (in Section 13.4.1). We'll develop the app as an Azure function (in Section 13.4.2), and we'll start and test it locally (in Section 13.4.3) before we prepare the cloud deployment (in Section 13.4.4) and deploy it there (in Section 13.4.5).

13.4.1   First Steps

First, you need a Microsoft Azure account. If you don't have one yet, you can sign up on the official Azure website (*https://azure.microsoft.com*) with a personal Microsoft account (such as Outlook or Hotmail) or a business email address.

Once you've created your account, the next step is to create an Azure subscription. A subscription serves as a billing unit and a logical container for the resources you've created, and you can have several subscriptions in one account (for example, to manage and bill different projects separately). To create a subscription, you must search for "Subscriptions" in the Azure Portal and then add a new subscription.

Figure 13.13 shows which parameters you need to set when creating your subscription. You need to have a billing account to cover the costs incurred, and you can make various advanced settings on the other tabs. You can also set the permissible monthly limit on the **Budget** tab, which will help you avoid unpleasant surprises when it comes to costs.

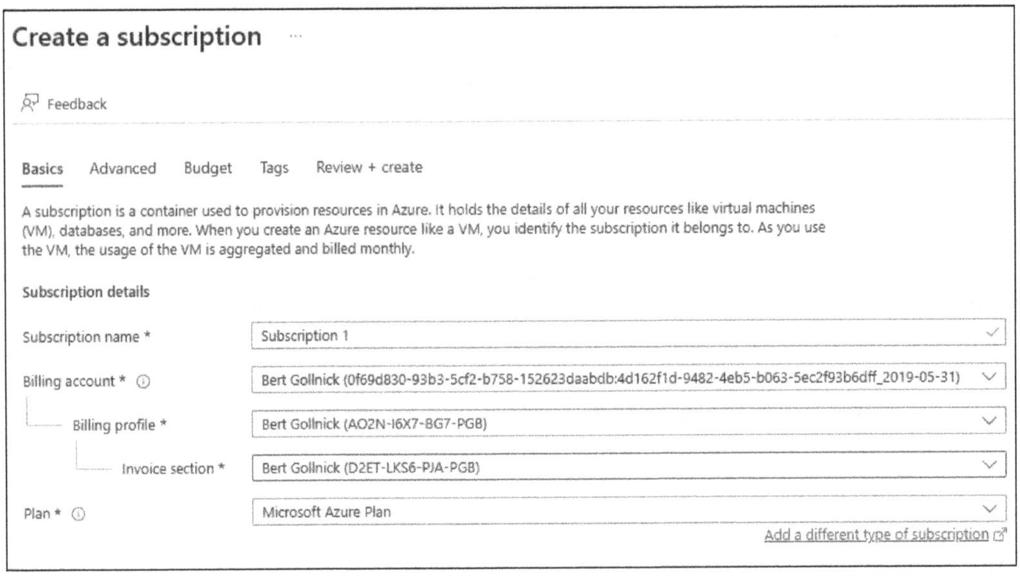

Create a subscription   ···

⟨⟩ Feedback

Basics   Advanced   Budget   Tags   Review + create

A subscription is a container used to provision resources in Azure. It holds the details of all your resources like virtual machines (VM), databases, and more. When you create an Azure resource like a VM, you identify the subscription it belongs to. As you use the VM, the usage of the VM is aggregated and billed monthly.

Subscription details

Subscription name *	Subscription 1
Billing account * ⓘ	Bert Gollnick (0f69d830-93b3-5cf2-b758-152623daabdb:4d162f1d-9482-4eb5-b063-5ec2f93b6dff_2019-05-31)
└── Billing profile *	Bert Gollnick (AO2N-I6X7-BG7-PGB)
└── Invoice section *	Bert Gollnick (D2ET-LKS6-PJA-PGB)
Plan * ⓘ	Microsoft Azure Plan

Add a different type of subscription ↗

Figure 13.13 Azure: Creating Subscription

To organize your resources, you should create a *resource group*, which is a logical container that holds related Azure resources together for an application or project. This makes it easier to manage and monitor resources. To create a resource group, you need to search for "resource group" in the Azure Portal.

Figure 13.14 shows the screen for creating a resource group. The resource group is always assigned to a subscription and has a unique name. The choice of region depends on various influencing parameters. For example, proximity to the app users may be important to ensure low latency. However, legal requirements can also play a role here. It should also be noted that the prices for the same Azure services can vary, depending on the region.

As soon as you've performed these steps, you can get down to the actual work and create the function app.

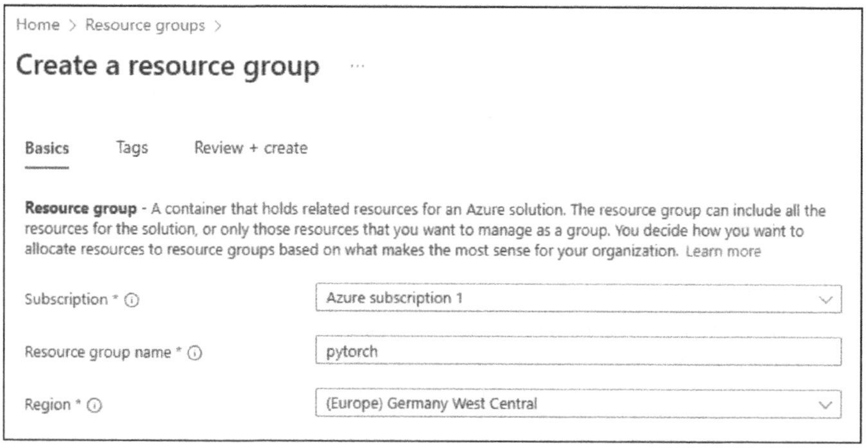

Figure 13.14 Azure: Creating Resource Group

13.4.2   App Development

To successfully deploy and run an Azure Function App, we need several essential files to define and configure the project. Essentially, this includes the following files, which can all be found in the working folder *140_Deployment\azure*:

```
azure
└───host.json
└───function.json
└───__init__.py
```

In our specific case, we also need these two additional files:

```
└───bin_class_model.onnx
└───api_app.py
```

This is the deep learning model (*bin_class_model.onnx*) and the actual program logic in *api_app.py*.

The *host.json* file from Listing 13.12 contains global configuration settings, and the behavior of the function app is also described there. What we set here will apply to all functions in the project, and typical settings relate to logging or extensions, for example.

```
{
  "version": "2.0",
  "extensions": {
    "http": {
      "routePrefix": ""
    }
  },
```

```
  "logging": {
    "applicationInsights": {
      "samplingSettings": {
        "isEnabled": true,
        "excludedTypes": "Request"
      }
    }
  },
  "extensionBundle": {
    "id": "Microsoft.Azure.Functions.ExtensionBundle",
    "version": "[4.*, 5.0.0)"
  }
}
```

Listing 13.12 Azure: host.json

Listing 13.13 shows the content of the *function.json* file, which is the configuration file for the Azure function. Here, the function uses the scriptFile parameter to find out which Python file is executed when someone calls the function. The authLevel parameter defines who is allowed to call the function, and for the sake of simplicity, it has been set to anonymous here so that anyone can call the function without having to log in.

The function responds to HTTP requests (of a parameter type with the httpTrigger value). We can define permitted methods via methods, and route specifies that all URL paths are forwarded to the function.

To summarize, the Azure file says, "Create an http API that forwards all requests to the __init__.py and sends the response back to the user."

```
{
  "scriptFile": "__init__.py",
  "bindings": [
    {
      "authLevel": "anonymous",
      "type": "httpTrigger",
      "direction": "in",
      "name": "req",
      "methods": ["get", "post", "put", "patch", "delete", "options"],
      "route": "{*route}"
    },
    {
      "type": "http",
      "direction": "out",
      "name": "$return"
```

```
    }
  ]
}
```

Listing 13.13 Azure: function.json

The request therefore comes via *__init__.py*, the contents of which are shown in Listing 13.14. This file connects the Azure function's HTTP trigger with the web app. It's important that the main function is async, because ASGI and Azure functions work asynchronously.

```
import azure.functions as func
from azure.functions import AsgiMiddleware
from api_app import asgi_app

async def main(req: func.HttpRequest, context: func.Context) -> func.HttpRe-
sponse:
    return await AsgiMiddleware(asgi_app).handle_async(req, context)
```

Listing 13.14 Azure: __init__.py

The api_app.py file is like the app.py file in the previous section.

Now, we can start the app by executing the following on the command line:

```
func start
```

```
Found Python version 3.12.9 (py).
Azure Functions Core Tools
Core Tools Version:
4.2.2+78afd8b84c8e31f0ddac570ba9e8128eefbd3d4a (64-bit)
Function Runtime Version: 4.1041.200.25360
[2025-09-17T07:17:23.443Z] Worker process started and initialized.

Functions:
HttpTrigger: [GET,POST,PUT,PATCH,DELETE,OPTIONS]
http://localhost:7071/{*route}
```

13.4.3 Local Test

We've started the server successfully, and it's waiting for requests. You can test it in the following ways:

- Open this URL directly in the browser:

  ```
  http://127.0.0.1:7071/predict?image_url=https://upload.wikimedia.org/wikipe-
  dia/commons/8/8a/Muffin_NIH.jpg
  ```

- Create a GET request with Postman. This procedure is identical to the procedure described in section Section 13.3.3, except that we need to adapt the URL.

- Work in the IDE, which is the preferred method of developers. You can simply open another terminal and use the `curl` command line tool to test the request:

  ```
  Curl "http://127.0.0.1:7071/predict?image_url=https://upload.wikimedia.org/
  wikipedia/commons/8/8a/Muffin_NIH.jpg"
  ```

  ```
  {"label":"chihuahua","prob_chihuahua":0.6972862305017932,"prob_
  muffin":0.30271376949820683}
  ```

Whichever method you choose, you should see that the server responds successfully to requests.

We can now move on to deploying the app in the cloud. To do this, we first need to create the app in the Azure portal.

13.4.4   Function App Development in the Azure Portal

The Azure Function App is the primary resource that provides the environment for executing your functions. To create a Function App, you need to search for "function" in the Azure Portal (*portal.azure.com*) as shown in Figure 13.15. You'll find the Function App service there.

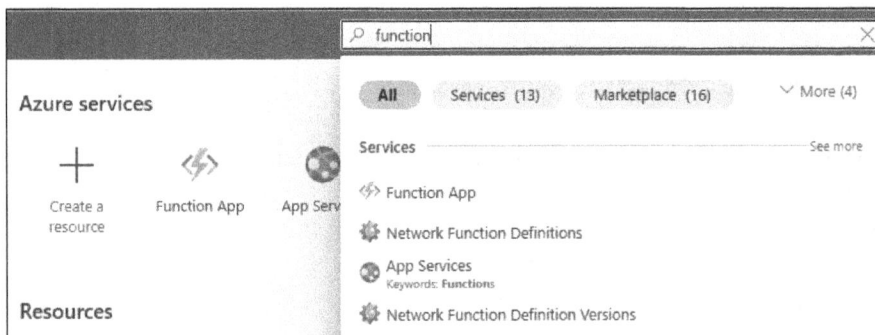

Figure 13.15 Azure: Function App

Then, you can create a new function app with **Create**. You can choose from various hosting options, and Figure 13.16 shows the selection options in which you select **Flex Consumption**. Costs are only incurred here for actual use, and clicking **Select** takes you to the next configuration setting.

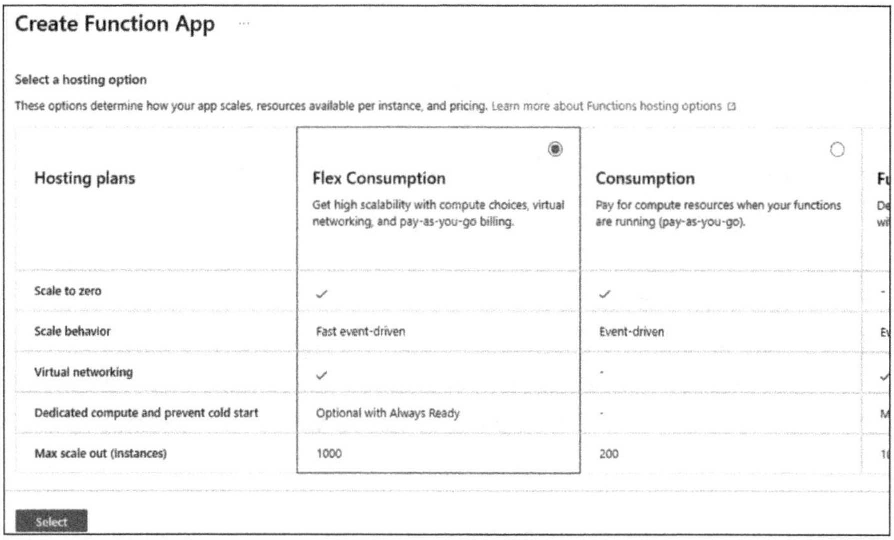

Figure 13.16 Azure: Function App Hosting

Figure 13.17 shows the other configuration settings. You must select the subscription and resource group in which you want to save the function app, and you must also enter the name of the app (which is pytorchmodel in our case).

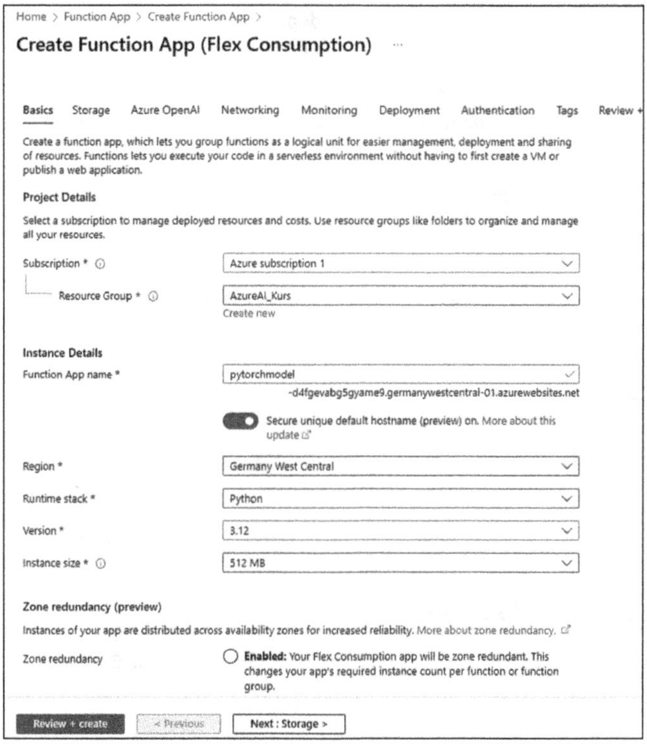

Figure 13.17 Azure: Function App Basics

You'll set the region to Germany West Central and set the runtime stack to Python version 3.12. Finally, you can define the instance size. You'll use the smallest size of 512MB since it's sufficient for this small model and the costs scale with the size. Then, leave all other details at the default settings so you can trigger the creation by clicking **Review + create**.

Now, the empty function app is ready and you must fill it with content. To do this, switch back to the IDE.

13.4.5   Cloud Deployment

You can close the running app in the terminal by pressing $\boxed{\text{CTRL}}$+$\boxed{\text{C}}$. There's a command line command that you can use for deployment, and it accesses the code in the current working directory and "fills" the app container with your content. You must then prepare the environment and start the app. You can do all of that with a single command.

You only need the app name that you defined in the previous section: pytorchmodel. Listing 13.15 shows the call of the Azure function with the publish parameter, followed by the app name.

```
func azure functionapp publish pytorchmodel
```

```
Getting site publishing info...
[2025-09-16T10:16:09.383Z] Starting the function app deployment...
[2025-09-16T10:16:09.391Z] Creating archive for current directory...
Performing remote build for functions project.
Uploading 148,98 KB [##############################################################
#######]
[2025-09-16T10:18:54.537Z] The deployment was successful!
Functions in pytorchmodel:
    HttpTrigger - [httpTrigger]
        Invoke url: https://pytorchmodel-d4fgevabg5gyame9.germanywestcentral-
01.azurewebsites.net/{*route}
```

Listing 13.15 Azure: App Deployment

The Invoke URL is also part of the output, and it's the endpoint that listens to the requests.

You can check whether the app has been successfully deployed in the Azure Portal, for example. Figure 13.18 shows the available function apps. Our pytorchmodel function app is listed in the overview with a status of **Running**.

Finally, let's test the app by using Postman again. Figure 13.19 shows the Postman request with the successful return.

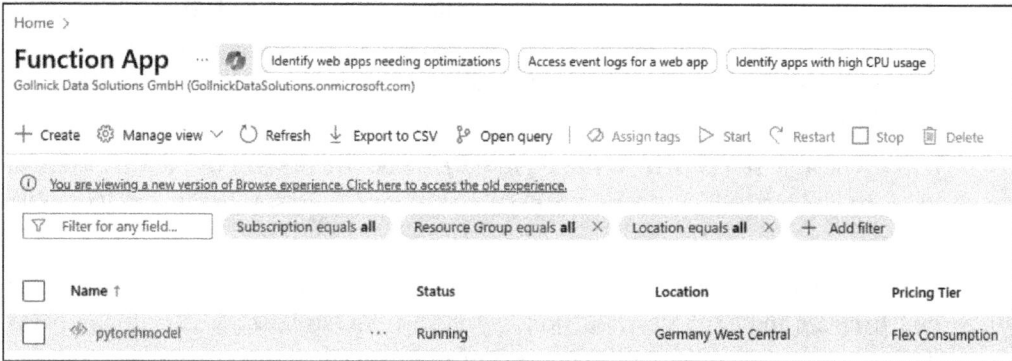

Figure 13.18 Azure: Function App Overview

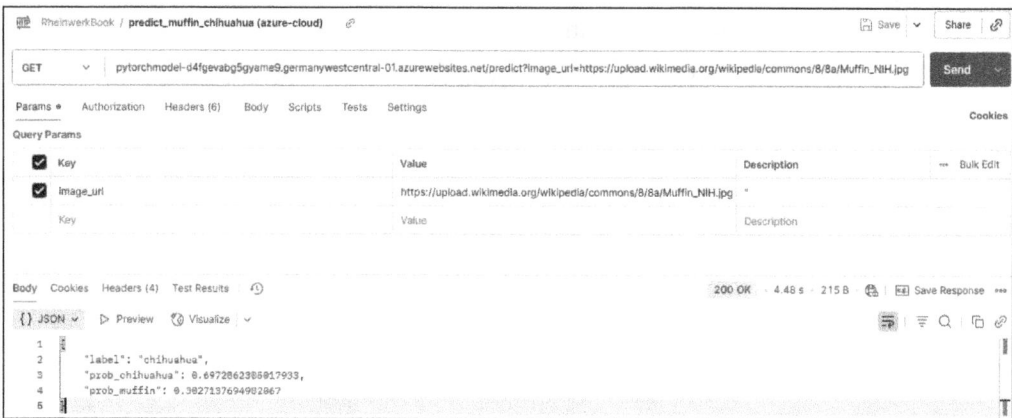

Figure 13.19 Azure: Deployment Test

The following parameters were used:

- HTTP method: GET
- URL: yourappname.azurewebsites.net/predict
- Parameter: pass an image URL to the image_url parameter (e.g., https://upload.wiki-media.org/wikipedia/commons/8/8a/Muffin_NIH.jpg)

This last successful test concludes this section on deployment with Azure.

13.5 Summary

This chapter provided comprehensive insights into the various aspects of model deployment. We started with a general overview of deployment strategies and looked at CI/CD and classic deployment using frontend and backend based on REST APIs.

We then turned our attention to local deployment and went through the entire process from API development to the actual deployment and final testing of the app in the local environment. This section laid the foundation for understanding the deployment workflow.

In the following sections, we ventured into the cloud. We first got to know the Heroku platform and learned how to use it to deploy, test, and (if necessary) stop or delete applications.

Lastly, we looked at Microsoft Azure. We started with the first steps in Azure and learned how to create and test a local Function App. Finally, we saw how this Function App was deployed in the Azure portal and successfully migrated to the cloud.

To summarize, in this chapter, you've learned about the most important concepts and practical steps of deployment—from local development to deployment on various cloud platforms such as Heroku and Azure.

You can now effectively manage and publish your own applications.

The Author

Bert Gollnick is a senior data scientist who specializes in renewable energies. For several years, he has taught courses about data science and machine learning, and more recently, about generative AI and natural language processing. Bert studied aeronautics at the Technical University of Berlin and economics at the University of Hagen. His main areas of interest are machine learning and data science.

Index

V

W

Y